Reactive Inkjet Printing
A Chemical Synthesis Tool

Smart Materials

Series editors:
Hans-Jörg Schneider, *Saarland University, Germany*
Mohsen Shahinpoor, *University of Maine, USA*

How to obtain future titles on publication:
A standing order plan is available for this series. A standing order will bring delivery of each new volume immediately on publication.

For further information please contact:
Book Sales Department, Royal Society of Chemistry, Thomas Graham House, Science Park, Milton Road, Cambridge, CB4 0WF, UK
Telephone: +44 (0)1223 420066, Fax: +44 (0)1223 420247
Email: booksales@rsc.org
Visit our website at www.rsc.org/books

Reactive Inkjet Printing
A Chemical Synthesis Tool

Edited by

Patrick J. Smith
University of Sheffield, UK
Email: patrick.smith@sheffield.ac.uk

and

Aoife Morrin
Dublin City University, Ireland
Email: aoife.morrin@dcu.ie

ROYAL SOCIETY
OF **CHEMISTRY**

THE QUEEN'S AWARDS
FOR ENTERPRISE:
INTERNATIONAL TRADE
2013

Smart Materials No. 32

Print ISBN: 978-1-78262-767-8
PDF ISBN: 978-1-78801-051-1
EPUB ISBN: 978-1-78801-350-5
ISSN: 2046-0066

A catalogue record for this book is available from the British Library

The Royal Society of Chemistry is a charity, registered in England and Wales, Number 207890, and a company incorporated in England by Royal Charter (Registered No. RC000524), registered office: Burlington House, Piccadilly, London W1J 0BA, UK, Telephone: +44 (0) 207 4378 6556.

For further information see our web site at www.rsc.org

Printed in the United Kingdom by CPI Group (UK) Ltd, Croydon, CR0 4YY, UK

Preface

Once upon a time, in the earliest years of the 21st century, I had just started my first post-doc which involved using inkjet printers. It was then that I started to ask questions about inkjet printing. At first, they were the questions we all ask, at the start, such as "Why doesn't it work?", "Why won't it do what I want it to do?" and so on, but as my career progressed the questions I asked became "What can I do with an inkjet printer that is unique?" and "What are inkjet's main strengths?"

This book forms the current answer to what I think is inkjet's principle strength, namely its ability to dispense more than one ink in a printing pass. We employ this ability every day when we print off an interesting journal paper, a tasty recipe or colourful invitations. With reactive inkjet printing the difference is in the inks. Instead of using suspensions of pigments or dyes, the inks we use are chemical reactants. Reactive inkjet printing extends the range of materials that can be patterned by inkjet. We are no longer bound to depositing materials that have to be made into ink form first.

The purpose of this book is to present a survey of the current research areas in which reactive inkjet printing either contributes to, or forms an essential foundation. As with many books, towards the end of the process one feels a sense of panic because the initial dreams of the book being a comprehensive, exhaustive and authoritative survey give way to realism and the external pressures for one's time that most modern day academics experience.

As such, it is with heartfelt sincerity that I and my fellow editor, Aoife, offer our thanks to all of our contributing authors. **Thank you!** We have been enriched and educated by your contributions. We are thrilled that our authors represent a broad range of disciplines and include those who wrote some of the first papers on reactive inkjet printing. The application topics range from printed electronics through additive manufacture to tissue engineering, which illustrates the versatility and appeal of both inkjet and reactive inkjet printing.

Smart Materials No. 32
Reactive Inkjet Printing: A Chemical Synthesis Tool
Edited by Patrick J. Smith and Aoife Morrin
© The Royal Society of Chemistry 2018
Published by the Royal Society of Chemistry, www.rsc.org

We have been honoured by the time, commitment and experience our contributing authors have been willing to share with us and with you, our reader. It is my hope and Aoife's that you find this book to be a source of information and inspiration.

Finally, Aoife and I would like to thank Leanne Marle for being at the start and end of this project. The initial idea for the book developed from a conversation in Boston in December 2013 between Leanne and me; along the way the discussion became serious and finally last year, 2016, Aoife and I began the project, the result of which you can now see. We'd also like to thank Cara Sutton and especially, Catriona Clarke for their guidance, assistance and encouragement.

It is my hope that you find this book to be a fascinating answer.

Patrick J. Smith, Sheffield, England

Contents

Smart Materials No. 32
Reactive Inkjet Printing: A Chemical Synthesis Tool
Edited by Patrick J. Smith and Aoife Morrin
© The Royal Society of Chemistry 2018
Published by the Royal Society of Chemistry, www.rsc.org

CHAPTER 1

Reactive Inkjet Printing— An Introduction

PATRICK J. SMITH*[a] AND AOIFE MORRIN*[b]

[a]University of Sheffield, UK; [b]School of Chemical Sciences, National Centre for Sensor Research, Insight Centre for Data Analytics, Dublin City University, Ireland
*E-mail: patrick.smith@sheffield.ac.uk, aoife.morrin@dcu.ie

1.1   Introduction

'Reactive inkjet (RIJ) printing', or 'reactive inkjet' is a type of inkjet technique whose history in the research literature has been documented for less than ten years. Despite its recent beginnings, it has attracted much attention as a cost-effective and highly controllable method to fabricate patterns, where it elegantly combines the processes of material deposition and chemical reaction. The combination of these two processes creates boundless possibilities for the fabrication of 2-dimensional and indeed 3-dimensional structures whereby it allows functional materials to be synthesized *in situ* at the same time as their final device geometries are patterned.

The concept of micro-dispensation of reactants actually dates back to the 1990s when the first report of a solenoid-based inkjet chemical dispenser, ChemJet, was made for combinatorial library synthesis.[1] Since then, the field has grown from this application to become a combinatorial library synthesis tool for rational coating design, and a tool used for patterning materials directly within devices. Materials that have been reactively printed for direct

Smart Materials No. 32
Reactive Inkjet Printing: A Chemical Synthesis Tool
Edited by Patrick J. Smith and Aoife Morrin
© The Royal Society of Chemistry 2018
Published by the Royal Society of Chemistry, www.rsc.org

device integration include conjugated polymers, fluorescent quantum dots, metallics and silk. It is the precursors to these materials that are dispensed (*e.g.*, monomer and oxidant) and subsequently react in solution droplets (micro- or pico-litre sized) on a solid support. While performing chemistry in patterned droplets using this printing approach promises high controllability, technical challenges do still remain. Fundamental droplet analysis in terms of size, shape and kinetics must be understood in each application in order to design optimal precursor ink formulations and selection of appropriate print parameters.

A wide spectrum of reactive inks that have been reported allows one to fabricate various patterns that perform different functionalities. In terms of print resolution, RIJ will theoretically have the same resolution as inkjet printing. However, in practice, given the multi-head combinatorial approach often used, maintaining a resolution of single-layer, single-head inkjet printing, is more challenging. Nonetheless, patterning 2D structures at resolutions at the low micron-scale are easily achievable using RIJ.

As this book is concerned with RIJ, it seems prudent that the first chapter introduce to the reader what is meant by this term. By its nature, RIJ requires the use of an inkjet printer, the first chapter will, therefore, survey the various types of inkjet printing (such as piezo and thermal), as well as describing droplet ejection mechanisms and ink considerations. The following chapters will discuss a number of fundamental inkjet areas that will help inform the reader.

The second chapter discusses droplet behaviour from the moment a droplet impacts a surface to where it coalesces with a second or more to form a feature. The various drying phenomena, such as 'coffee staining' are discussed, as well as possible phase changes. The third chapter is fascinating; typically, RIJ involves two dissimilar droplets being joined, with their contained reactants meeting to form a product. Chapter 3 discusses how two droplets interact; what is intriguing is the observation that two droplets don't appear to mix! What this chapter illustrates is the youth of the field of RIJ and that the early focus on applications has now opened up some fascinating physics to explore. The fourth chapter concludes the Fundamentals section with a discussion of the behaviour of high molecular weight polymers in ink. These polymers tend to break apart during jetting which offers the prospect that radical chemistry can be performed.

The third and largest section of the book is given over to a high level overview of many of the application areas that RIJ can be employed in. Alison Lennon illustrates the attraction of using RIJ in metallisation and the production of etchants for use in solar panel manufacture. The use of RIJ to generate HF *in situ* is a particularly compelling example. Ghassan Jabbour and co-authors continue the exploration of RIJ's use in inorganic chemistry with the synthesis of gold nanoparticles, zinc oxide nanostructures and lead sulphide quantum dots. They begin their chapter by illustrating that RIJ offers organic chemistry applications in their modification of the sheet resistivity of poly(3,4-ethylenedioxythiophene) polystyrene sulfonate (PEDOT:PSS) by selectively dispensing droplets of aqueous sodium hypochlorite onto the polymer, thereby tailoring the degree of oxidation.

The next two chapters move the focus on to biology with the preparation of silk by RIJ forming the core. The first of this duo looks at the preparation of dental barrier membranes. Here, the degradation rate of silk II can be tuned by RIJ. The second chapter looks at the preparation of Janus-particle like structures, silk micro-rockets, which could be employed in a range of applications such as environmental monitoring, Lab-on-a-Chip diagnostics, and *in vivo* drug delivery.

The final three chapters continue to portray the breadth of application areas where RIJ can make a contribution. Christopher Tuck and colleagues describe the use of RIJ in additive manufacture, whilst Paul Calvert looks at the use of RIJ to form conductive features of either copper or nickel. The use of copper is of particular interest since it is a cheaper alternative to silver. Finally, the focus returns to biology with a discussion of the use of RIJ in the production of alginate- and fibrin-based systems for tissue engineering.

1.2   Reactive Inkjet Printing—The Concept

In simple terms, the concept of reactive inkjet printing (RIJ) describes the process where an inkjet printer deposits a droplet of reactant that changes due to a chemical reaction with material on the substrate or with a subsequent printed second reactant (Figure 1.1). Reactive inkjet printing can be used to deposit systems that cannot be deposited in ink form, such as some polyurethanes or to prepare purer forms of systems, such as nanoparticles that no longer require surfactants to stabilise them in suspension.

Inkjet printing is an essential component of RIJ because it produces droplets of uniform, controllable size. The second feature of inkjet is that the produced droplets can be positioned accurately on a substrate in pre-determined locations. The ability to consistently produce droplets whose volume can be predicted allows the droplets to be treated as building blocks, thereby allowing inkjet to be employed as an additive manufacturing technique as well as a synthesis tool. Moreover, this faculty of uniform droplets that are accurately positioned allows inkjet to be widely employed in the field of graphics.

From an RIJ point of view it is the fact that an inkjet printer can be employed as a synthesis tool that is of the greatest interest. In traditional chemical synthesis, the chemist seeks to have a high degree of precision with the aliquots of solution that they mix, as this ensures that the correct stoichiometry is obtained. In RIJ, the inkjet printer is treated as a precise pipette.

Figure 1.1   Reactive inkjet printing involves adding one reactant to another to form a product.

RIJ can be compared to micro-fluidic chips, ('Lab-on-a-Chip'), in which networks of channels have been etched in to the substrate. Typically, two or more channels each containing a reactive species meet in a chamber where the reactants mix forming a product. In terms of comparison, the channels in micro-fluidic chips are in the region of 300 to 3000 micron wide whereas an inkjet printer typically produces droplets that have contact diameters on the substrate of 100 microns. Due to improved mixing, micro-fluidic chips offer increased reaction yields and decreased consumption of material; these advantages are also enjoyed by RIJ. However, a further advantage of RIJ is that reaction products can also be placed such that the final device geometry is obtained.

Two types of RIJ have been defined.[1] These types are 'Single RIJ' and 'Full RIJ'. Single RIJ involves an inkjet printer dispensing an ink, typically a solution, that reacts with a species that has been deposited on the substrate by another deposition technique, such as spin-coating or spraying. Although an argument could be made for expanding the concept of single RIJ to involve etching of the substrate, this approach isn't covered in this book. Etching, or chemical machining, is typically several orders of magnitude smaller than RIJ such as the etching of silicon to form integrated chips or several of orders of magnitude larger such as in bulk cleaning or smoothing. RIJ, whether Single or Full, requires a substrate that does not take part in the reaction.

The main advantage of Single RIJ is that most research laboratories possess inkjet printers that allow the jetting of one ink only. Although, inkjet printer manufacturers are making machines that dispense more than one ink these tend to be more expensive and complex than single ink systems. Another advantage of Single RIJ is that one reactant can be deposited in bulk which increases the speed of the overall process. The second patterning step forms the desired product and the unreacted material can be removed.

Full RIJ is where an inkjet printer deposits two or more reactants, usually in separate passes from independent printheads. The advantage of Full RIJ is that less waste is generated when compared to Single RIJ and direct patterning is obtained. However, Full RIJ can be more time consuming and in some cases, depending on the solvents used, the drying out of the first reactant may be an issue.

1.3   Types of Inkjet Printer

Inkjet printing can be divided into two types: continuous inkjet and drop-on-demand ink. A further division can be made with drop-on-demand (DoD) into piezo DoD and thermal DoD, with the divisor being the actual method of actuation employed. Regardless of the type of inkjet method used, the in-flight diameters of the produced droplets range from 150 micron down to 10 micron.[2]

1.3.1 Continuous Inkjet Printing

There are a number of ways droplets are produced in continuous inkjet (CIJ), however all methods involve the droplets passing between a set of electrodes. Domino's A-Series SureStart print head contains a vibrating drive rod, which creates ultrasonic pressure waves in the ink, breaking it up into individual droplets.[3] Linx CIJ technology involves the ink being pulsed inside the ink chamber by a piezoelectric crystal, which causes the ejected ink column to break up into droplets.[4] Other CIJ versions involve the droplets forming by the Rayleigh instability of the ejected ink column that has been forced out of the nozzle under pressure.[2]

In CIJ (Figure 1.2), the charged droplets pass through a set of deflector plates that direct the droplets either towards the substrate or towards a gutter, where they are collected and recycled. As its name implies droplet production in CIJ is continuous, droplets are either used or recycled. The predominant use of CIJ is in product marking, with bar-codes being an oft-cited example. Typically, a CIJ ink contains a pigment, a carrier and additives (*e.g.* humectant) to ensure ink lifetime and performance.

CIJ is fast, and droplet generation can reach up to 60 kHz; however, the main drawbacks from an RIJ point of view are a limit on the range of inks that can be printed due to the droplets having to be able to carry a charge, and more significantly the risk of contamination due to recycling. However, CIJ should not be discounted from RIJ, as Wheeler and Yeates discuss in Chapter 4.

1.3.2 Drop-on-demand Inkjet Printing

Unlike CIJ, in which a continuous stream of droplets is generated and either placed on the substrate or recycled, drop-on-demand (DoD) printers are so-called because droplets are only generated when required. As such, all droplets generated by DoD are positioned on the substrate.

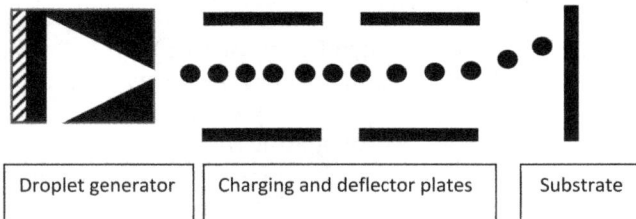

| Droplet generator | Charging and deflector plates | Substrate |

Figure 1.2   A simple schematic of a CIJ printhead, showing droplets generated due to the action of a transducer (diagonal hatching), passing through a set of charging plates before being directed by deflector plates onto a substrate. Droplets that are not deflected are collected by a gutter (not shown) and recycled.

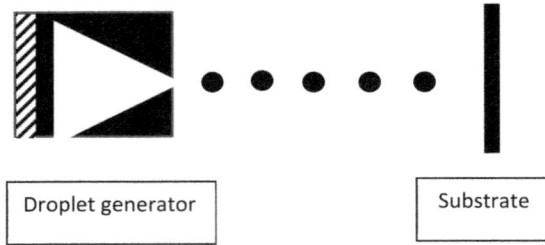

Figure 1.3 A simple schematic of a DoD printhead, showing droplets generated due to the action of a transducer (diagonal hatching), which can either be a heater or a piezoelectric material.

As mentioned earlier, DoD (Figure 1.3) can be divided into thermal DoD (often called TIJ – thermal inkjet) and piezo DoD. In TIJ, a thin film resistor is placed inside the ink chamber, a current is passed through the thin film causing it to rapidly heat up and inducing localised vaporisation of the ink in the chamber nearest to the heater. The vaporisation results in a bubble (which is why TIJ is also called bubble jet). The current passing through the resistor is then stopped, causing the thin film to cool down and triggering the collapse of the bubble. The rapid switching on and off of the current through the resistor and the subsequent rapid formation and collapse of the bubble causes pressure pulses throughout the ink in the chamber, which results in a droplet being generated.

Piezoelectric DoD works in a similar way to TIJ in that droplets are ejected due to an induced pressure pulse. However, in piezo DoD the actuator is based on a piezoelectric crystal, which changes shape as a consequence of a current being passed through it. In some printhead designs, the piezo component pushes into the ink chamber to create pressure pulses, in others it surrounds the chamber or forms a wall of the chamber.

1.4   Droplet Formation

An example of a piezo DoD printhead is shown in Figure 1.4(a). The operating principle is described as follows: pressure waves form as a result of a sudden volume change, which is caused by a voltage being applied to the piezoelectric actuator, and begin propagating throughout the capillary. When a positive pressure wave approaches the nozzle, a column of fluid is pushed outwards. The ink column, or droplet, is ejected when the transferred kinetic energy exceeds the surface tension required to form a droplet. The droplet's velocity is linked to the amount of kinetic energy transferred, with large actuating voltages producing droplets with higher velocities. The initial velocity of a droplet needs to be several metres per second to overcome the decelerating action of ambient air.[6,7]

The next few paragraphs describe the process of ejecting a droplet more fully. Figure 1.4(b) illustrates the pressure waves as boxes, below the line the boxes represent negative pressure waves, above positive. The arrows inside

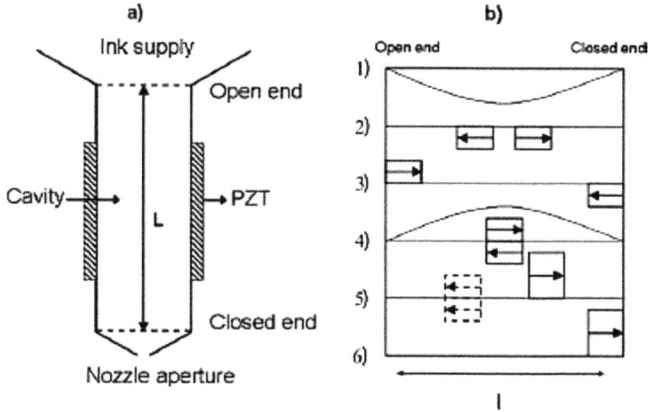

Figure 1.4 (a) Schematic diagram of a piezoelectric inkjet printhead. (b) Schematic representation of wave propagation and reflection in a piezoelectric tubular actuator. Reproduced from ref. 5 with permission from the Royal Society of Chemistry.

the boxes represent the direction of travel. At the start of the process, the piezoelectric actuator expands as a consequence of the applied voltage generating two negative waves (stage 2 in Figure 1.4(b)). These negative pressure waves mean that the ink is withdrawn to the centre; the ink at the nozzle retreats upwards, away from the nozzle aperture. The ink at the reservoir also retreats but as it does so it draws in more ink.

The nozzle is considered as closed, according to acoustic wave theory, as the size of the nozzle orifice is small compared to the overall cross-sectional area of the ink chamber. The end of the ink chamber that connects to the reservoir (ink supply in Figure 1.4(a)) is considered as open since the inside diameter of the ink supply is usually larger than the ink chamber. The pressure wave reflected from the nozzle keeps its phase, whereas the wave reflected from the reservoir—the open end—has its phase reversed. Another way of explaining this reversal is that the negative wave at the reservoir causes additional ink to be drawn in.

The two reflected waves travel back to the centre of the ink chamber (Stage 3). At this point, the voltage applied across the piezo is reset (or lowered below the starting value) causing the actuator to contract (Stage 4). The contraction of the actuator generates a positive pressure wave that travels towards the reservoir and the nozzle meeting the incoming initial waves. The negative wave that is returning from the nozzle is cancelled whilst the positive wave travelling from the reservoir is magnified (Stage 5). The magnified positive pressure wave travels to the nozzle, ejecting a column of ink.[8]

Bogy and Talke reported on a series of experimental observations using piezo DoD printheads. Their study examined the dependence of four operating characteristics on the length of the ink chamber. They concluded that DoD inkjet phenomena are related to the propagation and reflection of acoustic waves within the ink chamber.[9] The four measurable quantities that were found to be linearly dependent on length (*l*) and the speed of sound (*c*) in the ink were the optimum pulse width, which is equal to *l*/*c*; the delay time before the meniscus starts to protrude, which is equal to 3*l*/2*c*; the period of meniscus oscillation, which is equal to 4*l*/*c*; and the period of low frequency resonant and anti-resonant synchronous operation, which is equal to 4*l*/*c*.

As mentioned earlier, the velocity of a droplet is influenced by the actuating voltage, with larger voltages generating droplets with higher velocities. The volume of a droplet is also influenced by the actuating voltage – larger droplets are caused by larger voltages. Figure 1.5 shows how a droplet's volume and velocity are linearly determined by voltage. In terms of reactive inkjet printing (RIJ) the data shown in Figure 1.5 is of significant interest, as it demonstrates that the dispensed droplet volume can be adjusted.

Typically, in a laboratory when one generates a droplet three values can be set, which are actuating voltage, pulse width (when the two pressure waves recombine and the voltage across the piezo actuator is reset) and frequency. Although all three terms have some effect on droplet volume, that of voltage is most straightforward. A sine-wave type effect on droplet volume can be observed when pulse width is varied. From an RIJ point of view, the experimentalist is advised to optimise pulse width and then keep the value constant.

Piezo DoD printheads can operate at frequencies as high as 20 kHz.[2] However, operating piezo printheads at such high frequencies can result in chaotic droplet ejection events, which is because earlier pressure waves have not been damped out, but have instead interacted with subsequent pressure waves. From an RIJ point of view, high frequency droplet ejection is not recommended since a degree of control over, and confidence in, droplet volume is desired. In the early stages, the RIJ experimentalist is interested in

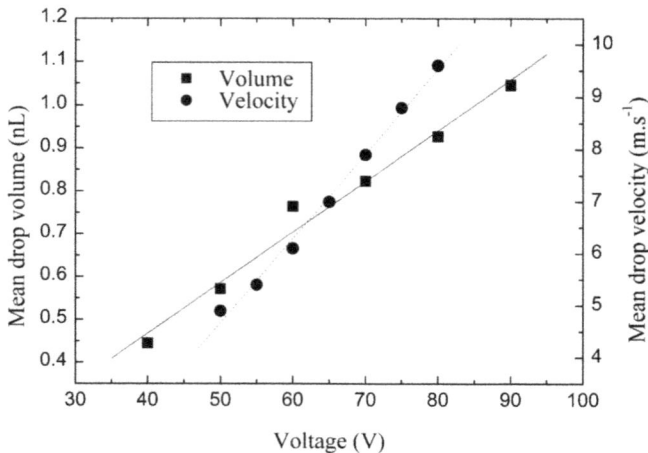

Figure 1.5   The influence of driving voltage on ejected droplet volume. Reprinted from N. Reis, C. Ainsley and B. Derby, *Journal of Applied Physics*, 2005, **97**, 094903, with the permission of AIP Publishing.[8]

determining if their target system can be synthesised by inkjet, in later stages high frequencies may be employed if speedier fabrication is a particular goal.

The viscosity of an ink, or solution (as is usual for RIJ), affects the residence time of the residual waves. A more viscous ink/solution eliminates pressure waves quicker. Wallace and Antohe compared ethylene glycol, which is more viscous, to water and found the decay time of residual waves was shorter than in water.[10]

1.5   Printability and *Z* Number

Fluid properties such as viscosity and surface tension have an influence on the formation of droplets from an inkjet printer. The Reynolds number, which is the ratio of inertia and viscosity, and Weber number (inertia and surface) are combined to form the Ohnesorge number.

The Ohnesorge number is given as $\sqrt{We}/Re$ or $\dfrac{\eta}{(\gamma \rho a)^{1/2}}$ where η equals dynamic viscosity, γ is surface tension, ρ is density and a is a characteristic length (usually the diameter of the printhead's nozzle). The issue is then complicated somewhat since the inverse of the Ohnesorge number is used to determine printability. The inverse is known as the *Z* number.

Fromm predicted that if *Z* was greater than 2, drop formation in DoD systems was possible.[7] Derby *et al.* refined this prediction saying that if *Z* lay between 1 and 10 an ink was printable.[8] If Z was less than 1 then viscosity would damp out the pressure pulse, whereas if *Z* was above 10 satellite droplets would form. Satellites are smaller droplets that form in a droplet ejection event. Typically, a column of liquid is ejected, which then, ideally, contracts into a sphere. However, in a number of cases the front of the column forms a spherical bulb and the rest of the column forms a long tail, as can be seen in Figure 1.6.[11] It can also be seen in Figure 1.6 that in some cases the droplet's tail has broken up into more than one droplet. From an RIJ point of view satellites are problematic if they do not re-combine with the main droplet either during flight or during impact.

Derby *et al.* also observed that *Z* has an influence on droplet volume too. As *Z* increases there is an increase in volume. However, the volume increase reaches a plateau once *Z* equals 10. Recently, Moon and co-workers have provided an additional refinement of what *Z* number means in terms of ink printability[12] by saying that if *Z* is between 4 and 14 then an ink is printable.

The picture is further complicated when one considers the work of Schubert *et al.* who found that a number of common solvents with low viscosities, ranging from 0.4 to 2 mPa.s, and surfaces tensions ranging from 23 to 73 mN/m were printable. The calculated *Z* numbers for these solvents varied from 21 to 91![13] In fact, they judged that the main factor affecting printability was vapour pressure. If values of vapour pressure were above 100 mm Hg then droplet formation was either unstable or non-existent.

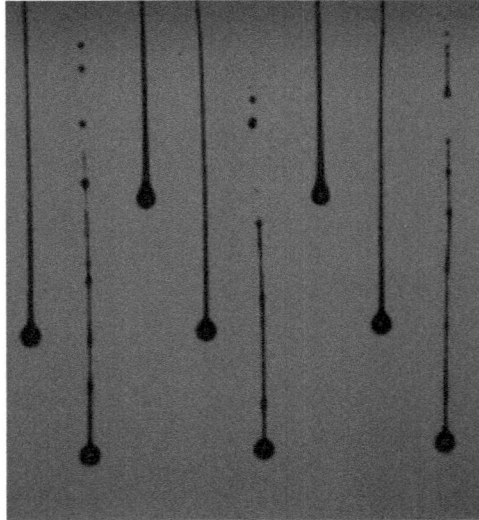

Figure 1.6 Jets formed from solutions of polymer in diethyl phthalate (0.4% monodisperse polystyrene with MW = 110 000) before and after detachment
from the nozzle plane. Reprinted from ref. 11 with permission of IS&T:
The Society for Imaging Science and Technology, sole copyright owners
of *NIP23: Twenty-third International Conference on Digital Printing Technologies and Digital Fabrication 2007*.

In his recent review of inkjet printing where he discussed fluid property
requirements, Derby identified where the region of printable fluid lies.
Although the preceding discussion has illustrated that there is a variety of
opinion as to what exact values of Z predict printability what can be said is
as follows. At low values of Z, viscosity dominates. If an ink or solution is too
viscous droplets cannot be ejected. At high values of Z satellites are, or may
be observed.

To summarise this section, the science behind inkjet printing regarding
droplet size and velocity is well understood. Key papers that have contributed
to this understanding are influenced, in piezo DoD printheads, by the voltage
applied to the piezoelectric actuator. Viscosity is the principal limiting factor
in terms of droplet ejection. If an ink is too viscous then the pressure waves
are nullified and the kinetic energy dissipated.

References

1. P. J. Smith and A. Morrin, Reactive Inkjet Printing, *J. Mater. Chem.*, 2012,
   22, 10965.
2. B. Derby, Inkjet Printing of Functional and Structural Materials: Fluid
   Property Requirements, Feature Stability, and Resolution, *Annu. Rev.
   Mater. Res.*, 2010, **40**, 395.

3. http://www.domino-printing.com/Global/en/Product-Range/Continuous-Inkjet/Continuous-Inkjet-Range.aspx, date accessed: 15th August 2016.

4. http://www.linxglobal.com/en-gb/technology-guide/cij, date accessed: 15th August 2016.

5. E. Tekin, P. J. Smith and U. S. Schubert, Inkjet printing as a deposition and patterning tool for polymers and inorganic particles, *Soft Matter*, 2008, **4**, 703.

6. J. F. Dijksman, Hydrodynamics of small tubular pumps, *J. Fluid Mech.*, 1984, **139**, 173.

7. J. E. Fromm, Numerical calculation of the fluid dynamics of drop-on-demand jets, *IBM J. Res. Dev.*, 1984, **28**, 322.

8. N. Reis, C. Ainsley and B. Derby, Ink-jet delivery of particle suspensions by piezoelectric droplet ejectors, *J. Appl. Phys.*, 2005, **97**, 094903.

9. D. B. Bogy and F. E. Talke, Experimental and theoretical study of wave propagation phenomena in drop-on-demand ink jet devices, *IBM J. Res. Dev.*, 1984, **28**, 314.

10. B. V. Antohe and D. B. Wallace, Acoustic Phenomena in a Demand-Mode Piezoelectric Ink-Jet Printer, *J. Imaging Sci. Technol.*, 2002, **46**, 409.

11. S. Hoath, G. Martin, R. Castrejón-Pita and I. Hutchings, Satellite formation in drop-on-demand printing of polymer solutions, *Proceedings of Digital Fabrication*, 2007, p. 331.

12. D. Jang, D. Kim and J. Moon, Influence of Fluid Physical Properties on Ink-Jet Printability, *Langmuir*, 2009, **25**, 2629.

13. B.-J. de Gans, E. Kazancioglu, W. Meyer and U. S. Schubert, Ink-jet Printing Polymers and Polymer Libraries Using Micropipettes, *Macromol. Rapid Commun.*, 2004, **25**, 292.

CHAPTER 2

From Inkjet Printed Droplets to Patterned Surfaces

JONATHAN STRINGER

Department of Mechanical Engineering, University of Auckland, Auckland, New Zealand
*E-mail: j.stringer@auckland.ac.nz

2.1 Introduction

In the vast majority of applications, inkjet printing is used as a means to selectively pattern a surface with a multitude of micron-sized droplets; with such an action typically performed so that material is transferred from the droplet to the surface in the desired pattern. Applications for this are numerous, from traditional graphical applications through to printed electronics and tissue engineering. For any of these applications, the success in how the desired pattern is accurately reproduced will come down to how the many individual droplets interact with the surface, neighbouring droplets and the environment around them.

This chapter will review the various physical processes that are encountered during the deposition of a droplet on a substrate, as well as the subsequent change of phase required to go from a liquid droplet to a solid deposit. While this chapter will not go into any specific issues associated with reactive inkjet printing, such as mixing of dissimilar droplets, the underlying mechanisms involved in droplet deposition remain, fundamentally, the same.

Smart Materials No. 32
Reactive Inkjet Printing: A Chemical Synthesis Tool
Edited by Patrick J. Smith and Aoife Morrin
© The Royal Society of Chemistry 2018
Published by the Royal Society of Chemistry, www.rsc.org

2.1.1 Dimensionless Numbers

As is apparent from the previous section, droplet impact and deposition on a surface is a complex phenomenon that is affected by a myriad of variables, such as the velocity of the droplet impacting the substrate (U_0), the size of the droplet (D_0), and the inherent properties of the fluid such as surface tension (σ), density (ρ) and dynamic viscosity (η). To help understand how each of the different variables influence the droplet deposition behaviour, it is appropriate to group the variables into a series of dimensionless numbers that represent the relative magnitudes of competing forces within the deposition process. Within this chapter, the following dimensionless number will be used:

The impact Reynolds number (Re) is representative of the ratio between inertial and viscous forces, and is defined as:

$$Re = \frac{\rho U_0 D_0}{\eta} \qquad (2.1)$$

The impact Weber number (We) is representative of the ratio between inertial and surface energies, and is defined as:

$$We = \frac{\rho U_0^2 D_0}{\sigma} \qquad (2.2)$$

The Bond number (Bo) is representative of the ratio between gravitational and surface energies, and is defined as:

$$Bo = \frac{\rho g D_0^2}{\sigma} \qquad (2.3)$$

where g is the acceleration due to gravity. The impact capillary number (Ca) is representative of the ratio between viscous and surface energies, and is defined as:

$$Ca = \frac{\rho U_0}{\sigma} \qquad (2.4)$$

The Ohnesorge number (Oh) is another dimensionless number that shows the ratio of viscous forces to inertia and surface forces independent of impact velocity, and is defined as:

$$Oh = \frac{\eta}{\sqrt{\rho \sigma D_0}} \qquad (2.5)$$

2.1.2 The Lifetime of a Deposited Droplet

The evolution of a droplet upon impact with a substrate is a subject that has been the subject of scientific interest and investigation for hundreds of years,[1,2] and such research has found application in fields such as aerospace coatings,[3] sprays for cooling,[4] self-cleaning of superhydrophobic structures,[5] all in addition to inkjet printing itself.[6,7] While both the applications and, more importantly in this chapter, the exact conditions of droplet impact can differ markedly between all the studies that has been carried out, each impact will share a broadly similar lifetime between initial impingement on the surface and the droplet (or deposit) reaching equilibrium with the environment. More detailed discussion of each stage will be given in the bulk of this chapter, but a brief overview and schematic diagram (Figure 2.1) will be given herein.

The first stage of the deposition process is known as the impact driven phase, in which the initial kinetic energy of the impinging droplet is dissipated (Figure 2.1, point 1). This stage will either lead to the droplet being maintained as a single body of fluid attached to the substrate (stable impact) or the droplet either detaches from the substrate or parts of the droplet are ejected from the main fluid body (unstable impact). If the impact is stable, the droplet will reach a maximum diameter (Figure 2.1, point 2). This stage of spreading typically occurs over an order of magnitude timescale of D_0/U_0, although can be far longer if there is significant recoiling of the droplet as a means to dissipate this energy (Figure 2.1, point 3).

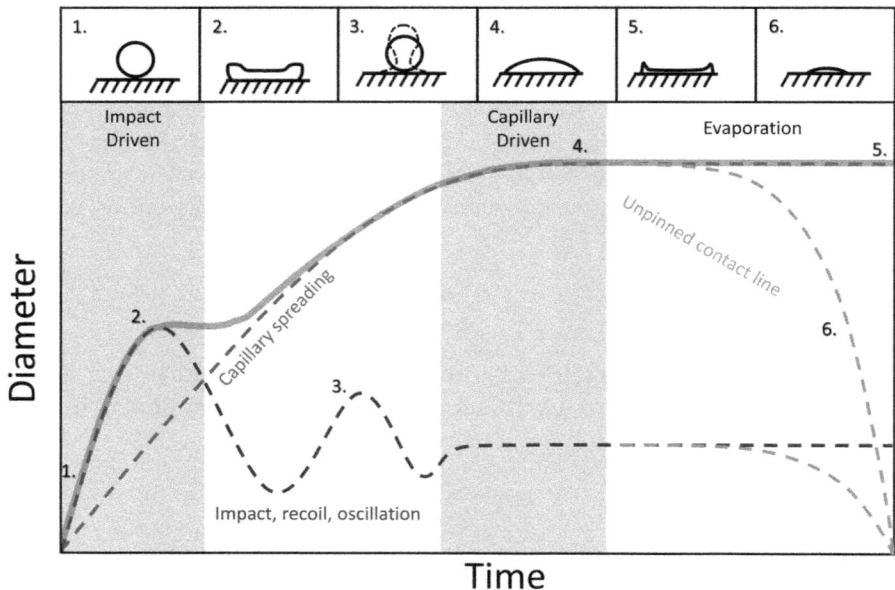

Figure 2.1 Evolution of droplet diameter as a function of time over the droplet deposition process. Typical spreading behaviour of a droplet during inkjet printing is represented by the solid line. Typical droplet and deposit morphologies observed during the deposition process are given at different points (1–6). Adapted from ref. 8.

The second stage of spreading is driven by the droplet and the substrate reaching a surface energy equilibrium, with the extent of spreading being dependent upon the surface energy interactions between the droplet, substrate and the environment. The speed with which spreading takes place is dependent upon the energetic driving force for spreading and the amount of spreading necessary to reach equilibrium, but will typically be orders of magnitude longer than impact driven spreading. Typically, this will lead to the droplet forming a spherical cap on the substrate (Figure 2.1, point 4).[9]

The other factor typically of importance for inkjet printing is how the droplet changes phase from a liquid to a solid. This can be by a number of different routes, such as evaporation, gelation, light-induced polymerisation and solidification, with the timescale over which this acts being dependent upon the nature of the phase change. The most common phase change found in inkjet printing is evaporation, and this is of particular relevance for reactive inkjet printing, as the reactants are typically carried by a solvent of some kind.

As evaporation proceeds, the droplet will reduce in volume; whether this leads to a reduction in the contact diameter of the droplet will be dependent upon whether the contact line of the droplet with the substrate is free to move, or whether it is pinned. If the contact line is pinned, the droplet evaporates with a constant contact area, leaving a deposit of the same size (Figure 2.1, point 5). If there is no pinning of the contact line, the contact area gradually gets smaller (Figure 2.1, point 6), and if it remains unpinned the droplet will eventually completely evaporate to leave no deposit. In the case of inkjet printing, in particular reactive inkjet printing, it is clearly preferable that a deposit of known size is produced, and as such a pinned contact line is preferred and, fortunately, far more likely to occur in most real-world ink systems.[10]

2.2 Droplet Spreading on a Surface

2.2.1 Impact-driven Spreading

It has been mentioned previously that studies into droplet impact have previously been performed for a wide range of applications and over a wide range of length and time scales. To better understand how the disparate studies interrelate, it is useful to construct a map based on the dimensionless parameters defined previously. This was first carried out by Schiaffino and Sonin,[11] who constructed a map based upon We (dependent upon impact velocity) and Oh (independent of impact velocity), a recreation of which is shown in Figure 2.2. Schiaffino and Sonin identified four primary regions of interest starting at the extremities of both axes, and deduced boundaries between them based on order of magnitude estimates of the balances between the different variables that influence flow. While the nature of the deposition map means that the boundary between the four different regions are defined as sharp transitions, they are in reality more gradual. This is due to both the nature of an order of magnitude analysis and due to the assumption that

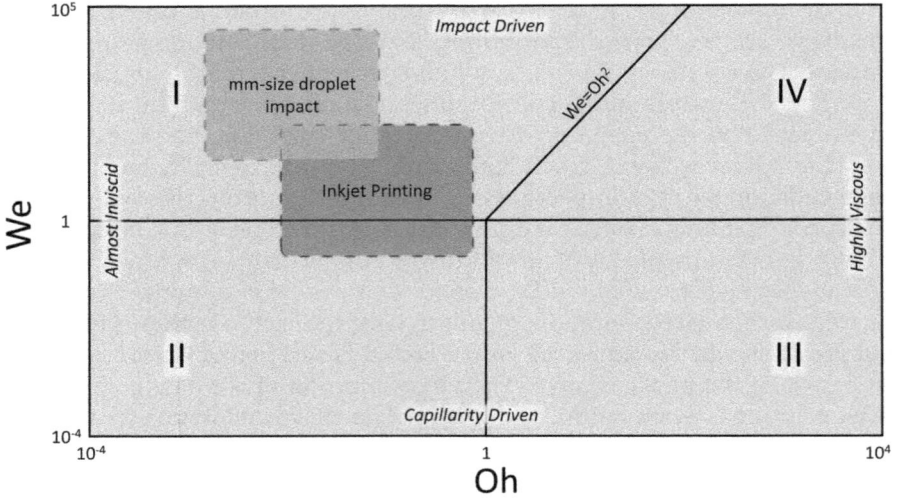

Figure 2.2   Deposition map based upon that initially constructed by Schiaffino and Sonin.[11] Superimposed upon the deposition map are regions representing the deposition regime for inkjet printing and for mm-size droplet impact experiments most commonly performed in the literature.

the initial impact conditions (which the dimensionless terms capture) are representative of the droplet impact behaviour throughout the deposition process. This is of particular significance for inkjet printing, as the region of interest potentially straddles regions 1 and 2, and is quite close to regions 3 and 4; this is in contrast to previous work with mm-size droplets that lies definitively in region 1. It is therefore apparent that previously derived models that predict the behaviour of mm-size droplets do not necessarily correspond with the regime experienced by an inkjet droplet, and should therefore be used with due caution.

2.2.1.1   Predicting Impact-driven Spreading

During a stable impact, the droplet deforms upon impact but stays as a single body of fluid attached to the substrate. If it is assumed that the impact energy is dissipated by deforming the droplet upon impact, a generic energy balance upon these assumptions can be constructed as follows:

$$E_{K0} + E_{S0} = E_{K1} + E_{S1} + E_{VD} \tag{2.6}$$

where E_{K0} and E_{S0} are the kinetic and surface energies at the point of impact, E_{K1} and E_{S1} are the kinetic and surface energies at the point in time after impact, and E_{VD} is the energy consumed by viscous dissipation.

The generic energy balance given in eqn (2.6) is most frequently used to predict the maximum extent to which impact drives droplet spreading over the substrate, referred to as β_{max}. The common assumption when doing so is that this will coincide with all the impact kinetic energy being converted into other forms and the droplet therefore having no kinetic energy (*i.e.* $E_{K1} = 0$).

If the substrate is both flat and homogeneous, the impact is orthogonal to the substrate, and the droplet is considered spherical, it is reasonable to assume that the shape formed by the impact droplet is axisymmetric around the orthogonal axis. Based upon these assumptions, it is possible to derive two extreme cases based upon regions 1 and 4 in Figure 2.2 that dictate how the impact energy is dissipated. In addition to the assumptions above, it is also necessary to assume that gravity does not play a significant role in the impact process. For inkjet printed droplets, this is reasonable due to the small size of the droplets, as shown by Bo $\ll$ 1.

In the case of region 1, the fluid is considered almost inviscid and therefore energy losses due to viscous dissipation can be ignored ($E_{VD} = 0$). This means that at the point of maximum extension upon the substrate, the kinetic energy of impact must be converted into surface energy. Using the assumptions discussed previously, this leads to the following equation for β_{max}, similar to that initially derived by Collings *et al.*:[12]

$$\beta_{max} = \sqrt[2]{\frac{We + 12}{3\left(1 - \cos\theta_{eqm}\right)}} \qquad (2.7)$$

In the case of region 4, the highly viscous nature of the fluid means that the energy change due to the change in surface energy is negligible compared to the that lost through viscous dissipation (*i.e.* $E_{S0} = E_{S1} = 0$). A generalised expression for β_{max}, similar to that derived by Madejski[3] can be derived:

$$\beta_{max} \sim Re^{1/5} \qquad (2.8)$$

The expressions eqn (2.7) and (2.8) represent extreme cases of droplet impact, with their validity in real world situations being limited to droplet impacts at the extremes of experimental conditions. It is therefore unsurprising that a large number of studies have looked at refining these models to make them more applicable to real world situations. While some of these revised derivations are of particular relevance to inkjet printing, a large number have a different focus and will therefore not be discussed in this chapter. The reader is referred to the following review article for a more in-depth discussion of impact-driven spreading more generally.[13]

2.2.1.2 Impact-driven Spreading of Inkjet Droplets

As previously discussed, droplet impact during inkjet printing may lie within all four impact regimes (Figure 2.2). It is therefore instructive to look at the experimental studies that have looked at droplet impact to verify what the salient factors that affect how the droplet spreads are. The number of experimental studies, however, are not as numerous as perhaps would be desired.

This is due to the high degree of precision required to obtain quantitative information about the droplet impact, primarily the ability to obtain a series of sharp images of micron-sized droplets travelling at 1–10 ms^{-1}. This requires sub-µs light pulses of sufficient intensity to fully expose the microscopic droplet. This has limited the number of studies, and the studies that have been possible will now be discussed further.

The first extensive study that looked at inkjet printed droplet impact was conducted by van Dam and Le Clerc.[6] They looked at how individual water droplets impact a smooth glass surface, and compared these results to a number of previously derived models of spreading. An example of the images obtained are given in Figure 2.3. In comparison with previously derived models of droplet deposition for mm-size droplets given in eqn (2.7) and (2.8), as well as one derived by Pasandideh-Fard *et al.*,[4] which is similar to eqn (2.7) but allows for viscous dissipation. It was found that there was poor agreement between experiment and models, and this was mostly attributed to the models not taking into account the dissipation of energy during oscillation of the droplet surface. Dong *et al.*[14] saw similar oscillations upon impacts with surfaces with higher contact angles, but found that lower contact angle surfaces did not exhibit oscillations upon droplet impact.

Work by Son *et al.*[15] used an adaptation of the Weber number (We') that accounted for surface energy interactions with regards to both shape and spreading of the droplet and found that this adapted Weber number had an empirical correlation with β_{max} as given by:

$$\ln \beta_{\mathrm{max}} = 0.090 \ln \mathrm{We}' + 0.151 \qquad (2.9)$$

While showing relatively good agreement, the use of an empirical correlation is not ideal from a theoretical viewpoint. Further work by Dong *et al.*[16] showed that two developed models, those of Roisman *et al.*[17] and Park *et al.*[18] showed good agreement with experimental data; this was also found by Jung and Hutchings.[19]

The adaptations to the assumptions used in these models meant that the shape, contact angle dynamics and viscous dissipation were more appropriate to the inkjet printing regime. The assumptions inherent within any of the analytical models will mean that there is bound to be a level of imprecision. This is exacerbated when there is increased complexity within the fluid, such as polymer additives that give a non-Newtonian viscosity or suspended particulates.

One of the primary assumptions made is that the impact is stable as defined earlier. If this condition is not met, the impact can be deemed as unstable, resulting in either complete recoil and bouncing[5] of the droplet upon impact or separation of fluid from the main droplet, referred to as splashing. While splashing is not generally considered an issue in inkjet printing due to the small size of the droplets (and therefore lower kinetic energy in comparison to surface energy), in the area of reactive inkjet printing it has the potential

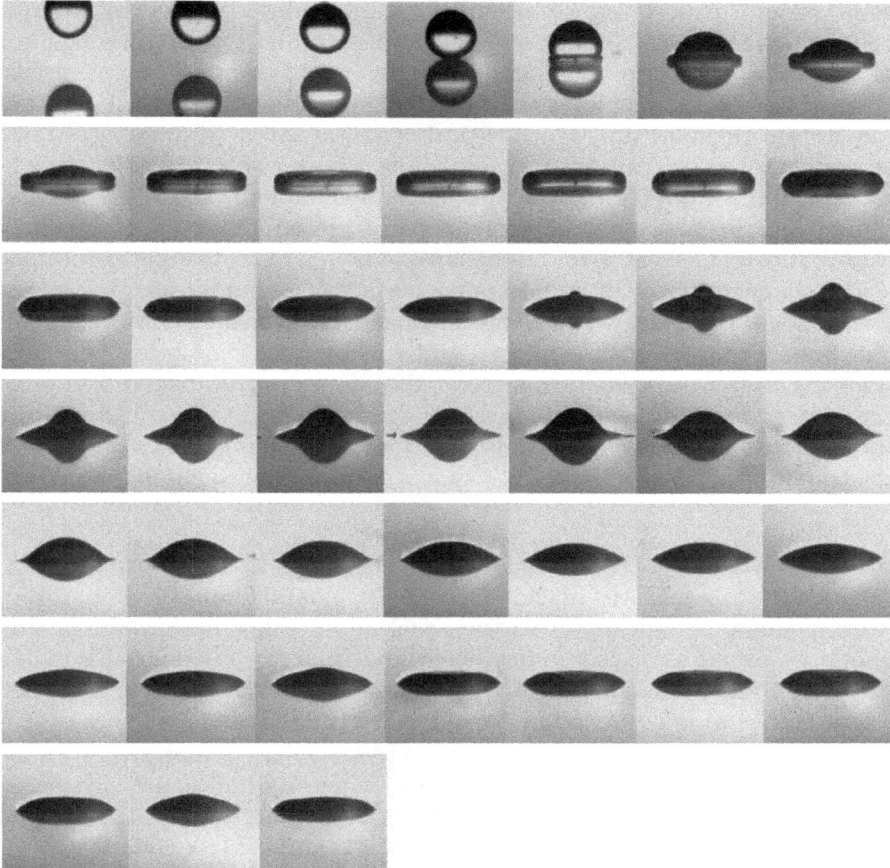

Figure 2.3 Sequential images of an 85 μm water droplet impacting onto a smooth glass substrate at 5.1 ms⁻¹. Each image is taken with a time delay of 3 μs. Reprinted from D. B. van Dam and C. Le Clerc, Experimental study of the impact of an ink-jet printed droplet on a solid substrate, *Physics of Fluids*, 2004. **16**(9), 3403–3414, with the permission of AIP Publishing.[6]

to be an issue. This is due to some of the parameters that have been found to increase the likelihood of an unstable impact, primarily a rough surface and a pre-existing film of liquid upon the surface.

2.2.2 Surface Energy-driven Spreading

As was shown previously, the dissipation of impact kinetic energy is not sufficient to explain the final spreading behaviour of an inkjet printed droplet upon the substrate. This was perhaps shown most clearly in the work by van Dam and Le Clerc, where the final diameter of a deposit made from a silver colloidal solution was largely invariant to impact conditions.[6] The driving force behind the spreading to the final diameter is in fact minimisation of surface energy between the droplet, substrate, and environment, referred to as capillary-driven spreading in Figure 2.1. During the transition from impact-driven spreading to capillary driven spreading, it is necessary for the contact line to move, with this motion dissipating energy. This section will give some background into the theory behind surface energy equilibrium and the contact line, with particular focus on elements of substantial relevance to inkjet printing.

The shape of a body of liquid upon a surface will tend towards the lowest energy state. This state will be determined solely by a balance of surface energies between the droplet, substrate, and environment if no other forces act upon the body in a significant manner or there is any impediment to motion of the contact line between the phases (see Figure 2.4). Following these assumptions, it is possible to derive a simple balance of these forces as a function of the equilibrium contact angle, θ_{eqm}, known as the Young equation:[1]

$$\sigma_{\text{sv}} = \sigma_{\text{ls}} + \sigma_{\text{lv}} \cos \theta_{\text{eqm}} \qquad (2.10)$$

Under the assumptions made, this equation is valid; however, the assumptions made do not always reflect the situation in reality. Besides the kinetic energy of impact as discussed previously, the other force that is liable to disturb this contact line equilibrium and distort the surface is that of gravity. The extent to which this has an impact is determined by the balance of surface and gravitational forces acting upon the droplet, as represented by Bo (eqn (2.3)). If Bo > 1, this indicates that gravitational forces are significant and this will alter the surface curvature of the droplet. In this situation, the surface curvature can be calculated using the Young-Laplace equation:[20]

$$\Delta P_0 + \Delta \rho g z = \sigma_{\text{lv}} \left(\frac{1}{R_1} + \frac{1}{R_2} \right) \qquad (2.11)$$

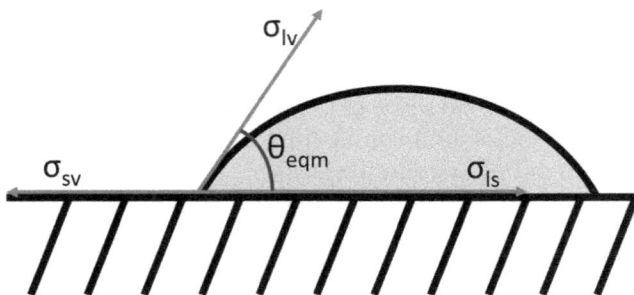

Figure 2.4 Schematic diagram showing a droplet at equilibrium on a surface assuming no external forces or constraints to contact line motion. The constituent surface energies and contact angle are also marked.

where P_0 is the pressure at a fixed reference point, $\Delta\rho$ is the density difference between the liquid and the surrounding vapour, z is the distance from a fixed reference point and R_1 and R_2 are the two principal radii of curvature of the liquid. In the case of inkjet printing, it is found that Bo $\ll 1$, therefore there is negligible influence of gravity and at rest a droplet will therefore form a shape dictated solely by surface energy minimisation. By assuming that this shape is a spherical cap, it is possible to construct a volume conservation model that will give the ratio of initial and spread diameters solely as a function of contact angle, β_{eqm}:[6]

$$\beta_{\text{eqm}} = \sqrt[3]{\frac{8}{\tan{\theta_{\text{eqm}}}/_2 \left(3 + \tan^2{\theta_{\text{eqm}}}/_2\right)}} \qquad (2.12)$$

In a similar manner to how eqn (2.7) and (2.8) can be thought of as extreme examples of group 1 and group 4 droplet depositions (Figure 2.2), this equation can be thought of as an extreme example for cases 2 and 3, as the impact energy is considered irrelevant to the spreading process. Due to the lack of inertia-driven flow, this process typically takes place over significantly longer timescales, and corresponds to inset 4 in Figure 2.1. This was observed experimentally by Jung and Hutchings,[19] with the spreading ratio over large timescales being dictated by the respective surface energies of the droplet and substrate.

The assumption of a flat homogenous substrate is not always appropriate, in the case of reactive inkjet printing this is likely to occur due to previous deposition of one of the reactive inks. For a rough surface, there are two possible outcomes, either the contact line penetrates each asperity of the rough surface (Wenzel state),[21] or the asperities are not penetrated and the contact line bridges between asperities (Cassie–Baxter state).[22] The Wenzel state effectively increases the length of the contact line, and leads to the following variation of eqn (2.10):

$$\sigma_{\text{LV}} \cos \theta_{\text{W}} = r_{\text{W}}(\sigma_{\text{SV}} - \sigma_{\text{LS}}) \qquad (2.13)$$

where θ_{W} is the apparent equilibrium contact angle observed in a Wenzel state and r_{W} is the ratio between actual surface length and projected surface length at the contact line. The term apparent equilibrium contact angle is used as the equilibrium contact angle is a thermodynamic property of the interaction at the contact line. The Cassie–Baxter state was initially derived for wetting on surfaces with heterogeneity of surface energy, and takes the form:

$$\cos \theta_{\text{CB}} = f_1 \cos \theta_{\text{e1}} + f_2 \cos \theta_{\text{e2}} \qquad (2.14)$$

where θ_{CB} is the apparent equilibrium contact angle observed in the Cassie–Baxter state, f_1 and f_2 are the fraction of components 1 and 2 at the contact line, and θ_{e1} and θ_{e2} are the equilibrium contact angles for component 1 and 2 respectively. By treating one of the two components as air (*i.e.* θ_{e2} = 180°), this can be used in situations where the liquid does not penetrate the asperities of the rough surface. By maximising the contact made with the gaseous phase (and thereby minimising the contact with the solid substrate) it is possible to make a superhydrophobic surface exhibiting a contact angle that approaches 180°. Such a structure has been found to occur naturally in foliage such as lotus leaves, and has also been made by various synthetic routes.[23]

As well as affecting the apparent equilibrium contact angle, roughness and surface energy heterogeneities can also be obstacles to contact line motion. This was elegantly demonstrated by Meiron *et al.*[24] by measuring the contact angle of a droplet on a rough surface both before and after supply of additional energy to overcome an energetic barrier to movement. It was found that the additional energy enabled the droplet to spread and form the typical spherical cap geometry, indicating that the droplet had previously been in a metastable state. This indicates that the assumption of a freely moving contact line is not always the case. The lack of free movement means that rather than there being one contact angle that is always observed, there are in fact a range of contact angles that a droplet can have without motion of the contact line. Overcoming the energetic barrier requires additional driving force for contact line motion, it is therefore found that for the contact line to advance across the solid surface a contact angle greater than equilibrium is required, and for the contact line to recede from the solid surface back into the liquid a contact angle less than equilibrium is required. These two angles are most commonly referred to as the advancing and receding contact angles, respectively, and the difference between the two is referred to as the contact angle hysteresis.[25]

One of the most relevant mechanisms that generates an energetic barrier and leads to contact angle hysteresis in inkjet printing is the deposition of solid material from within the ink at the contact line. As will be discussed later, this is commonplace with droplets containing any solid material and is referred to as contact line pinning.[10] As the evaporation of a droplet proceeds, both the driving force for contact line recession and the energetic barrier increase. If the energy barrier is always maintained to be greater than the driving force for recession, the contact angle will be pinned throughout the entirety of droplet drying. The receding contact angle is therefore effectively zero, and it means that the contact area between the droplet and substrate (and therefore deposit diameter) is constant throughout drying. This has been verified experimentally to commonly occur with inkjet printing by comparing the deposit diameter to that predicted using eqn (2.12) and variations thereof, with good agreement found.[6,9,26]

2.3 Phase Change

For there to be a permanent pattern formed by the inkjet printing process, it is at some point necessary for the liquid ink to change phase and leave a solid deposit. The mechanism by which this phase change occurs, and the timescale over which it occurs, will determine to what extent this phase change controls how the final solid deposit is formed. These phase changes include melt solidification, gelation, photopolymerisation, and evaporation.

For very rapid phase changes, such as solidification on an undercooled substrate, this phase change can disrupt the spreading processes discussed above, leading to a non-equilibrium deposit shape. This can take the form of either a series of oscillations on the surface that correspond to the oscillations of the droplet, or a temperature controlled deposit diameter.[11] Such behaviour has also been observed with a gelation phase change ink,[27] and an evaporating ink.[28] Due to the potential thermal fluctuations during the phase change and spreading process, as well as more mundane issues such as elevated temperatures possibly leading to nozzle clogging, there is limited scope for this to be used as a method to control feature size.

Perhaps the most significant phase change mechanism, in particular for reactive inkjet printing, is that of evaporation. Most reactive inkjet printing ink systems are likely to consist of two or more reactants in separate carrier liquids, with the carrier liquids forming no substantive part within the reaction beyond being a medium in which the reaction can take place. Evaporation is also the most prevalent phase change encountered within inkjet printing more generally, and it is therefore instructive to understand the evaporation process of a droplet in more detail.

2.3.1 Evaporation

On a molecular level, evaporation is a process whereby a molecule in the liquid state has sufficient energy to overcome any cohesive forces within the liquid body to escape across the liquid vapour interface and enter the gaseous state. After evaporation of the molecule, the average kinetic energy of the liquid body will be reduced, meaning that evaporation is inherently an endothermic process. For the molecule to leave the liquid phase, it is necessary for it to traverse the interface between the liquid and the surrounding vapour; the driving force for which will be governed by the concentration gradient of molecules between the liquid and vapour phases. Taken together, this means that the net flux of molecules between vapour and liquid, or evaporation rate, is dependent upon the surface area over which evaporation occurs, the temperature, and the local vapour concentration.

As previously discussed, a droplet at equilibrium on a substrate when Bo $\ll$ 1 will form a spherical cap, the shape and surface of which is controlled by the contact angle. As evaporation from this surface will result in a reduction in volume of the droplet over time, conceptually it is worth considering whether this spherical cap shape will be maintained throughout the evaporation process.

For this to be the case, it is necessary that surface tension forces are greater than viscous forces, so that the droplet shape can successfully relax to the equilibrium shape. This is shown using a variation on the impact capillary number (eqn (2.4)), with the impact velocity being replaced by a velocity characteristic of the evaporation process. While this value is likely to change during the evaporation process in all but pure single solvent systems, typical inks will have a value significantly less than 1, indicating that a spherical cap is maintained through the vast majority of the evaporation process.

To maintain a spherical cap geometry it is possible for the droplet to either maintain a constant contact angle and reduce contact area, or maintain a constant contact area and reduce contact angle.[29] The prevalence for which of the two modes will occur is determined by the amount of contact angle hysteresis between the droplet and the substrate: for a droplet and substrate that exhibit little or no hysteresis the constant contact angle mode will be observed, for a droplet and substrate with a zero receding contact angle the constant contact area mode will be observed, and for intermediate contact angle hysteresis a mixed mode stick-slip recession of the contact line is often observed.[30] While it is possible to observe this behaviour in pure solvent droplets, it is perhaps most easily observed in dilute particle suspension. In this case, small deposits of material at the contact line act as an energetic barrier that must be overcome before the contact line can recede. When the energy barrier is overcome, the contact line is able to recede significantly, until further accumulation at the new contact line acts as a further barrier. This typically exhibits itself as a series of concentric deposits within the footprint of the droplet upon complete evaporation.

As most inkjet printed liquids aim to deposit a solid phase, it is likely that there will be some kind of accumulation of material at the contact line during the evaporation process. If this accumulation causes the energetic barrier to recession to rise more quickly than the driving force due to reduction in contact angle, the contact line is effectively pinned throughout the entire evaporation process; thus the condition of a zero receding contact angle is effectively achieved. As discussed earlier, it is common to observe such behaviour in inkjet printed inks, as shown by the good agreement between deposit size and the spherical cap model of droplet spreading. In addition to even very small solid particle additions leading to a zero receding contact angle, it is also possible to observe such behaviour due to contamination of the substrate and surface roughness.[31]

2.3.1.1   *Solute Segregation*

Over the surface of a droplet, there is a variation in the ratio of surface area to volume directly beneath that surface, with this ratio being lowest in the centre and highest at the periphery of the droplet. As evaporation, all else being equal, is proportional to the surface area over which it takes place, it is therefore a reasonable assumption that over a given timeframe a greater proportion of liquid will be lost to evaporation at the edge of

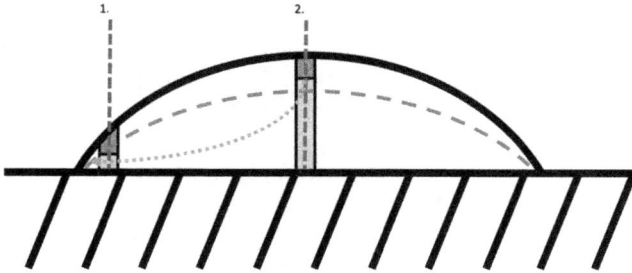

Figure 2.5   Schematic diagram demonstrating how the differential relative evaporation rate over the surface at the droplet generates outward flow towards the periphery.

the droplet when compared to the centre (see Figure 2.5). As has been previously stated, for inkjet printed droplets Ca ≪ 1. This means that to maintain the spherical cap shape, the liquid within the droplet must redistribute from the region where there is 'excess' liquid (the centre, point 2 in Figure 2.5) to the 'deficient' region of the droplet (the edge, point 1 in Figure 2.5). This redistribution of liquid means that there is a flow from the centre of the droplet towards the periphery. If there is a solid phase within the liquid, whether that be a solution or a particulate suspension, this flow will tend to carry material towards the edge of the droplet and preferentially deposit there. This segregation of solute towards the edge of the droplet is commonly referred to as a 'coffee stain' due to the resemblance to the coffee remnants left by a drying coffee droplet, which has the same physical basis.[10] The shape of the liquid vapour interface also contributes to an increased evaporation rate at the edges. Upon escaping from the liquid body and entering the vapour phase, a molecule at the periphery of the droplet undergoing Brownian motion is statistically less likely to re-enter the body of the liquid in comparison to a molecule leaving the droplet centre.[31]

In addition to the flow during evaporation inherent to the droplet geometry, additional flows may also act upon the droplet during evaporation. These flows, referred to as Marangoni flows, originate due to surface tension gradients upon the droplet surface, with flow occurring from low surface tension regions to high surface tension regions. The origins of a surface tension gradient are typically either due to variations in temperature upon the surface[32] or variations in composition or surfactant concentration.[33,34]

For thermal Marangoni flow to occur, it is necessary for there to be a temperature gradient to be present across the surface of the droplet. Hu and Larson[35] calculated using a finite element method that a temperature gradient develops over the droplet surface during evaporation, with the region at the periphery having a higher temperature (and therefore lower surface tension) than at the centre of the droplet. This leads to an inward flow that it has been shown can counteract the outward flow caused by replenishment of evaporative losses.[32] The direction of the temperature gradient modelled is due to the shorter conductive pathway between the droplet surface and the substrate at the edge of the droplet when compared to that at the centre. This was investigated in more detail by Ristenpart *et al.*,[36] who found that the ratio of liquid and substrate thermal conductivity had a critical influence on the direction of the Marangoni flow. If the thermal conductivity of the substrate is over twice that of the liquid then the Marangoni flow will be inward, as modelled and observed by Hu and Larson. If the substrate thermal conductivity is less than 1.45 times that of the liquid, then the substrate is unable to effectively equilibrate the temperature of the droplet and the extra evaporative cooling at the edge of the droplet will lead to a reversal in surface tension gradient and the Marangoni flow being outward.

For compositional based Marangoni flow, the gradient in surface tension is achieved by either the use of a binary solvent mixture or the addition of surfactant. In a binary solvent mixture, a miscible solvent of different surface tension and volatility is used. As one of the components more readily evaporates, and evaporation is greater at the edge of the droplet, the binary solvent mixture will become depleted of the volatile component as evaporation proceeds, with this exacerbated at the edge of the droplet. This variation in composition means that there is also a variation in surface tension across the droplet surface, which generates a Marangoni flow towards the high surface tension region.

Due to the interaction between the geometry induced flow and Marangoni flow, there are a number of varying morphologies that can be observed besides the classical coffee stain. The other extreme to the coffee stain is that the outward flow is completely inverted due to Marangoni effects, more complex interactions between the solvent and solute,[37] or localised heating.[38] This leads to a concentration of material in the centre of the droplet surrounded by a thin film, and is often referred to as a 'sombrero' morphology. Observed deposit morphologies will typically fall somewhere between these two extremes, with their character dependent upon the nature of the generated flow field.

While some applications have been found that exploit the coffee stain morphology, such as alignment of nanotubes,[39] close packing of metallic nanoparticles,[40] and stretching of DNA,[41] the morphology typically most desired for inkjet printing is a flat film. Soltman and Subramanian[42] demonstrated that a flat film could be achieved, and coffee staining eliminated, by lowering the temperature of the substrate. It was reasoned that due to the shorter distance between surface and substrate at the edge of the droplet, the droplet edge will be significantly cooled and therefore the evaporation rate lowered sufficiently to eliminate the need for an outward flow. Conversely, Wang *et al.*[43] found that raising the temperature of the substrate helped to eliminate coffee staining for a carbon nanotube-based ink droplet. This was hypothesised to be due to the increased inward Marangoni flow, as discussed earlier.

Besides use of varying substrate temperature, one of the most used methods to obtain a uniform deposit is the use of additional solvents and surfactants to the ink. Park and Moon[44] first demonstrated this using addition of a humectant (either formamide or dielthylene glycol) to an aqueous silica suspension. Tekin *et al.*[45] demonstrated similar behaviour with a polymer solution. In both cases this is attributable to compositionally derived surface tension gradients as described above.

2.4 Interaction of Multiple Droplets

While an understanding of how an individual printed droplet impacts, spreads, and changes phase upon a substrate is of fundamental importance, in most cases of interest to reactive inkjet printing, it is not all that takes place in achieving the desired printed pattern. For an understanding of how individual droplets proceed to form a pattern on a substrate, it is necessary to understand how each of these droplets interact with each other in the liquid state.

2.4.1 Coalescence

Coalescence is the process by which two or more droplets of liquid will join together to form a single larger body of liquid. The driving force for this is primarily minimisation of surface energy, as the larger body of liquid will have a lower surface energy to volume ratio than the constituent droplets. In inkjet printing, it is possible that a large number of droplets will coalesce within a given pattern, it is therefore instructive to understand how coalescence proceeds and the timescale over which it takes place.

In the case of binary droplet coalescence, the mechanism can be divided into three distinct stages. In the first stage, known as bridge formation and broadening, two droplets come into contact with each other and form a liquid bridge between them that becomes wider over time. In the second stage, known as relaxation, the formed neck reduces curvature and the contact line of the formed single liquid droplet moves. In the third stage, here known as equilibration, the surface profile relaxes to the minimal energy state, typically a truncated spherical cap geometry. While similar mechanisms occur whether coalescence is between sessile droplets on a substrate, or spherical droplets in free space, the discussion contained herein will focus on the coalescence of sessile droplets.

Upon initial contact between two sessile droplets, a bridge between the two droplets is nucleated. This bridge can be formed either due to an energetic impulse (such as impact kinetic energy), but will form without such impulse due to the van der Waals force between the neighbouring droplets when in close enough proximity. The formation of the neck region creates a small region in which there is a large change in curvature. As previously discussed for surface energy-driven spreading, this dramatic change in surface curvature leads to a high surface stress (eqn (2.11)), the lowering of which acts as the driving force for further broadening of the bridge.

Upon initial formation of the liquid bridge, the liquid bridge is very small, typically on the nanometre scale for most common liquids. Due to this small size, Re is significantly below 1, and the flow is dominated by the balance between the capillary forces driving the coalescence and the viscous forces that impede it. This type of flow will continue at a flow velocity equal to σ/η until the bridge has broadened to a point where Re = 1.[46] Observation of this period of bridge formation is understandably challenging, due to the short time and length scales over which it occurs. Efforts to observe it have therefore focused on trying to slow it down by either using highly viscous fluids[47] or lowering the surface tension experienced.[48] It was found that by using phase separated binary colloidal suspension with minimal effective surface tension gradients, the growth rate was found to be constant with time, which agrees with models previously proposed.[49]

After this period of bridge broadening, the nature of the flow within the bridge changes and can be assumed to be inviscid. During this period it is found that the bridge width grows in a manner proportional to the square root of the time since bridge formation. This square root dependence was initially derived for spherical droplets in contact,[50] and similar power law relationships have since been derived for sessile droplets.[51] Experimental observations with differing fluids such as mercury,[52] water and silicone oil,[48] water,[53] and silicone oil[54] have been found to agree well with this power law relationship, with exponents found between 0.41 and 0.57.

Once bridge formation is complete, there is a driving force for the single body of coalesced liquid to minimise surface energy. In the case of a sessile droplet on a substrate, this will mean that the droplet will tend towards a spherical cap geometry, which requires the contact line to either advance or retract. At this point, it should be expected that the factors influencing contact line motion for single droplets also influence coalescence, such as the magnitude of the driving force, the topology of the substrate, and contact line hysteresis and pinning. This is indeed found to be the case.[55,56] One notable factor that influences this period of relaxation and equilibration is the vapour pressure in the surrounding atmosphere, with a higher vapour pressure slowing down any reconfiguration of the contact line. This is most likely due to the need for evaporation to take place to overcome any hysteresis to contact line retraction.

Of perhaps greater interest for inkjet printing is whether this period of relaxation, and associated contact line motion, is influenced by the presence of additional energy as would be the case of sequentially deposited droplets. Initial analysis of this form of coalescence were carried out with mm-size droplets travelling at a high velocity; this meant that impact, unlike inkjet printed droplets, were primarily dominated by the inertia of the droplets and were therefore concerned with a splashing instability.[57] Subsequent studies have looked at mm-size droplet impact in impact regimes that more closely match that experienced by inkjet printing.[58] It was found that a retraction of the contact line (known as drawback) was possible, with there being a greater propensity for the second deposited droplet to drawback towards the first droplet. This was attributed to local changes in the dynamic contact angle brought about by the oscillations upon impact of the droplet. The initial coalescence of inkjet printed droplets pairs has also been studied.[59] This work showed that not only does drawback occur with inkjet printed droplets, it also demonstrated that this drawback was exacerbated as further offset droplets were printed.

2.4.2 Bead Formation

While the coalescence of binary droplets is of academic interest, the need to understand the coalescence process is of greater interest to applications that require more complex patterns of droplets to coalesce. Perhaps the simplest of these shapes, both geometrically and in terms of the underlying physics that control it, is that of a liquid bead. This structure is of fundamental importance to a number of potential applications for inkjet printing, such as a conductive trace in printed electronics, and the ability to reliably form such a structure is critical for the successful function of any fabricated structures; with the example of a conductive trace, a gap in the track means that it ceases to function as a continuous conductive trace.

2.4.2.1 Stable Bead Formation

Perhaps the ideal shape of a linear feature formed by a bead is that of a parallel sided rectangle of known width. As previously discussed with regards to single droplets, surface energy minimisation means that a body of liquid will tend towards a spherical geometry if unconstrained, and when the liquid is on a surface this results in a spherical cap. It would therefore be expected that the creation of a liquid bead by deposition of multiple offset droplets would be inherently unstable if the contact line is free to move. This is analogous to the classical Rayleigh instability that is the basis for continuous inkjet printing,[60] and has also shown to be the case if the contact line is free to move over the surface both theoretically[61,62] and experimentally.[63] As discussed previously in relation to solute segregation, however, fluids relevant to inkjet printing do not often have freely movable contact lines, with there being some degree of contact angle hysteresis or pinning observed. This pinning means that the liquid bead formed by the coalescence of droplets can be maintained over the whole drying period of the bead, resulting in a rectangular shaped deposit with parallel sides.

Based upon the assumption of the pinned contact line in conjunction with volume conservation, it is possible to construct a simple geometrical model that relates the spacing and size of the deposited drop to the bead formed upon the substrate (Figure 2.6):[64]

$$\frac{\pi D_0^3}{3p} = \frac{w^2}{2}\left(\frac{\theta}{\sin^2\theta} - \frac{\cos\theta}{\sin\theta}\right) \tag{2.15}$$

where w is the width of the bead and p is the spacing between each deposited droplet. From this volume balance (eqn (2.10)) it is possible to obtain a prediction of the bead width as a function of printing parameters and contact angle:

$$w = \sqrt{\frac{2\pi D_0^3}{3p\left(\dfrac{\theta}{\sin^2\theta} - \dfrac{\cos\theta}{\sin\theta}\right)}} \tag{2.16}$$

Due to the zero-receding contact line condition necessary for this volume balance, it is inherent that the minimum feature size attainable will be that of a single deposited droplet. This therefore necessitates that the minimum width of a bead for which eqn (2.16) is valid is equal to $\beta_{eqm}D_0$ as defined in eqn (2.12). Using this condition, it is possible to define a critical droplet

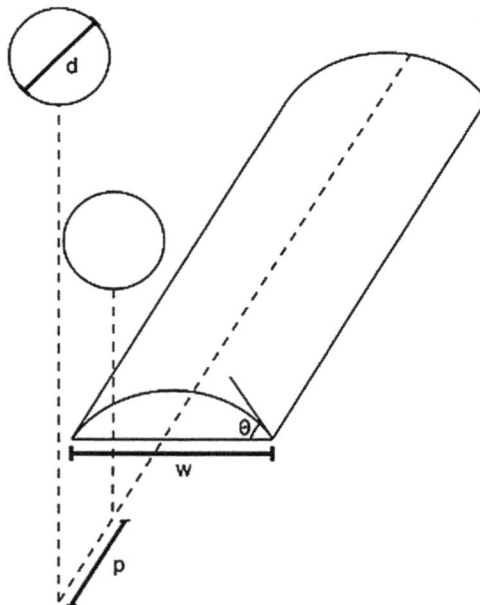

Figure 2.6 Diagram showing the parameters that are used in modelling the relationship between final track width and contact angle. *Journal of Materials Science*, Direct ink-jet printing and low temperature conversion of conductive silver patterns, **41**, 2006, 4153–4158, Smith, P. J., *et al.* 2006 Springer Science + Business Media, Inc. With permission of Springer.[64]

spacing, p_{max}, above which there is insufficient deposited liquid to form a stable bead with parallel contact lines:

$$p_{max} = \frac{2\pi D_0}{3\beta_{eqm}^2 \left(\dfrac{\theta}{\sin^2 \theta} - \dfrac{\cos \theta}{\sin \theta} \right)}$$ (2.17)

A droplet spacing greater than the value of p_{max}, but smaller than $\beta_{eqm}D_0$ will result in a periodic curvature to the contact line due to the insufficient volume of liquid, which is referred to as a 'scalloped' morphology.[42]

Predictions made using the above and similar models have been found to show good agreement with experimental observations.[9,26,42,64,65] This simple model can therefore be used to select appropriate printing parameters to attain the desired track morphology.

2.4.2.2 Unstable Bead Formation

In addition to the 'scalloping' morphology discussed above, another important unstable bead morphology is sometimes encountered. This morphology is generally observed when droplet spacing is significantly reduced, contact angles are relatively large, and the printing speed is low. The morphology consists of a series of ridges of width $\beta_{eqm}D_0$ that connect together a number of regularly spaced bulges of appreciably greater dimensions. This has become known as bulging.

Bulging was first observed and explained by Duineveld,[26] the explanation focusing on the relative driving forces for flow of deposited droplets. A droplet deposited next to a pre-existing bead will have a smaller radius of curvature than the pre-existing bead with which it coalesces. As shown in eqn (2.11), this difference in curvature leads to a difference in pressure between the newly deposited droplet and the bead. This pressure difference acts as a driving force for flow, with liquid tending to flow from the newly deposited droplet into the bulk of the pre-existing bead. This flow induced by the pressure difference is competing with the capillary flow as the contact line moves from the initial droplet diameter to the stable width of the bead. If the magnitude of the capillary force is significantly greater than the pressure driven flow, the droplet will spread to the width of the pre-existing bead and a stable track will be formed. If the magnitude of the pressure driven flow is sufficient to overcome the capillary flow, the liquid will flow down the bead before the contact line is able to move appreciably. This pressure driven flow will lead to a reduction in the contact angle at the head of the bead and therefore further reduce the driving force for capillary flow, leading to a ridge of width equal to the initial width of the droplet.

Duineveld used the above arguments to construct a mathematical model of the bulging process that showed good agreement with experimental results. The primary conclusions of both the model and associated experiments were that the occurrence of bulging was exacerbated by decreasing deposition frequency (reducing the applied flow rate), increased ink/substrate contact angle, and reducing droplet spacing (both increasing the driving force for axial flow). This model was subsequently adapted and coupled with the equation for stable track width given in eqn (2.16). This adaptation managed to decouple the fluid properties and printing speed from the interactions between the bead and the surface, which in turn made it possible to construct a stability map.[66] This enabled the prediction of line morphology based upon ink properties, printing conditions and ink/substrate interactions (Figure 2.7). This map has subsequently shown good agreement with experimental results for a wide variety of ink/substrate combinations.[43,65,66]

Based upon this understanding of the bulging instability, other methods have been devised that eliminate bulging and produce a stable track morphology. These methods typically rely on the creation of time-dependent fluid properties, in that either the viscosity or surface tension of the fluid changes after deposition to either remove the driving force for bulging or inhibit the pressure-driven flow. The first work to investigate this, by van den Berg et al.,[67] used a polymer that underwent a thermally triggered gelation. The sharp increase in viscosity upon gelation effectively inhibited any axial flow along the bead when printing on heated substrates, and led to stable track morphologies when unstable morphologies were observed at room temperature. Other work has looked at reducing the driving force for pressure-driven flow, this has typically been done by using a two solvent system of which one is more volatile than the other in a manner similar to coffee staining prevention strategies mentioned previously.[68]

2.4.2.3   Film Formation

In addition to single tracks, perhaps one of the most common shapes desired in inkjet printing is a planar film. While the same underlying mechanisms occur in film formation when compared to bead formation, there is added

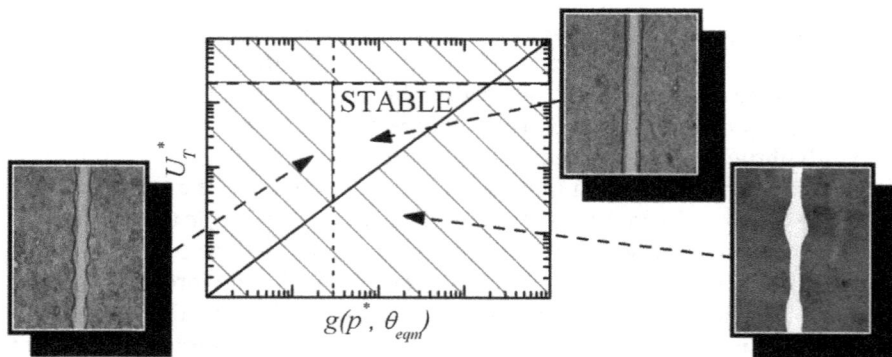

Figure 2.7   A stability map for inkjet printed tracks, showing the triangular region of stability in the centre, the 'scalloped' region to the left and the 'bulging' region in the bottom right. Reprinted (adapted) from ref. 66. Copyright (2010) American Chemical Society.

complexity due to the possibility of flow between adjacent beads. It should be noted that the work on film formation is comparatively limited compared to bead formation and coalescence in general, and has so far been focused on the sequential deposition of adjacent beads with a single printing nozzle. Due caution should therefore be taken when trying to interpret behaviours of films generated with multiple nozzle systems, as the relevant timescales of interaction between adjacent beads will likely significantly alter the observed behaviour.

Some of the first work to look at film formation by inkjet printing was conducted by Tekin *et al.*,[69] where it was found that the homogeneity of produced films was improved by printing droplets in a quasi-random order and at low speed. This is in agreement with observed behaviour of beads in that the quasi-random order means that each deposited droplet is limited in how it can interact with neighbouring droplets before it evaporates, and therefore eliminating any fluid flow caused by differential surface curvature. This work was, however, limited due to the nature of quasi-random droplet deposition, which eliminates a number of the advantages inherent with inkjet printing such as potential for reel-to-reel processing and ease of parallelisation.

To this end, further work was carried out by Kang *et al.*,[70] where the variation in the contact angle of the printed film due to flow and evaporation was numerically modelled. This led to the use of variable line spacing so that the contact angle was kept between the advancing and receding contact angle. The need to keep the contact angle of the liquid film with the substrate between the advancing and receding contact angle has also led to the use of two solvent mixtures[71] and controlled surface roughness.[72]

Due to the complexity of the interactions between droplets, environment, and substrate, the fabrication of non-circular patterns such as films is often templated. This is potentially beneficial for two primary reasons; it eliminates variation due to the aforementioned instabilities between films, and it makes the resolution of the feature dictated by the templating process rather than the droplet size. The templating, also referred to as substrate patterning, typically involves the fabrication of a barrier (for example, by lithography) that impedes spreading of the liquid past that point by means of increasing the advancing contact angle at that point. This barrier can either be purely down to changes in surface energy,[73] or can also be a physical barrier.[74] By exploiting the differential wetting behaviour of two different inks, it has even been possible to inkjet print such defining barriers, with patterns as fine as 100 nm achieved.[75] Such techniques are particularly useful in printed electronics, where consistency and resolution of pattern define device performance and reliability.

2.5 Summary

This chapter has given an introduction into the many physical process that underpin how a series of liquid droplets can form an arbitrary solid pattern upon a substrate. These include the initial impact process, capillary-induced spreading of the liquid contact line, phase change of droplets to a solid deposit, and how individual droplets interact and coalesce to form larger patterns. While the majority of investigation into this area up to now has concerned non-reactive systems, by looking at the underlying mechanisms it is possible to infer how reactive inkjet systems will behave.

References

1. T. Young, An essay on the cohesion of fluids, *Philos. Trans. R. Soc., London*, 1805, **95**, 65–87.
2. A. M. Worthington, On the forms assumed by drops of liquids falling vertically on a horizontal plate, *Proc. R. Soc. London*, 1876, **25**, 261–271.
3. J. Madejski, Solidification of droplets on a cold surface, *Int. J. Heat Mass Transfer*, 1976, **19**(9), 1009–1013.
4. M. Pasandideh-Fard, *et al.*, Capillary effects during droplet impact on a solid surface, *Phys. Fluids*, 1996, **8**(3), 650–659.
5. C. Clanet, *et al.*, Maximal deformation of an impacting drop, *J. Fluid Mech.*, 2004, **517**, 199–208.
6. D. B. van Dam and C. Le Clerc, Experimental study of the impact of an ink-jet printed droplet on a solid substrate, *Phys. Fluids*, 2004, **16**(9), 3403–3414.
7. C. W. Visser, *et al.*, Dynamics of high-speed micro-drop impact: Numerical simulations and experiments at frame-to-frame times below 100 ns, *Soft Matter*, 2015, **11**(9), 1708–1722.
8. B. Derby, Inkjet printing of functional and structural materials: Fluid property requirements, feature stability, and resolution, *Annu. Rev. Mater. Res.*, 2010, 395–414.
9. J. Stringer and B. Derby, Limits to feature size and resolution in ink jet printing, *J. Eur. Ceram. Soc.*, 2009, **29**(5), 913–918.
10. R. D. Deegan, *et al.*, Capillary flow as the cause of ring stains from dried liquid drops, *Nature*, 1997, **389**(6653), 827–829.
11. S. Schiaffino and A. A. Sonin, Molten droplet deposition and solidification at low Weber numbers, *Phys. Fluids*, 1997, **9**(11), 3172–3187.
12. E. Collings, *et al.*, Splat-quench solidification of freely falling liquid-metal drops by impact on a planar substrate, *J. Mater. Sci.*, 1990, **25**(8), 3677–3682.
13. A. L. Yarin, Drop impact dynamics: Splashing, spreading, receding, bouncing, *Annu. Rev. Fluid Mech.*, 2006, 159–192.
14. H. Dong, W. W. Carr and J. F. Morris, Visualization of drop-on-demand inkjet: Drop formation and deposition, *Rev. Sci. Instrum.*, 2006, **77**(8), 085101.
15. Y. Son, *et al.*, Spreading of an inkjet droplet on a solid surface with a controlled contact angle at low Weber and Reynolds numbers, *Langmuir*, 2008, **24**(6), 2900–2907.
16. H. Dong, *et al.*, Temporally-resolved inkjet drop impaction on surfaces, *AIChE J.*, 2007, **53**(10), 2606–2617.

17. I. V. Roisman, R. Rioboo and C. Tropea, Normal impact of a liquid drop on a dry surface: Model for spreading and receding, *Proc. R. Soc. A*, 2002, **458**(2022), 1411–1430.

18. H. Park, *et al.*, Single drop impaction on a solid surface, *AIChE J.*, 2003, **49**(10), 2461–2471.

19. S. Jung and I. M. Hutchings, The impact and spreading of a small liquid drop on a non-porous substrate over an extended time scale, *Soft Matter*, 2012, **8**(9), 2686–2696.

20. P. S. Laplace, *Mecanique Celeste*, 1806.

21. R. N. Wenzel, Resistance of solid surfaces to wetting by water, *Ind. Eng. Chem.*, 1936, **28**(8), 988–994.

22. A. B. D. Cassie and S. Baxter, Wettability of porous surfaces, *Trans. Faraday Soc.*, 1944, **40**, 546–551.

23. L. Feng, *et al.*, Super-Hydrophobic Surfaces: From Natural to Artificial, *Adv. Mater.*, 2002, **14**(24), 1857–1860.

24. T. S. Meiron, A. Marmur and I. S. Saguy, Contact angle measurement on rough surfaces, *J. Colloid Interface Sci.*, 2004, **274**(2), 637–644.

25. P. G. De Gennes, Wetting: Statics and dynamics, *Rev. Mod. Phys.*, 1985, **57**(3), 827–863.

26. P. C. Duineveld, The stability of ink-jet printed lines of liquid with zero receding contact angle on a homogeneous substrate, *J. Fluid Mech.*, 2003, **477**, 175–200.

27. M. Di Biase, *et al.*, Inkjet printing and cell seeding thermoreversible photocurable gel structures, *Soft Matter*, 2011, **7**(6), 2639–2646.

28. T. Lim, *et al.*, Deposit pattern of inkjet printed pico-liter droplet, *Int. J. Precis. Eng. Manuf.*, 2012, **13**(6), 827–833.

29. R. G. Picknett and R. Bexon, The evaporation of sessile or pendant drops in still air, *J. Colloid Interface Sci.*, 1977, **61**(2), 336–350.

30. J. R. Moffat, K. Sefiane and M. E. R. Shanahan, Effect of TiO2 Nanoparticles on Contact Line Stick–Slip Behavior of Volatile Drops, *J. Phys. Chem. B*, 2009, **113**(26), 8860–8866.

31. R. D. Deegan, *et al.*, Contact line deposits in an evaporating drop, *Phys. Rev. E: Stat. Phys., Plasmas, Fluids, Relat. Interdiscip. Top.*, 2000, **62**(1 pt B), 756–765.

32. H. Hu and R. G. Larson, Marangoni effect reverses coffee-ring depositions, *J. Phys. Chem. B*, 2006, **110**(14), 7090–7094.

33. Y. Hamamoto, J. R. E. Christy and K. Sefiane, The Flow Characteristics of an Evaporating Ethanol Water Mixture Droplet on a Glass Substrate, *J. Therm. Sci. Technol.*, 2012, **7**(3), 425–436.

34. T. Still, P. J. Yunker and A. G. Yodh, Surfactant-induced Marangoni eddies alter the coffee-rings of evaporating colloidal drops, *Langmuir*, 2012, **28**(11), 4984–4988.

35. H. Hu and R. G. Larson, Analysis of the Effects of Marangoni Stresses on the Microflow in an Evaporating Sessile Droplet, *Langmuir*, 2005, **21**(9), 3972–3980.

36. W. D. Ristenpart, *et al.*, Influence of substrate conductivity on circulation reversal in evaporating drops, *Phys. Rev. Lett.*, 2007, **99**, 23.

37. K. A. Baldwin, *et al.*, Drying and deposition of poly(ethylene oxide) droplets determined by Peclet number, *Soft Matter*, 2011, **7**(17), 7819–7826.

38. D. Ta, *et al.*, Dynamically controlled deposition of colloidal nanoparticles suspension in evaporating drops using laser radiation, *Soft Matter*, 2016, **12**, 4530–4536.

39. Q. Li, *et al.*, Self-Organization of Carbon Nanotubes in Evaporating Droplets, *J. Phys. Chem. B*, 2006, **110**(28), 13926–13930.

40. M. Layani, *et al.*, Transparent conductive coatings by printing coffee ring arrays obtained at room temperature, *ACS Nano*, 2009, **3**(11), 3537–3542.

41. S. S. Abramchuk, *et al.*, Direct observation of DNA molecules in a convection flow of a drying droplet, *EPL (Europhys. Lett.)*, 2001, **55**(2), 294.

42. D. Soltman and V. Subramanian, Inkjet-printed line morphologies and temperature control of the coffee ring effect, *Langmuir*, 2008, **24**(5), 2224–2231.

43. W. Tianming, *et al.*, Inkjet printed carbon nanotube networks: the influence of drop spacing and drying on electrical properties, *J. Phys. D: Appl. Phys.*, 2012, **45**(31), 315304.

44. J. Park and J. Moon, Control of Colloidal Particle Deposit Patterns within Picoliter Droplets Ejected by Ink-Jet Printing, *Langmuir*, 2006, **22**(8), 3506–3513.

45. E. Tekin, *et al.*, InkJet printing of luminescent CdTe nanocrystal-polymer composites, *Adv. Funct. Mater.*, 2007, **17**(1), 23–28.

46. S. T. Thoroddsen, K. Takehara and T. G. Etoh, The coalescence speed of a pendent and a sessile drop, *J. Fluid Mech.*, 2005, **527**, 85–114.

47. W. Yao, *et al.*, Coalescence of viscous liquid drops, *Phys. Rev. E: Stat., Nonlinear, Soft Matter Phys.*, 2005, **71**(1), 016309.

48. D. G. A. L. Aarts, *et al.*, Hydrodynamics of droplet coalescence, *Phys. Rev. Lett.*, 2005, **95**(16), 164503.

49. R. W. Hopper, Coalescence of Two Viscous Cylinders by Capillarity: Part I, Theory, *J. Am. Ceram. Soc.*, 1993, **76**(12), 2947–2952.

50. L. Duchemin, J. Eggers and C. Josserand, Inviscid coalescence of drops, *J. Fluid Mech.*, 2003, **487**, 167–178.

51. S. C. Case and S. R. Nagel, Coalescence in Low-Viscosity Liquids, *Phys. Rev. Lett.*, 2008, **100**(8), 084503.

52. A. Menchaca-Rocha, *et al.*, Coalescence of liquid drops by surface tension, *Phys. Rev. E: Stat., Nonlinear, Soft Matter Phys.*, 2001, **63**(4 pt II), 463091–463095.

53. N. Kapur and P. H. Gaskell, Morphology and dynamics of droplet coalescence on a surface, *Phys. Rev. E: Stat., Nonlinear, Soft Matter Phys.*, 2007, **75**(5), 056315.

54. W. D. Ristenpart, *et al.*, Coalescence of spreading droplets on a wettable substrate, *Phys. Rev. Lett.*, 2006, **97**(6), 064501.

55. C. Andrieu, *et al.*, Coalescence of sessile drops, *J. Fluid Mech.*, 2002, **453**, 427–438.

56. D. A. Beysens and R. D. Narhe, Contact line dynamics in the late-stage coalescence of diethylene glycol drops, *J. Phys. Chem. B*, 2006, **110**(44), 22133–22135.

57. I. Roisman, *et al.*, Multiple drop impact onto a dry solid substrate, *J. Colloid Interface Sci.*, 2002, **256**(2), 396–410.

58. R. Li, *et al.*, Coalescence of two droplets impacting a solid surface, *Exp. Fluids*, 2010, **48**(6), 1025–1035.

59. W.-K. Hsiao, G. D. Martin and I. M. Hutchings, Printing Stable Liquid Tracks on a Surface with Finite Receding Contact Angle, *Langmuir*, 2014, **30**(41), 12447–12455.

60. L. Rayleigh, On the capillary phenomena of jets, *Proc. R. Soc. London*, 1879, **29**, 71–97.

61. S. H. Davis, Moving contact lines and rivulet instabilities. Part 1. The static rivulet, *J. Fluid Mech.*, 1980, **98**(2), 225–242.

62. K. Sekimot, R. Oguma and K. Kawasaki, Morphological stability analysis of partial wetting, *Ann. Phys.*, 1987, **176**(2), 359–392.

63. S. Schiaffino and A. A. Sonin, Formation and stability of liquid and molten beads on a solid surface, *J. Fluid Mech.*, 1997, **343**, 95–110.

64. P. J. Smith, *et al.*, Direct ink-jet printing and low temperature conversion of conductive silver patterns, *J. Mater. Sci.*, 2006, **41**(13), 4153–4158.

65. R. Dou, *et al.*, Ink-jet printing of zirconia: Coffee staining and line stability, *J. Am. Ceram. Soc.*, 2011, **94**(11), 3787–3792.

66. J. Stringer and B. Derby, Formation and stability of lines produced by inkjet printing, *Langmuir*, 2010, **26**(12), 10365–10372.

67. A. M. J. Van Den Berg, *et al.*, Geometric control of inkjet printed features using a gelating polymer, *J. Mater. Chem.*, 2007, **17**(7), 677–683.

68. M. Liu, *et al.*, Inkjet printing controllable footprint lines by regulating the dynamic wettability of coalescing ink droplets, *ACS Appl. Mater. Interfaces*, 2014, **6**(16), 13344–13348.

69. E. Tekin, B. J. De Gans and U. S. Schubert, Ink-jet printing of polymers - From single dots to thin film libraries, *J. Mater. Chem.*, 2004, **14**(17), 2627–2632.

70. H. Kang, D. Soltman and V. Subramanian, Hydrostatic Optimization of Inkjet-Printed Films, *Langmuir*, 2010, **26**(13), 11568–11573.

71. J.-L. Lin, Z.-K. Kao and Y.-C. Liao, Preserving Precision of Inkjet-Printed Features with Solvents of Different Volatilities, *Langmuir*, 2013, **29**(36), 11330–11336.

72. Y. V. Kalinin, V. Berejnov and R. E. Thorne, Contact Line Pinning by Microfabricated Patterns: Effects of Microscale Topography, *Langmuir*, 2009, **25**(9), 5391–5397.

73. J. Léopoldès, *et al.*, Jetting Micron-Scale Droplets onto Chemically Heterogeneous Surfaces, *Langmuir*, 2003, **19**(23), 9818–9822.

74. C. Chin-Tai, *et al.*, An inkjet printed stripe-type color filter of liquid crystal display, *J. Micromech. Microeng.*, 2010, **20**(5), 055004.

75. C. W. Sele, *et al.*, Lithography-free, self-aligned inkjet printing with sub-hundred-nanometer resolution, *Adv. Mater.*, 2005, **17**(8), 997–1001.

CHAPTER 3

Droplet Mixing

MARK C. T. WILSON*[a], J. RAFAEL CASTREJÓN-PITA[b] AND
ALFONSO A. CASTREJÓN-PITA[c]

[a]School of Mechanical Engineering, University of Leeds, Leeds LS2 9JT, UK;
[b]School of Engineering and Materials Science, Queen Mary, University of
London, London E1 4NS, UK; [c]Department of Engineering Science,
University of Oxford, Oxford OX1 3PS, UK
*E-mail: M.Wilson@leeds.ac.uk

3.1   Introduction

This chapter explores a somewhat paradoxical issue. There are many examples of successful reactive inkjet printing illustrated in this book, and a key requirement for the relevant chemical reactions to occur is sufficient mixing of the reagents within the droplets. However, there are also several independent studies in the literature, both experimental and computational, which indicate that when two droplets coalesce—especially on a surface—there is little or no mixing!

To make sense of this observation we will begin by reviewing the fundamental mechanisms by which mixing can take place in liquids. Armed with this, we will examine available studies of droplet mixing, particularly in the case of two droplets impacting and coalescing on a surface. It will be seen that the size of the droplets is an essential consideration in reconciling the findings of these studies and the practical success of reactive inkjet printing. Note that we will focus almost exclusively on mixing in the absence of chemical reactions, though the latter will be discussed briefly at the end.

Smart Materials No. 32
Reactive Inkjet Printing: A Chemical Synthesis Tool
Edited by Patrick J. Smith and Aoife Morrin
© The Royal Society of Chemistry 2018
Published by the Royal Society of Chemistry, www.rsc.org

3.1.1 Mechanisms of Mixing

When two different but miscible liquids are brought together—for example by adding cream or milk to coffee, or printing a droplet of one liquid on top of another liquid—the initially inhomogeneous combined liquid will eventually homogenise into a uniform mixture in which the molecules of one species are uniformly dispersed among those of the other. At the molecular level, this homogenisation is driven by molecular diffusion, which acts to smooth out gradients in the concentrations of the component species. However, diffusion is a slow process arising from the random motion of molecules over very small inter-molecular distances. It is much slower in liquids than in gases, since gas molecules are typically more energetic and have a longer mean free path.

The homogenisation process can be speeded up dramatically by setting the liquid in motion. In the example of the coffee this is simply achieved by stirring with a spoon. This clearly moves parts of the liquid rapidly over much larger distances and helps in bringing different sections of the liquid together for diffusion to smooth out into a uniform blend. In the case of a droplet landing on another, the impact produces a significant disturbance resulting in substantial movement of the combined liquid, albeit over a rather brief period. Indeed, the effectiveness of this disturbance in assisting mixing within droplets will be discussed in detail later in this chapter.

Fluid motion that is turbulent is particularly effective in enhancing mixing since, in addition to the bulk movement and redistribution of material, the random fluctuations in fluid velocity enable transport of material perpendicular to the average flow direction. In contrast, laminar flows feature smooth fluid motion in parallel layers with no transport between layers except that due to molecular diffusion, which is commonly insignificant compared to the transport by advection along the flow path. Hence laminar flows are less effective in mixing than turbulent flows.

Whether or not a flow is turbulent is indicated by the Reynolds number, $Re = UL/v$, where U and L are respectively the characteristic speed and length scale of the flow, and v is the kinematic viscosity of the fluid (with SI units of $m^2 \ s^{-1}$). This number expresses the importance of inertial forces compared to viscous forces. Reynolds numbers below about 2300 indicate laminar flow, so for inkjet printing applications, where typical values might be $U \sim 5 \ m \ s^{-1}$, $L \sim 50 \ \mu m$, $v \sim 10^{-5} \ m^2 \ s^{-1}$ and hence $Re \sim 25$, the flow in the droplets is very much laminar. This is also true of microfluidic 'lab-on-a-chip' devices, which also rely on the manipulation and mixing of small droplets, and has led to much research into methods of enhancing mixing at these length scales by chaotic advection. Some of these are mentioned in Section 3.2.3. The next section explores the interplay of advection and diffusion in more detail.

3.2 Diffusion and Advection

As mentioned above, diffusion acts to smooth out inhomogeneities in the concentration of a substance (it has the same effect on differences in temperature or momentum *etc.* also). The flux, $\mathbf{j}_d$, of material transported by diffusion, *i.e.* the amount of the substance of interest passing through a unit area per unit time, is proportional to the concentration gradient, as expressed in Fick's law:

$$\mathbf{j}_d = -D\nabla c \tag{3.1}$$

where $c(\mathbf{x},t)$ is the concentration at position $\mathbf{x}$ and time t and D is the diffusion coefficient, which varies with the substances being mixed but for liquids is of the order 10^{-9} m^2 s^{-1}.[1] The minus sign appears since material diffuses from an area of higher concentration to one with lower concentration.

Advection[†] is the transport of material by a fluid flow. For example, a concentrated blob of dye injected into a slow, laminar stream will be carried away downstream. The flux of material carried by advection is given by $\mathbf{j}_a = c\mathbf{u}$, where $\mathbf{u}(\mathbf{x},t)$ is the velocity of the flow. By imagining a fixed volume within a fluid flow and considering the fluxes of material into and out of the volume, and the rate of change of the amount of material in the volume, one arrives at a partial differential equation for the concentration combining the effects of advection and diffusion:

$$\frac{\partial c}{\partial t} + \nabla \cdot (c\mathbf{u}) = \nabla \cdot (D\nabla c). \tag{3.2}$$

If the fluid is incompressible and the diffusion coefficient is constant, this advection–diffusion equation simplifies to

$$\frac{\partial c}{\partial t} + \mathbf{u} \cdot \nabla c = D\nabla^2 c. \tag{3.3}$$

By comparing the relative magnitudes of the advective and diffusive fluxes using characteristic velocity and length scales U and L, one can obtain the Péclet number,

$$\mathrm{Pe} = \frac{UL}{D} \tag{3.4}$$

as a useful measure of the significance of advection compared to diffusion. Notice the similarity in its form to that of the Reynolds number. For large Pe, advection dominates. The next two sub-sections explore solutions of eqn (3.3) under different conditions to illustrate the roles of these mixing mechanisms.

3.2.1   Pure Diffusion in a Sessile Droplet

Ignoring the advection term in eqn (3.3), we can estimate the time required for material to diffuse over a distance L as

---

[†]Note that advection is also commonly referred to as convection. However, the latter term—particularly in the context of heat transfer—can be used to mean transport by advection together with diffusion. To be precise we will use the term advection in this chapter.

$$T_d \sim L^2/D. \tag{3.5}$$

For an inkjet droplet of say 50 μm diameter mixing with another droplet of the same size, with a diffusion coefficient ~10^{-9} m^2 s^{-1}, this gives a time of about 0.6 seconds to diffuse over a distance of one droplet radius, *i.e. L* ~ 25 μm.

To give a more visual picture of the effect of mixing by diffusion alone, Figure 3.1 shows snapshots from a numerical solution of an axisymmetric version of eqn (3.3) with **u** = **0** calculated using COMSOL Multiphysics® (version 5.2a). The domain is a hemispherical sessile droplet whose volume is equal to that of two 50 μm-diameter droplets combined. The corresponding base radius of the hemisphere is approximately 40 μm. The top half of the volume is initialised with a normalised concentration $c = 1$ (shown as black), while the bottom half has $c = 0$ (shown as white). This simulation represents the effect of placing one droplet on top of another, but without considering any of the dynamics of the coalescence process itself. The diffusion coefficient was set to 10^{-9} m^2 s^{-1}. As can be seen, with the particular geometry now considered, the concentration field is essentially uniform after half a second. Note, however, that if the liquids being mixed are more complex, with larger molecules, the diffusion coefficient is likely to be lower than the one used

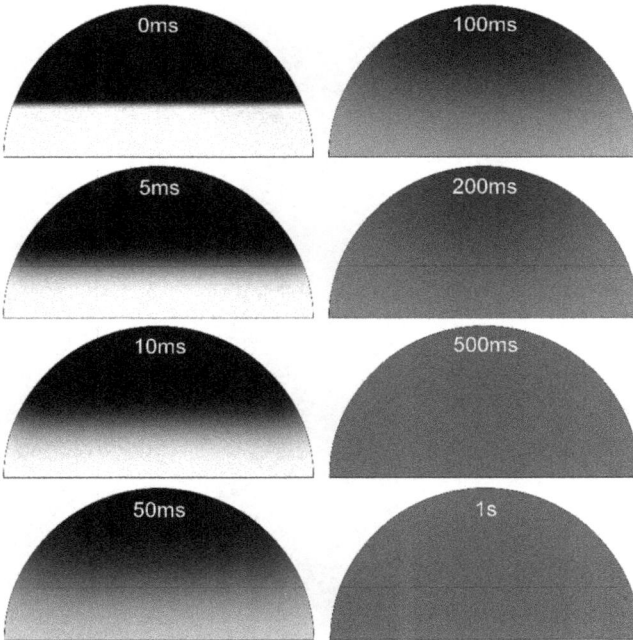

Figure 3.1 Simulation of the mixing of dye by diffusion alone in a hemispherical sessile droplet with base radius 40 μm. The images show the concentration field at the indicated times, with black corresponding to a normalised concentration of one and white representing zero concentration.

here, and the time taken to reach a given state of homogenisation will be correspondingly longer.

In stark contrast to the previous case, Figure 3.2 shows a repeat of the simulation in Figure 3.1, but with a droplet volume equivalent to that of two droplets of diameter 2.4 mm combined. The base diameter of the corresponding hemisphere is 3.8 mm. Whereas the droplet in Figure 3.1 was fully mixed after half a second, after a full second the droplet in Figure 3.2 looks almost unchanged except for a slight smoothing of the 'interface' between the black and white fluid. Simply because of the much bigger distances involved, this composite droplet takes 10 minutes to reach the same level of mixing that the smaller drop achieved after half a second. We will return to this in Section 3.3.

3.2.2   Advection and Diffusion in a Cavity Flow

To illustrate the role of advection in enhancing the rate of mixing, Figures 3.3 and 3.4 show a simple 2D rectangular cavity flow, in which the fluid motion is driven by moving the upper and lower solid boundaries (referred to as 'lids') in opposite directions, as indicated in Figure 3.3. The reason for considering this system, in this chapter on droplet mixing, is that it provides a way of controlling the fluid motion and hence observing its effect. The side walls are

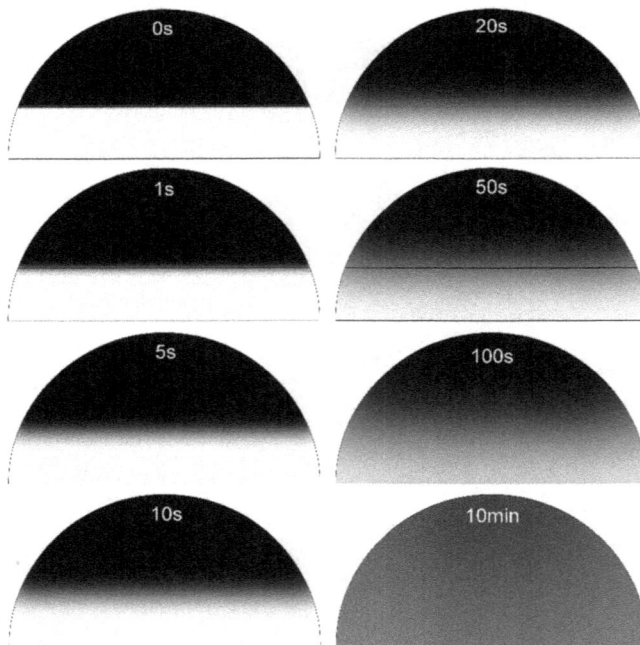

Figure 3.2   Diffusion of dye in a hemispherical sessile droplet with base radius of 1.9 mm.

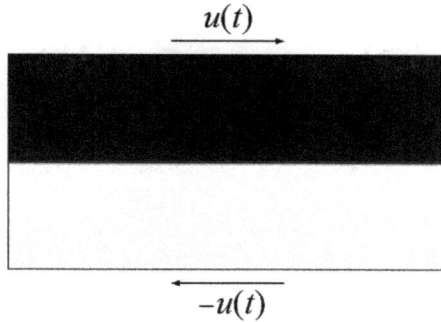

$$\overrightarrow{u(t)}$$

$$\overleftarrow{-u(t)}$$

Figure 3.3 Initial condition of the lid-driven cavity mixing example shown in Figure 3.4. The upper half of the domain is dyed black, represented by a concentration of one, with white indicating a zero concentration of dye. The domain size is 1 × 0.5 mm.

stationary, and the size of the flow domain is 1.0 × 0.5 mm. Initially the upper half of the domain is dyed black, represented by a normalised concentration value $c = 1$, while the lower half has an initial concentration $c = 0$ – see Figure 3.3. Figure 3.4 shows snapshots at different times from numerical solutions of eqn (3.3) under different conditions. The first column of images shows the effect of diffusion alone in smoothing out the concentration field—just as in Figures 3.1 and 3.2. The diffusion coefficient is again 10^{-9} m^2 s^{-1}. From eqn (3.5), we can see that it will take about a minute to smooth out the concentration field across the domain, and the thickness of the diffuse band between the black and white regions grows as $2\sqrt{Dt}$ (the factor of two appears since diffusion occurs in both directions).

In the images in the middle and right-hand columns of Figure 3.4, the lids are moving at a speed given by $u(t) = U\sin\omega t$, where $U = 0.1$ m s^{-1} and $\omega = 10\pi$, meaning that they switch direction every 0.1 s, creating an alternating clockwise and anticlockwise stirring pattern. The middle column of Figure 3.4 shows what happens when there is no diffusion. These images were created by directly advecting 100 000 passive tracer particles according to the equation

$$\frac{d\mathbf{x}_p}{dt} = \mathbf{u}(\mathbf{x}_p, t) \tag{3.6}$$

where $\mathbf{x}_p(t)$ is the position of a particle following the flow field. This is common practice in studies of mixing by chaotic advection,[2,3] and helps to reveal the finely detailed structure of the mixing process which is not so easily seen when a concentration field and eqn (3.3) are used. Initially the tracer particles are uniformly distributed in the top half of the domain, and the middle column of Figure 3.4 shows later positions of those particles as black dots.

For the (anti-)symmetrical movement of the lids used here, the centre point of the domain is always a stagnation point, *i.e.* where the fluid

pure diffusion   advection, no diffusion   advection and diffusion

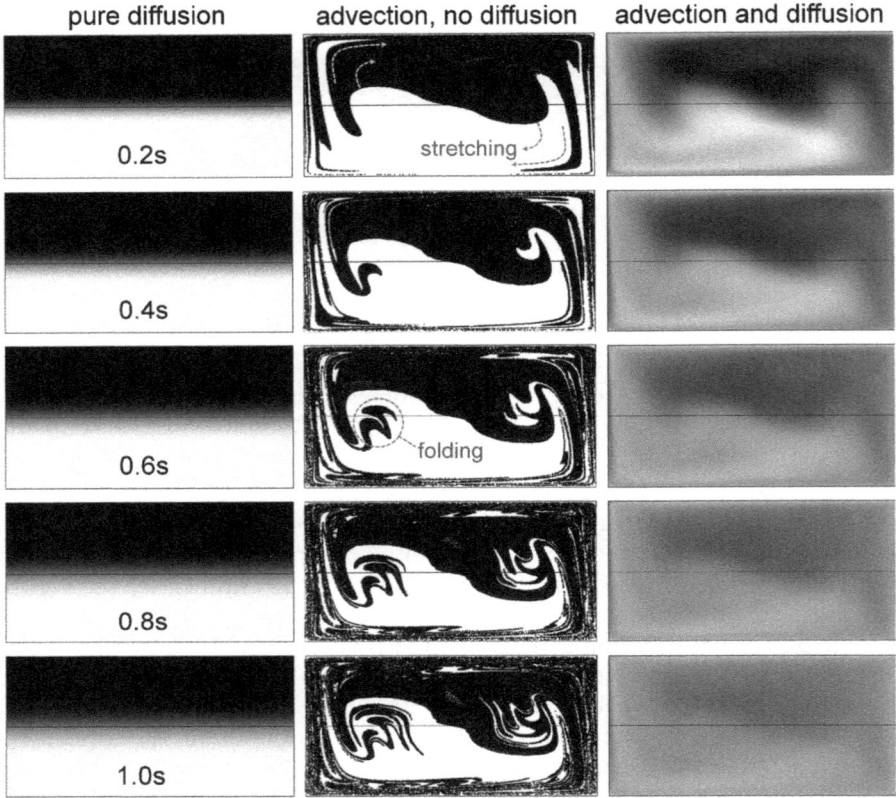

stretching

0.2s

0.4s

folding

0.6s

0.8s

1.0s

Figure 3.4   Simulation of mixing in a lid-driven cavity showing the effects of advection and diffusion. The snapshots in the left- and right-hand columns show the concentration of a dye, with black indicating a normalised concentration of one and white a concentration of 0. The middle column shows the positions of 100 000 passive tracer particles advected by the flow in the absence of diffusion. The initial condition is shown in Figure 3.3. The time indicated in each row applies to all images on the row.

velocity vanishes. The fluid velocity is also zero along the side walls. Therefore, as the lids start to move (initially in a clockwise direction), the initially horizontal 'interface' between the black and white fluid bows downwards to the right of the centre point and upwards to the left, while remaining pinned at its ends. This action *stretches* the 'interface' and increases its length, and this stretching is substantial as the fluid continues to rotate clockwise several times until the lids switch direction at $t = 0.1$ s. This switch in direction produces a *folding* of parts of the stretched 'interface' back on itself, and further stretching in the opposite direction. This folding action is most clearly seen in the $t = 0.6$ s and later images (see the dashed grey circle in Figure 3.4). The five images in the middle column of the figure show the state of mixing after one, two, three, four and five full cycles of movement of the lids, and illustrate that this continuous *stretching and folding* rapidly creates fine striations and dramatically increases the length of the black–white boundary. This has the dual effect of vastly increasing the length of the 'interface' along which molecular diffusion can take place as well as greatly reducing the distance over which diffusion needs to act to homogenise the liquid.

The right-hand column in Figure 3.4 shows the evolution of the concentration field calculated from eqn (3.3) with the same movement of the lids but with diffusion now also acting (with diffusion coefficient 10^{-9} m^2 s^{-1}). The corresponding Péclet number, eqn (2.4), is 10^5. Again, the initial condition is as shown in Figure 3.3, and it is clear that the action of stretching and folding combined with diffusion is very effective in mixing the black and white sections of fluid. After a further second the concentration field is essentially uniform; much faster than the minute required by diffusion alone.

Before considering the actions of advection and diffusion in the context of droplets, as an aside we make a few remarks on the specific stirring flow described above. The Reynolds number for the flow is approximately 100, which is very much in the laminar range but is sufficiently large for the inertia of the fluid to play an important role. In the limit Re $\rightarrow$ 0, the symmetric sinusoidal stirring protocol would not be effective in mixing the fluid (in the absence of diffusion). This is because Stokes flows (*i.e.* where Re = 0) are reversible and so the stretching of the black–white 'interface' caused by the clockwise motion would be undone by the ensuing anticlockwise motion which is precisely the opposite. This situation can be avoided by introducing a phase difference between the lid speeds, which prevents any part of the cycle being a reversal of a preceding part. This would also make the mixing in Figure 3.4 more effective, since the changing asymmetry in lid speeds would mean that the stagnation point in the centre of the cavity would move up and down over the course of the stirring.

The lack of inertia in a Stokes flow means that the flow field responds instantaneously to changes in the boundary conditions imposed. In the context of the cavity problem, this means that if Re were zero, the whole fluid would start (or stop) moving as soon as the boundaries started (or stopped) moving (hence the reversibility of such flows). In the Re = 100 case shown in Figure 3.4, if starting from rest, the inertia of the fluid results in the fluid near the centre remaining stationary while fluid near the moving boundaries is set in motion. Over time, the momentum of the moving fluid diffuses towards the centre of the cavity until all the fluid is rotating clockwise (say) inside the cavity. If the moving boundaries then switch direction, the fluid close to those boundaries will quickly switch to moving anticlockwise also, but the fluid away from the boundaries will continue to rotate clockwise for a while, before the momentum imparted by the switched boundaries propagates throughout the cavity and the whole flow then rotates anticlockwise.

Hence under these conditions the flow is not reversible, despite the symmetry of the boundary movement, and this is the mechanism by which the stretching and folding continues in the cavity flow in Figure 3.4, even when the motion of the lids follows a cycle where half of the cycle is the reverse of the other half. This also explains why the fluid near the centre of the cavity takes longer to mix in this case.

3.2.3 Droplets as Micro-reactors

In addition to their fundamental importance in reactive inkjet printing, droplets also feature in other applications where small-scale chemical reactions can be advantageous. Examples include droplet (or 'digital') microfluidic systems,[4,5] which offer benefits such as reduced sample sizes, faster reaction and heat transfer rates, and the associated 'micro total analysis systems'[6] and 'lab-on-a-chip'[7] concepts. Even at these small scales (microns to millimetres), however, diffusion can still be too slow for practical purposes, so various means of exploiting chaotic advection to stretch and fold material interfaces, and hence speed up diffusion, have been developed. These include 'passive' systems such as the transport of droplets along serpentine channels,[8] where two miscible reagents are injected simultaneously into an immiscible carrier liquid flowing along the channel in such a way that they come together to form a droplet with each reagent occupying separate halves of the droplet. The movement along the winding channel then distorts the droplet and stretches and folds the boundary between the reagents rather like in Figure 3.4. Other mixing systems use 'active' agitation of droplets for example through electrowetting[9] phenomena.

The above applications are clearly somewhat removed from reactive inkjet printing systems, but they illustrate the usefulness and interest in exploiting droplets as microreactors, and also the difficulties in mixing.

3.3 Mixing in Impacting and Coalescing Droplets

We consider now a system more relevant to reactive inkjet printing, namely a droplet impacting on another (sessile) droplet, and explore the internal mixing process as the two droplets coalesce. The impact of a jetted droplet onto a previously deposited sessile droplet results in a substantial disturbance in the free-surface shapes of both droplets; for similar sized droplets, there is typically a flattening of the impacting droplet into a disc or 'pancake' shape, and a recoil due to the surface tension of the free surfaces. Hence there are strong movements of the liquid inside the droplets, which could potentially result in some advection-driven mixing of the droplets. However, unlike the advection examples given in Sections 3.2.2 and 3.2.3, this liquid movement occurs over a rather short, finite period, so it is not clear *a priori* how effective it will be in promoting mixing and this is something we will now explore.

3.3.1 Impact and Coalescence

For typical inkjet-sized droplets, the small size poses a significant challenge for visualising the internal dynamics. In their demonstration of reactive ink-jet printing of micron-scale polyurethane features, Kröber *et al.*[10] added fluorescent dye to one of the droplets to give an indication of the level of mixing upon coalescence. Fluorescence images of the combined drop, with surface contact diameter about 240 μm, showed greater fluorescence towards the top of the droplets, consistent with the greater depth of fluorophore there, but otherwise no concentration gradients were observed, suggesting that homogeneous mixing had occurred. However, this was based purely on views from above, so no detail of the vertical intermixing or stratification could be revealed.

Fluorescence was also used by Fathi and Dickens[11] in their exploration of nylon 6 polymerization *via* reactive inkjet printing. In this case, multiple droplets were jetted to form larger droplets on the surface, following difficulties in achieving sufficiently accurate individual drop-on-drop placements. Again, the composite droplets (in this case approximately 5 mm in diameter) were viewed from directly above, and the lateral extent of the dyed liquid was examined *via* fluorescent imaging. Fluorescence was seen across about 80% of the top view area. Now, it is not clear from a top view whether the dyed liquid is simply sitting on top of the undyed liquid or if mixing has occurred. However, it is clear that across 20% of the top view area, no mixing occurred. These unmixed regions lay mainly around the perimeter of the contact area, suggesting that no dyed droplets landed on those parts. Unfortunately, it is not clear how much time elapsed between the droplet deposition and the fluorescence observations, but from eqn (3.5) and the size of the droplets it is clear that diffusion of dye into those unmixed regions would take several minutes.

Fathi and Dickens[11] assumed that the liquids within the fluorescing area of their droplets mixed upon impact of the small dyed droplets. There is some justification for a certain degree of local mixing, given the classical studies[12] illustrating that when a droplet falls into a liquid pool, it can form a vortex ring. The same phenomenon has been shown to occur when a small droplet is drawn into a large droplet by surface tension if the viscosity is not too large.[13] Figure 3.5 shows a sketch of what happens: the small dyed droplet penetrates the large droplet, leaving a tail behind it. At its leading edge, resistance from the surrounding liquid causes a flattening of the dyed droplet and a slowing of its outer edges, and the droplet rolls up into a toroidal vortex ring. Recalling the action of advection described in Section 3.2.2, creation of this vortex ring stretches the 'interface' between the dyed and undyed liquid and creates a similar action to the folding of this interface, thus speeding up the process of mixing by molecular diffusion and therefore enhancing local mixing.

Other methods for analysing the internal dynamics of consecutively printed droplets include observations from below, through a transparent substrate, of particles seeded in the fluids.[14,15] Particle image velocimetry

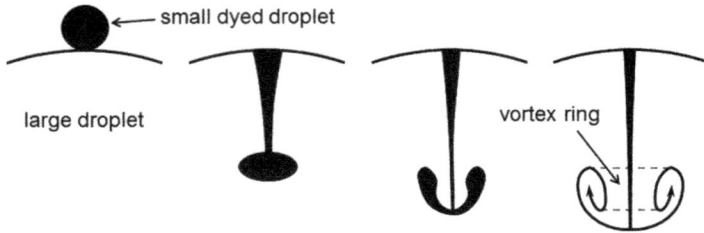

Figure 3.5   Sketch of vortex ring formation when a small droplet enters a large one.

(PIV) has been used to obtain the velocity field within millimetre-size droplets during impact and coalescence.[14] However, this technique is not particularly useful for direct observation of mixing within the droplets, since PIV algorithms are statistical in nature and do not track the movement of individual particles. The flow visualization is also limited to one plane within the flow, hence providing a limited picture of the flow. Yang *et al.*[15] tracked the movement of 1.1 μm-diameter fluorescent beads during coalescence and evaporation of consecutively printed droplets, primarily to understand the particle deposition dynamics.

The lack of a clear picture of the mixing during impact and coalescence of two droplets motivated us to observe this directly from an oblique viewpoint using coloured dye—a technique which had previously been used effectively to study mid-air collisions of free droplets.[16] However, this is very difficult to do with inkjet-scale droplets, so millimetre-scale droplets were used,[17] but with liquid properties selected to ensure that the associated Reynolds number and Weber number,[‡] We, lay in the same range that arises in typical inkjet printing systems. This means that at least the hydrodynamics observed in the millimetre-scale system should be the same as at the inkjet scale. Figure 3.6 gives an example from this work and shows the impact of a dyed droplet directly onto a previously deposited undyed sessile droplet of approximately the same volume. Time zero is taken as the moment at which contact is first made between the droplets.

The impact of the falling droplet clearly creates a substantial disturbance to the sessile droplet, as the combined droplet flattens out to a pancake shape. At this stage of the coalescence (the middle row of Figure 3.6), the liquid from the two droplets appears to be completely mixed together. However, this is not the case; the apparent mixing is an illusion. As the combined droplet starts to recoil under the action of surface tension, it becomes clear that the liquid has not significantly mixed and dyed liquid can be seen sitting on top of the undyed liquid.

While the droplet impact clearly creates a large 'interface' between the dyed and undyed liquid, there is no folding of that interface to cause striations like those in Figure 3.4. Note also that since the droplets are of approximately

---

‡The Weber number measures the relative importance of inertial *versus* surface tension forces.

Figure 3.6 Experimental observation of a 1.2 mm radius dyed droplet impacting on a sessile droplet of similar volume. The impact speed is approximately 1 m s^{-1} and the liquid is a glycerol-water mixture with kinematic viscosity $v = 7 \times 10^{-4}$ m^2 s^{-1}. Reprinted with permission from ref. 17. Copyright (2013) by the American Physical Society.

equal volume, there is no space within the sessile droplet for a vortex ring to be formed. Hence the action of advection here is limited to a single expansion of the dyed/undyed liquid 'interface', and the only mechanism for mixing the liquids is the slow process of diffusion. The hemispherical droplet shown in Figure 3.2 has an equivalent volume to the combined droplet in Figure 3.6. In fact in the experimental system, diffusion will be even slower than indicated in Figure 3.2, because the diffusion coefficient for the glycerol-water mixture used in the experiments is closer to 8×10^{-11} m^2 s^{-1}[18] rather than the 10^{-9} m^2 s^{-1} used to generate Figure 3.2.

The millimetre-scale droplet experiments based on matching inkjet-scale Reynolds and Weber numbers are helpful in revealing the effects of the flow on the mixing of the two droplets, but a key parameter that is not matched is the Péclet number, eqn (3.4). Using the glycerol–water diffusion coefficient above, together with the radius and speed of the impacting drop as length and velocity scales, the Péclet number for the system in Figure 3.6 is roughly 1.5×10^7, and confirms that diffusion is negligible over the timescale of the droplet impact and coalescence.

What about the Péclet number and the anticipated effect of diffusion for a typical inkjet-sized droplet impact? Taking a representative inkjet droplet radius of 25 μm as the length scale, a typical impact speed of 5 m s^{-1} for the velocity scale, and using a more generous diffusion coefficient of 10^{-9} m^2 s^{-1}, we have a value Pe ~ 1.25×10^5. Although this is two orders of magnitude lower, it is still quite large (of the same order as in the right-hand column of Figure 3.4), suggesting that diffusion still has only a small role to play (particularly as the diffusion coefficient is likely to be smaller, and Pe larger, for complex *e.g.* polymer molecules).

However, it is perhaps more helpful to estimate the distance over which diffusion can reasonably act. In the 20 ms impact and coalescence period shown in Figure 3.6, the dyed/undyed liquid 'interface' can be expected to diffuse to a thickness of about $2\sqrt{Dt}$ ~ 9 μm, and about half this distance after just 5 ms. This would be of the order of the thickness of the disc/pancake shape formed during impact and coalescence (see Figure 3.6) if the droplet radius were about 25 μm. Hence, for inkjet-sized droplets it is reasonable to expect that substantial intermixing of the dyed and undyed liquid could occur during the period when the combined droplet (and the 'interface' between its components) is stretched out—see the sketch in Figure 3.7. The recoil stage shown in the bottom row of Figure 3.6 would then look quite different and show a more homogeneous state.

Figure 3.8 shows further observations from the same set of experiments, and illustrates the effect of having the impacting droplet offset from the sessile one, so that the impact is no longer axisymmetric. As before, the two components of the combined droplet emerge as separate parts after the disturbance of the impact. For small offset (*e.g.* the 1 mm row in Figure 3.8), the combined droplet again flattens into a pancake with the dyed liquid almost covering completely the undyed liquid. This is the opportunity for molecular diffusion to act to start homogenising the mixture before the droplet recoils and the dyed/undyed 'interface' shrinks again. In fact, in the 20 ms and 100 ms images for the 1 mm offset (Figure 3.8) there is evidence of some blurring of the 'interface' between the dyed and undyed liquid and hence mixing of the liquids. Scaling the droplets down to typical inkjet droplet size, this blurring would be expected to be much more pronounced by the argument above.

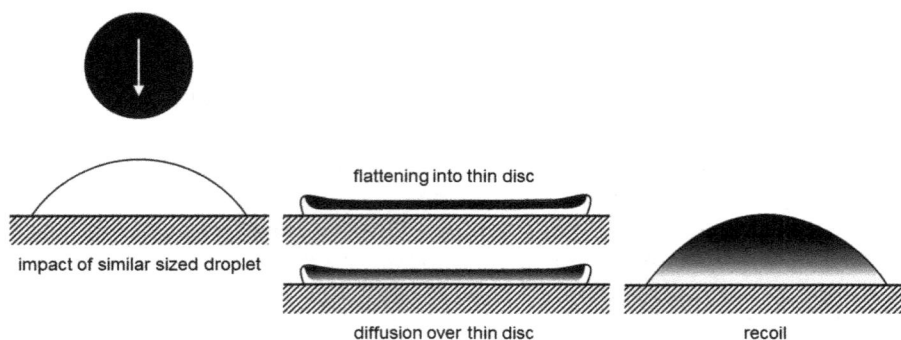

Figure 3.7   Sketch of the impact of a droplet upon another of similar size.

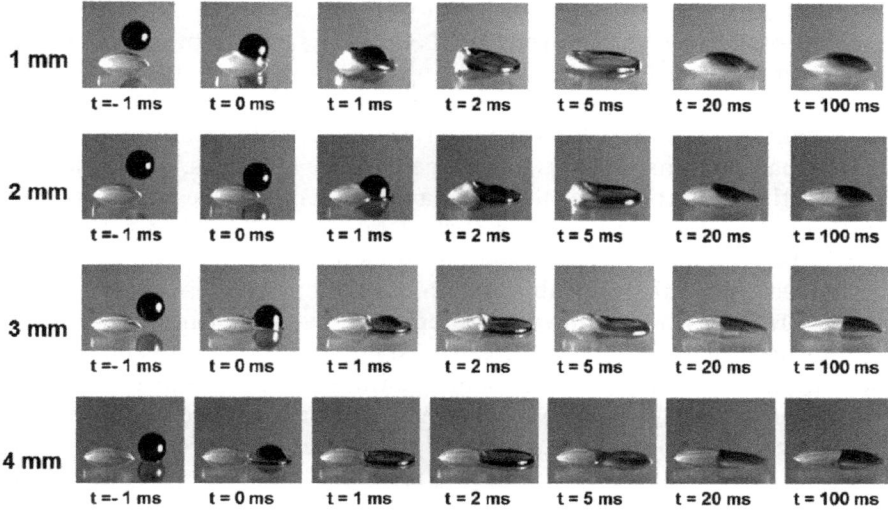

Figure 3.8 Offset impact of a dyed droplet of radius 1.2 mm on a sessile droplet of similar volume. The dyed droplet is released at the lateral offset between the centres of the droplets indicated in each row. The impact speed is approximately 1 m s^{-1} and the liquid is a glycerol–water mixture with kinematic viscosity $v = 7 \times 10^{-4}$ m^2 s^{-1}. Reprinted with permission from ref. 17. Copyright (2013) by the American Physical Society.

As the offset between droplets increases, and especially when the first contact of the dyed droplet is with the solid rather than the sessile droplet (bottom two rows of Figure 3.8), the dyed/undyed 'interface' becomes much smaller and is less stretched by the disturbance of the impact. The dyed droplet no longer spreads over the top of the sessile droplet; it still forms a pancake shape, but this is to the side of the sessile droplet, and the 'interface' now arises only at the edge of the pancake, where it pushes sideways into the sessile droplet. This clearly reduces the opportunity for mixing by diffusion during the impact, and in the 20 ms and 100 ms images for the larger offsets the boundary of the dyed part of the droplet appears much sharper than for the 1 mm offset.

The behaviour seen in Figures 3.6 and 3.8 has also been seen in several numerical simulations, both in our work[17] and more recently in that of others.[19,20] These simulations use different 3D lattice Boltzmann methods to model the two-phase flow of the droplets and surrounding medium, and examine the mixing of the two droplets by advecting large collections of passive tracer particles *via* eqn (3.6)—as was done in the cavity simulation of Section 3.2.2. Hence no diffusion effects were captured. The simulations show that the liquid originating in each droplet remains in two distinct adjacent regions within the combined droplet, with no significant inter-penetration, though oblique droplet impacts appear to help extend and distort the boundary between the two liquids.[20]

Note that in the experiments shown in Figures 3.6 and 3.8, the surface tensions of the dyed and undyed liquids were the same. In practical reactive inkjet printing, it is likely that the mixing liquids will have surface tensions that differ to some degree. An important mechanism that could help to spread out the boundary between the two liquids in the combined droplet after impact is transport due to surface tension gradients, *i.e.* Marangoni flows. If the two droplets coalescing have different surface tensions, upon contact the liquid with the lower surface tension will be driven along the free surface in order to reduce the overall surface energy, and the resulting movement would inevitably stretch out the boundary between the two liquids. This has been shown to be effective in promoting internal mixing during coalescence of two freely suspended droplets (*i.e.* not in contact with a solid).[21] For droplets on a substrate, the same effect would also occur, though it is likely to be somewhat more limited because of the viscous drag of the solid surface and the restricted dynamics of the contact line. Indeed this is seen to be the case for coalescence of two sessile droplets discussed below.

3.3.2   Coalescence of Two Sessile Droplets

As the offset of the two droplets in Figure 3.8 is increased, eventually a limit is reached beyond which coalescence of the droplets does not occur because the spreading of the impacting droplet is insufficient to reach the sessile droplet. Near this limit, the coalescence of the droplets essentially becomes a coalescence of two static droplets. The coalescence of two sessile droplets has been more widely studied than the impacting and coalescing case, though it is perhaps less important for most reactive inkjet purposes, so only a few examples are discussed here.

van Dongen *et al.*[22] inkjet printed two droplets containing different coloured fluorescent dyes onto a glass substrate at a separation large enough to prevent them from coalescing. Coalescence was then triggered by irradiating the substrate with UV light to lower the contact angle and hence cause spreading of the droplets. The droplets (with contact radius about 100 μm) were then observed from above. Upon contact between the droplets, a narrow bridge forms and rapidly expands due to surface tension (see the sketch in Figure 3.9), drawing in liquid from both droplets. However, for droplets of equal size, the symmetry of the system means that no mixing is possible except by diffusion. Indeed, van Dongen *et al.*[22] found the relative positions of the dyes to advance at a rate consistent with Fick's law of diffusion, with a diffusion coefficient of 2×10^{-9} m^2 s^{-1}, when the droplets were of equal size. For unequal droplets, they used deviations from the theoretical pure-diffusion rate of advancement to identify the contribution of advection (due to higher Laplace pressure in the smaller droplet) to the transport of the dye. This was also visible in the overhead views of the dyes in the composite droplet, which showed transport of dye from the smaller droplet into the larger one along the outer edges of the neck region between the droplets.

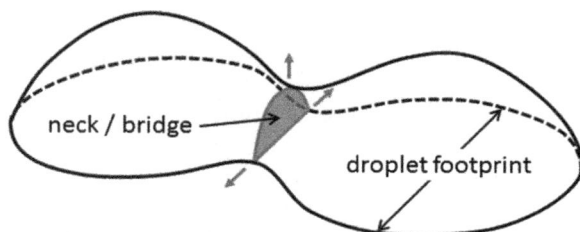

Figure 3.9   Sketch of two sessile droplets coalescing.

The degree of penetration is rather small, however, and the main mechanism of mixing is still diffusion.

Lai *et al.*[23] and Yeh *et al.*[24] explored the mixing dynamics of two differently dyed sessile droplets coalescing on a self-assembled monolayer surface with a wettability gradient, which drove one droplet into the other. They used micro-PIV to observe the rapid but short-lived flows between the droplets over the first 100 ms after first contact. These advective currents transport material between the droplets. The mixing patterns over much longer times-cales were visualised using micro laser induced fluorescence and confocal microscopy to build up 3D images of the regions of the combined droplet. These images showed the stretching and distortion of the boundary between the two dyes caused by the earlier advection, which assists the eventual homogenisation *via* diffusion. In the later work,[24] droplets with different sur-face tensions were also considered, and these did produce more pronounced advective mixing, as a result of the low-surface tension liquid moving around the outside of the composite drop to minimise the surface energy. However, the effect was not as extensive as in freely suspended droplets[21] since, as remarked above, the presence of the solid surface damps out the movement within the combined droplet.

3.4   Reactions Within Coalescing Droplets

So far this chapter has explored mixing within droplets without explicitly considering the effect of reactions taking place simultaneously. Here we point out some interesting findings in the recent literature. Yeh *et al.*[25] have extended their previous work on mixing dyes in sessile droplets on a wettability gradient[23,24] to examine the progress of a chemical reaction. Specifically, a droplet of sodium hydroxide solution was mobilised by a wettability gradi-ent to coalesce with a stationary droplet of pH indicator phenolphthalein dissolved in ethanol. As a control, on the same surface, a droplet containing yellow dye was coalesced with one containing blue dye. The droplets used each had a diameter of contact with the surface of roughly 1 mm prior to coalescence, though the size of the stationary droplet was varied to change the radius of curvature of the neck (see Figure 3.9) between the droplets. The colour change in the droplets was observed from directly above.

For the reacting droplets, two features were made visible by the colour change: a smooth circular front and a fingered front, both of which penetrated into the stationary droplet. For droplets of equal size, both fronts advanced rapidly at the same rate such that within 2 ms more than half of the stationary droplet was covered by the smooth front. In the same time, the yellow and blue dyed droplets mixing purely by diffusion had produced only a narrow green band still within the neck region. Based on achieving a given area exhibiting colour change, the reactive coalescence was 100 times more efficient than the unreactive mixing. For larger stationary droplets, the smooth circular front still advances rapidly, but the fingered front is slower, and the fingers become less jagged. The fingering was attributed to a thermal instability of the chemical front, though no measurements of heat generation were reported.

Although Yeh *et al.*[25] did not seem to consider them, there are in fact several competing or cooperating effects at work in the coalescence process they studied—aside from the chemical reaction. These arise from the differences in material properties of the coalescing solutions: the phenolphthalein/ethanol in the stationary droplet has a lower density, lower surface tension, and slightly higher dynamic viscosity than the sodium hydroxide solution in the moving droplet. The higher surface tension of the NaOH will have a twofold effect. As already noted in Section 3.3.1, movement along the free surface towards areas with higher surface tension should be expected. The higher surface tension will also lead to a higher Laplace pressure within the NaOH droplet than in the stationary droplet, even when the droplets are the same size; this pressure difference will be increased in the cases where the stationary droplet is larger, and hence has a smaller curvature. This pressure difference would act to drive liquid from the NaOH into the stationary droplet, and this would explain the rapidly advancing smooth circular front. The effect of the pressure gradient would be further assisted by the higher density of the NaOH, which would be expected to produce a gravity-driven current.[26]

Though the observations of Yeh *et al.*[25] are revealing, they do suffer the same limitations of other studies based on views from directly overhead:[10,11] they do not show the full detail of what is happening through the depth of the droplet. However, they do illustrate that coalescence and mixing of reactive droplets is a complex process involving several phenomena in addition to the chemical reaction(s) going on. Essentially the same reaction has also been explored in mid-air collisions of much smaller (36–38 μm) droplets fired at each other from opposing nozzles and visualised using a stroboscopic technique.[27] Based on visual images and absorbance measurements, it takes about 2 ms to reach a homogeneous state, but the small size of the droplets means that details of reaction fronts cannot be seen. A later development of this system to detect fluorescence arising in the amidation reaction of dansyl chloride has been applied[28] under conditions where the colliding droplets flatten to form a disc, rather like Figure 3.7 but without the solid surface. This provides a side view of the coalescence but, as dye was not used, the visual images do not show the internal dynamics. By symmetry, though, a planar contact surface is expected in the middle of the disc and, indeed, fluorescence was detected in this region, around the outer rim of the disc, showing that the reaction was indeed taking place there.[28]

3.5 Conclusions

Molecular diffusion is ultimately responsible for complete homogenisation of miscible liquid systems, and acts along gradients in the concentration of the species being mixed. However, mixing can be accelerated by advective currents within the liquid, and deformations of the liquid domain, that stretch the material interface between the liquids being mixed and (ideally) fold this interface upon itself to create thin striations in the concentration field. This mechanism for reducing the distance over which diffusion needs to take place, while expanding the area over which it *can* take place, is the process by which mixing is achieved in *e.g.* electrowetting- or microchannel-based microfluidic droplet systems, despite the dominant viscous effects.

Isolating the role of flow and droplet deformation in the mixing of ink-jet-printed droplets can be achieved through experiments and numerical simulations with non-reactive millimetre-scale droplets with Reynolds and Weber numbers matched to typical inkjet conditions. The larger length scale means that the Péclet number is much larger than for inkjet-sized droplets, and hence diffusion effects become negligible compared to advection. Observations of impact and coalescence of these larger scale droplets reveals that, unlike in active microfluidic systems, the drop-on-drop impact offers only limited scope for stretching and folding of the interface between the liquids being combined. Hence, in such experiments, the droplets can appear not to mix.

There are, however, two mechanisms by which drop-on-drop impact can help achieve more rapid mixing than diffusion alone. When a small droplet lands on and enters a much larger one, a vortex ring can be created within the larger drop (see Figure 3.5). This stretches out the liquid from the smaller droplet and rolls up the leading edge of that liquid, thus creating striations and extended areas over which diffusion can be effective. The other mechanism applies when a droplet lands on another of similar size. If the impacting droplet has sufficient momentum, and the offset between the droplet positions is not too large, the combined droplet will be flattened into a thin disc or pancake shape, with the impacting liquid sitting in a thin layer above a thin layer of the pre-deposited liquid (see Figure 3.7). This action both stretches out the boundary between the two liquids and simultaneously reduces the distance over which diffusion needs to act to homogenise the liquid mixture, albeit for a limited time before the droplet recoils into a more spherical shape. For a large (*i.e.* millimetre scale) droplet, there will not be time for mixing by diffusion to advance very far across the layers, and after the droplet regains its shape, the liquids from each droplet will appear almost completely unmixed. However, when this mechanism is scaled down to inkjet-sized droplets, the flattened layers could in principle be only a few microns thick and so, even in the few milliseconds for which the droplet is flattened, substantial diffusive mixing across the layers could occur.

By its very nature, reactive inkjet printing involves bringing together liquids that are different in composition and hence are likely to have different material properties, *i.e.* density, viscosity and surface tension. These differences can all contribute to enhancing or modifying transport and hence mixing between coalescing droplets. Surface tension gradients produce Marangoni flows along the liquid surfaces from areas of low surface tension towards those with high surface tension, while higher surface tension also leads to a higher Laplace pressure within a given sized droplet, and this can lead to flow away from the higher pressure regions. Density differences can potentially lead to gravity-driven movement, though at inkjet droplet scales this is not likely to be significant. Differences in viscosity can lead to viscous fingering of advancing fronts driven by the above phenomena. When chemical reactions are also taking place, there are added complications of reaction fronts and heat generation, which will change local material properties that depend on temperature, as well as the creation of reaction products with potentially different intrinsic material properties from the initial reagents. In reactive inkjet printing, therefore, the humble droplet hosts many simultaneous interacting effects. Since these occur across the depth and breadth of a very small, confined three-dimensional space, and over multiple time scales, obtaining a complete picture of precise droplet mixing dynamics still offers an interesting area for further exploration.

Acknowledgements

The experiments presented in Figures 3.6 and 3.8 were conducted as part of the Innovation in Industrial Inkjet Technology (I⁴T) Programme Grant (EP/H018913/1) funded by the Engineering and Physical Sciences Research Council.

References

1. E. L. Cussler, *Diffusion Mass Transfer in Fluid Systems*, Cambridge University Press, 2nd edn, 1997.
2. H. Aref, Stirring by chaotic advection, *J. Fluid Mech.*, 1984, **143**, 1–21.
3. M. C. T. Wilson, J. L. Summers, N. Kapur and P. H. Gaskell, Stirring and transport enhancement in a continuously modulated free-surface flow, *J. Fluid Mech.*, 2006, **565**, 319–351.
4. S.-Y. Teh, R. Lin, L.-H. Hung and A. P. Lee, Droplet microfluidics, *Lab Chip*, 2008, **8**, 198–220.
5. M. Abdelgawad and A. R. Wheeler, The digital revolution: a new paradigm for microfluidics, *Adv. Mater.*, 2009, **21**, 920–925.
6. D. R. Reyes, D. Iossifidis, P.-A. Auroux and A. Manz, Micro total analysis systems. 1. Introduction, theory, and technology, *Anal. Chem.*, 2002, **74**, 2623–2636.

7. H. A. Stone, A. D. Stroock and A. Ajdari, Engineering flows in small devices: microfluidics toward a lab-on-a-chip, *Annu. Rev. Fluid Mech.*, 2004, **36**, 381–411.
8. H. Song, M. R. Bringer, J. D. Tice, C. J. Gerdts and R. F. Ismagilov, Experimental test of scaling of mixing by chaotic advection in droplets moving through microfluidic channels, *Appl. Phys. Lett.*, 2003, **83**(22), 4664–4666.
9. P. Paik, V. K. Pamula and R. B. Fair, Rapid droplet mixers for digital microfluidic systems, *Lab Chip*, 2003, **3**, 253–259.
10. P. Kröber, J. T. Delaney, J. Perelaer and U. S. Schubert, Reactive inkjet printing of polyurethanes, *J. Mater. Chem.*, 2009, **19**, 5234–5238.
11. S. Fathi and P. Dickens, Challenges in drop-on-drop deposition of reactive molten nylon materials for additive manufacturing, *J. Mater. Process. Technol.*, 2013, **213**, 84–93.
12. J. J. Thomson and H. F. Newall, On the formation of vortex rings by drops falling into liquids, and some allied phenomena, *Proc. R. Soc. London*, 1885, **39**, 417–436.
13. A. V. Anilkumar, C. P. Lee and T. G. Wang, Surface-tension-induced mixing following coalescence of initially stationary drops, *Phys. Fluids A*, 1991, **11**, 2587–2591.
14. J. R. Castrejón-Pita, E. S. Betton, K. J. Kubiak, M. C. T. Wilson and I. M. Hutchings, The dynamics of the impact and coalescence of droplets on a solid surface, *Biomicrofluidics*, 2011, **5**, 014112.
15. X. Yang, V. H. Chhasatia, J. Shah and Y. Sun, Coalescence, evaporation and particle deposition of consecutively printed colloidal drops, *Soft Matter*, 2012, **8**, 9205–9213.
16. N. Ashgriz and J. Y. Poo, Coalescence and separation in binary collisions of liquid drops, *J. Fluid Mech.*, 1990, **221**, 183–204.
17. J. R. Castrejón-Pita, K. J. Kubiak, A. A. Castrejón-Pita, M. C. T. Wilson and I. M. Hutchings, Mixing and internal dynamics of droplets impacting and coalescing on a solid surface, *Phys. Rev. E*, 2013, **88**, 023023.
18. G. D'Errico, O. Ortona, F. Capuano and V. Vitagliano, Diffusion coefficients for the binary system glycerol+water at 25 °C. A velocity correlation study, *J. Chem. Eng. Data*, 2004, **49**, 1665–1670.
19. K. A. Raman, R. K. Jaiman, T. S. Lee and H. T. Low, Lattice Boltzmann study on the dynamics of successive droplets impact on a solid surface, *Chem. Eng. Sci.*, 2016, **145**, 181–195.
20. K. A. Raman, R. K. Jaiman, T. S. Lee and H. T. Low, Dynamics of simultaneously impinging drops on a dry surface: Role of impact velocity and air inertia, *J. Colloid Interface Sci.*, 2017, **486**, 265–276.
21. F. Blanchette, Simulation of mixing within drops due to surface tension variations, *Phys. Rev. Lett.*, 2010, **105**, 074501.
22. M. H. A. van Dongen, A. van Loon, R. J. Vrancken, J. P. C. Bernards and J. F. Dijksman, UV-mediated coalescence and mixing of inkjet printed drops, *Exp. Fluids*, 2014, **55**, 1744.
23. Y.-H. Lai, M.-H. Hsu and J.-T. Yang, Enhanced mixing of droplets during coalescence on a surface with a wettability gradient, *Lab Chip*, 2010, **10**, 3149–3156.

24. S.-I. Yeh, W.-F. Fang, H.-J. Sheen and J.-T. Yang, Droplets coalescence and mixing with identical and distinct surface tension on a wettability gradient surface, *Microfluid. Nanofluid.*, 2013, **14**, 785–795.

25. S.-I. Yeh, H.-J. Sheen and J.-T. Yang, Chemical reaction and mixing inside a coalesced droplet after a head-on collision, *Microfluid. Nanofluid.*, 2015, **18**, 1355–1363.

26. Y. Zhang, S. D. Oberdick, E. R. Swanson, S. L. Anna and S. Garoff, Gravity driven current during the coalescence of two sessile drops, *Phys. Fluids*, 2015, **27**, 022101.

27. Y. Takano, S. Kikkawa, T. Suzuki and J. Kohno, Coloring rate of phenolphthalein by reaction with alkaline solution observed by liquid-droplet collision, *J. Phys. Chem. B*, 2015, **119**, 7062–7067.

28. T. Suzuki and J. Kohno, Collisional reaction of liquid droplets: amidation of dansyl chloride observed by fluorescence enhancement, *Chem. Lett.*, 2015, **44**, 1575–1577.

CHAPTER 4

Unwanted Reactions of Polymers During the Inkjet Printing Process

JOSEPH S. R. WHEELER[†a] AND STEPHEN G. YEATES[*b]

[a]Infineum UK Ltd, Milton Hill Business & Technology Centre, Abingdon OX13 6BB, United Kingdom; [b]Organic Materials Innovation Centre, School of Chemistry, University of Manchester, Oxford Road, Manchester, M13 9PL, United Kingdom
*E-mail: stephen.yeates@manchester.ac.uk

4.1 Introduction

Polymeric additives are used for many applications at all stages of the inkjet process in both drop on demand (DOD) and continuous inkjet (CIJ) formulations.[1] Polymeric dispersants[2] and stabilisers ensure the ink has a long shelf life and operating lifetime, high molecular weight polymers control the ink viscosity to optimise the drop formation[3] and branched polymeric additives improve image permanence on the substrate post printing.[4] In the field of printed electronics polymeric material can be included as synergists in graphene inks[5] and conducting polymers can be printed by inkjet methods as the functional material itself.[6] Furthermore, biopolymers,[7] DNA[8] and

---

[†]JSRW's inkjet research was carried out as doctoral candidate at The University of Manchester and current employment address is provided for contact purposes.

Smart Materials No. 32
Reactive Inkjet Printing: A Chemical Synthesis Tool
Edited by Patrick J. Smith and Aoife Morrin
© The Royal Society of Chemistry 2018
Published by the Royal Society of Chemistry, www.rsc.org

59

even cells[9] can be printed by inkjet methods. However, there are several for-mulation challenges associated with high molecular weight polymeric addi-tives in inkjet printing.[10] This is due to the non-Newtonian behaviour flow behaviour of polymer solutions in the high extension rates found in an inkjet head[11] and can manifest as problems with printhead blockage,[12] inconsis-tent drop formation and instability of ink components.[13]

Prior to work carried out by al-Alamry *et al.* the assumption was the ink-jet printing process was benign with respect to components of ink formula-tion.[13] For inkjet to find further high technology applications it is vital that functional materials are not degraded by the inkjet printing process. This chapter will introduce key concepts of polymer molecular weights[14] and polymer solutions under flow both in rheological studies[15] and in the inkjet head.[16] Unwanted degradation reactions of polymers during inkjet printing will be introduced, explanations given on their occurrence in both DOD and CIJ systems and methods presented to reduce these effects through the use of branched materials in inkjet formulations.[13,17,18]

4.2   Polymer Definitions

Polymers are long chain molecules composed of many repeating sub units (monomers).[19] They are widely used in many consumer and industrial prod-ucts such as: inks,[1] coatings,[20] adhesives,[21] clothing,[22] personal care prod-ucts[23] and food.[24] This gives an incomplete list of the many uses of these materials but goes some way to demonstrate the ubiquity of polymeric mate-rials in modern life. Polymers can be divided into two broad classes: ther-moplastics[25] and thermosets.[26] Thermosets are cross-linked, do not melt and cannot be solutions processed or recycled. Thermoplastics are polymers which are not cross-linked, can usually be dissolved in a solvent, and in most instances will melt and flow.[25,26]

Polymers can be obtained both from natural sources[27] and by synthetic methods.[28] The properties of polymeric materials are determined by their constituent monomer and the molecular weight of the resulting polymer. Synthetic polymers can broadly be divided into two classes: step growth[28] and chain growth polymers.[29] Step growth polymers are formed from the reaction of difunctional monomers or multifunctional monomers, which react to form first dimers, then trimers, longer oligomers and eventually long chain polymers. Chain growth polymerisation involves the reaction of unsaturated monomers (generally alkenes) with an active site of the growing chain.[29] This active site is generated through the use of an initiator which can either be the decomposition product of a (relatively) unstable molecule, a highly reactive electrophile or nucleophile or by the formation of radical species through irradiation. The active site of a growing polymer chain can be radical,[29] cationic[30] or anionic.[31]

Polymer architecture is not limited to linear polymers chains. Poly-mers with exquisitely controlled architectures such as combs,[32] brushes,[33] controlled numbers of branches and stars[34] can be synthesised with a plethora of methods such as RAFT,[35] ATRP[36] and ROMP.[37]

4.3 Polymer Molecular Weights

When considering polymers for inkjet formulation the most important factor is the molecular weight of the polymer in question. Unlike small molecules in which it is simple to calculate the molecular mass from the chemical formula; polymeric samples will contain a distribution of different chain lengths and are therefore referred to as polydisperse. This is due to the random nature of the addition of monomers onto growing polymer chains. Some naturally occurring polymers have a single defined molecular weight and are referred to as monodisperse.

Therefore, it is necessary to consider how the number of polymer chains n_i with mass M_i varies with $\sum_i M_i$, *i.e.* the distribution of molar mass. Strictly, polymer molar mass distributions are discontinuous, but for a high molecular weight polymers the number of species becomes very large so distributions can be represented as continuous curves, as shown in Figure 4.1.

In the continuous number-fraction distribution, $n(M)$ *versus* M, the area of a slice under the curve, $n(M)dM$, is the number-fraction with molar mass in the range M to $M + dM$. Similarly, for the continuous weight-fraction distribution, $w(M)$ *versus* M, the area of a slice under the curve, $w(M)dM$, is the weight-fraction with molar mass in the range M to $M + dM$. The example given in Figure 4.1 above is an example of the 'most probable' distribution, which is predicted by theory for radical-chain polymerisation or linear condensation polymerisation in certain situations. Many of the techniques available for the determination of molar mass provide a single value which, for a

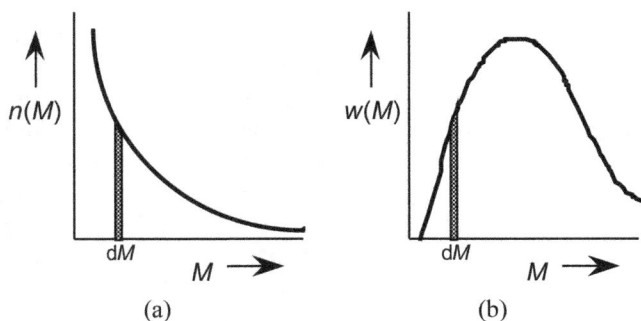

Figure 4.1   Illustration of (a) a number-fraction distribution and (b) the equivalent weight fraction distribution. Reproduced from S. D. Hoath, *Fundamentals of Inkjet Printing: The Science of Inkjet and Droplets,* John Wiley and Sons. © 2016, John Wiley and Sons.

polydisperse polymer sample, must therefore, be an average value. Important averages are:

$$\text{number-average } (\bar{M}_\mathrm{n}), \ \bar{M}_\mathrm{n} = \frac{\sum_i n_i M_i}{\sum_i n_i}$$

$$\text{weight-average } (\bar{M}_\mathrm{w}), \ \bar{M}_\mathrm{w} = \frac{\sum_i W_i M_i}{\sum_i W_i} = \frac{\sum_i n_i M_i^2}{\sum_i n_i M_i},$$

$$z\text{-average } (\bar{M}_z), \ \bar{M}_z = \frac{\sum_i W_i M_i^2}{\sum_i W_i M_i} = \frac{\sum_i n_i M_i^3}{\sum_i n_i M_i^2}.$$

Several important relationships between the different averages are given below:

polydisperse polymers, $\bar{M}_\mathrm{n} < \bar{M}_\mathrm{w} < \bar{M}_z$

uniform polymers, $\bar{M}_\mathrm{n} = \bar{M}_\mathrm{w} = \bar{M}_z$

most probable distribution, $\bar{M}_\mathrm{n} : \bar{M}_\mathrm{w} : \bar{M}_z = 1 : 2 : 3$.

The ratio $\bar{M}_\mathrm{w}/\bar{M}_\mathrm{n}$ is a measure of the width of the number distribution. This is strictly defined as the dispersity ($Đ$). This value is commonly referred to in the polymer literature as the polydispersity index (PDI). This is not strictly correct as the value of $Đ$ is not indexed to another value.[38]

The ratio $\bar{M}_z/\bar{M}_\mathrm{w}$ is a measure of the width of the weight distribution.

Analytical techniques to measure polymer molecular weight can either be classed as primary techniques (no calibration against standards of known molecular weight required) or secondary techniques, which require calibration. As has been previously stated polymer molecular weights do not have one single value. The value ($\bar{M}_\mathrm{n}$, $\bar{M}_\mathrm{w}$ or $\bar{M}_z$) provided by a certain analytical technique depends on the physical basis of the experiment. Techniques based on colligative properties (such as membrane osmometry or end group analysis) will calculate the $\bar{M}_\mathrm{n}$ as these methods are based on counting polymer chains without weighting them on the basis of their size.[39] Techniques which are sensitive to molecular size, such as light scattering, provide the $\bar{M}_\mathrm{w}$. The $\bar{M}_z$ can be measured by centrifugation experiments. These primary experiments are experimentally demanding, involving measurements on dilute polymer solutions[40] and have largely been superseded by chromatographic techniques such as gel permeation chromatography (GPC).[41]

GPC involves eluting polymer solutions through a column packed with beads of varying sizes. Higher molecular weight chains do not fit into the smaller void volumes of the packing material and therefore elute at the lowest elution times. Polymer chains of lower molecular weight will be impeded by the packing materials and progress through the column at a slower rate. A concentration sensitive detector (most commonly a refractive index detector) is placed at the end of the column. This produces a chromatogram of elution time *versus* concentration. Strictly speaking this technique separates polymer samples on the basis of different hydrodynamic volumes. However, it can be shown that hydrodynamic volume is proportional to molecule weight. GPC is also commonly referred to as size exclusion chromatography (SEC).[42,43]

This chromatogram is converted to a molecular weight distribution through calibration against the elution times of a set of monodisperse (in practise Đ < 1.1) polymer standards. These standards are generally polystyrene (PS) or poly methyl methacrylate (PMMA). This method of calculating the molecular weight distribution gives a value relative to the standard used. Workers in the field of polymer science must be aware that polymers of different chemistry may well adopt different solutions conformations relative to the standards. Therefore, samples with equivalent hydrodynamic volumes to the calibrants may well have very different molecular weights.[42]

More modern GPC technology solves these issues through the use of multiple detectors. The most common experimental set-up is a refractive index detector, light scattering detector and a viscometer. This method allows the measurement of the absolute molecular weight without comparing to standards. Triple detection GPC can also be used to characterise molecular architecture such as branching.[43]

4.4 Polymer Solutions

Here we consider inkjet printing inks which are deposited from an appropriate solvent. Due to the size of macromolecules the conformations of the polymer in solution is a key consideration; this is in contrast to considering molecules as non-interacting points as in ideal solution theory. The properties of polymer solutions are strongly determined by the interactions between the polymer and solvent. This impacts the physical properties of the solution such as viscosity and stability of the formulation with respect to temperature which are critically important in inkjet applications.[44]

When considering polymers in solution a single chain orientation is not representative of the dynamic nature of polymer chains in solution. Chain dimensions must be represented by an average size. There are various mathematical models for polymer chains in solution. These range in complexity from phantom chains in which monomer units are considered as points with zero dimensions, with all chain orientations equally allowed and monomers able to occupy the same point in space;[45] to energetically weighted distributions which do not allow for monomer units to occupy the same point in space with the range of monomer–monomer dihedral angles weighted by their energetic favourability and therefore, the likelihood of this orientation occurring in the polymer chain.[46]

If the interactions between segments of the polymer chain and solvent are unfavourable energetically (*i.e.* it is a poor solvent for the polymer) compared to polymer–polymer interactions, the polymers will adopt a tight coil conformation to maximise favourable polymer–polymer interactions. When polymer–solvent interactions are favourable, (a good solvent for the polymer), compared to polymer–polymer interactions the polymer will adopt an expanded self-avoiding walk conformation.[44] A special case is the theta (θ) condition in which the polymer adopts its ideal random walk orientation; in which polymer solvent interactions can be considered to be zero.[47] Therefore, when this condition is met the properties of the polymer chain itself can be investigated when unencumbered by any solvation effects.[47]

The nature of polymer–solvent interactions can be determined by the measurement of the intrinsic viscosity, $[\eta]$. This is the volume occupied by a unit mass of a polymer at infinite dilution. The value measured is not a viscosity, but rather an inverse concentration. The traditional method of measuring this quantity is by viscometry hence the term. Furthermore, the term reflects that the higher the intrinsic viscosity of a polymer the viscometric contribution of the polymer (the increase in solution viscosity per addition of polymer unit volume). The viscosity of a solution containing polymer of a known intrinsic viscosity can be calculated to a first approximation using the Einstein equation:

$$\eta = \eta_{\text{medium}}(1 + [\eta]c) \tag{4.1}$$

Where η is the predicted viscosity. η_{medium} is the solvent viscosity and c is the concentration of polymer.

The intrinsic viscosity can be measured using a triple detection GPC system or by viscosity measurements on dilute polymer solutions using an Ubbelohde viscometer. In the case of measurement of intrinsic viscosity by Ubbelohde viscometry the correct value is only calculated through extrapolation to an infinitely dilute solution to correct for polymer–polymer interaction. For a given polymer in a variety of solvents the larger the intrinsic viscosity, the more favourable the polymer–solvent interactions. The relationship between molecular weight and intrinsic viscosity is given by the phenomenological Mark–Houwink–Sakurada equation:

$$[\eta] = KM^a \tag{4.2}$$

where M is the molecular weight and K and a are empirical constants which are tabulated for most linear polymers in common solvent systems or determined experimentally. K and a are dependent upon the specific polymer-solvent system whilst a gives additional information on polymer architecture. For $a = 0.5$ the polymer is either in its Θ state or highly branched, 0.6 for a branched system, 0.6–0.8 for a coil conformation in a good solvent with an increase in α reflecting an increasingly inflexible linear conformation up to 2, which is an absolute rigid rod type polymer.[48] It is common to see α used interchangeably with a, but strictly speaking, a is the correct IUPAC notation for the Mark–Howink–Sakurada exponent.[49]

The intrinsic viscosity can be used to calculate the concentration at which polymers overlap in solution ($c*$). This was defined by Flory as $c* = 1/[\eta]$. The extent of chain overlap in solution is described by the reduced concentration (a unitless quantity), $\dfrac{c}{c^*} = c[\eta]$. The reduced concentration is an extremely useful description of polymer solutions as it allows solutions of widely differing molecular weight, molecular structure and solvent composition to be compared directly in terms of their chain overlap in solution. Inkjet formulation rules for polymer solutions to be discussed later in this chapter are described in terms of the reduced concentration.[50]

Polymer solutions as defined by de Gennes can be classified as being part of three concentration regimes which are shown qualitatively in Figure 4.2.[51] In a dilute polymer solution the reduced concentration is ≪1 and there are no frictional interactions between macromolecules in solution as the average separation between polymer chains is much greater than the radius of gyration (R_g) of the polymer chain. At very low concentrations polymer solutions show ideal solution behaviour; characterisation of molecular weight by primary physical methods often requires measurements to be carried out in this regime. High molecular weight polymer solutions suitable for DOD printing are typically in the dilute regime.[52]

As the concentration of polymer chains increases, frictional interactions between polymer chains are observed manifesting as a transition in solution properties. An example of this is the increase in the rate of increase of solution viscosity with increased polymer concentration. This transition occurs when the polymer concentration is of the same order of magnitude of $c*$. This is referred to as the semi-dilute region as the physical properties of the solution are dominated by that of the polymer chains but the volume fraction of polymer is low. The polymer solution in the semi-dilute regime is still a discernible mixture of polymer and solvent molecules.[52] Due to the higher viscosities required by CIJ formulations compared to DOD formulations, polymer solutions for CIJ printing are often in the semi-dilute concentration.[17]

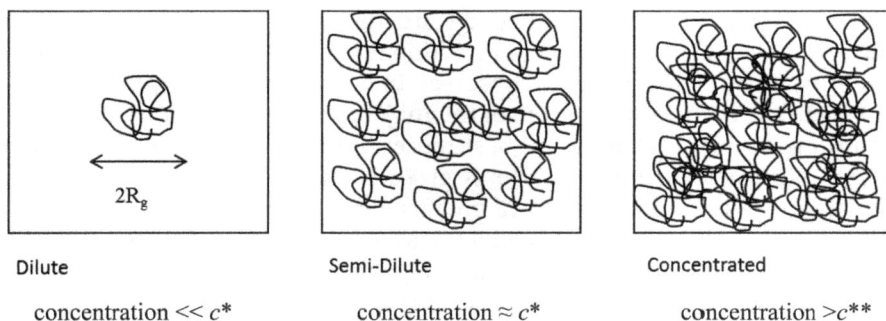

Dilute	Semi-Dilute	Concentrated
concentration ≪ $c*$	concentration ≈ $c*$	concentration > $c**$

Figure 4.2 A qualitative representation of the three regimes of polymer solutions.

As polymer concentration increases, such that the volume fraction of polymer becomes sufficiently high that it is not possible to discern between individual polymer chains; a second transition of solution properties is observed (c^{**}). On a molecular level the solution above this second transition is a continuum of overlapping chains. The movement of individual polymer chains through this continuum of polymer molecules and the prediction of the dynamics of concentrated polymer solutions is addressed by reptation theory. The diffusion of polymers through this viscous media is considered akin to a reptile (hence repatation) slithering through a tube of other polymer molecules.[53,54] Due to the high viscosities and elastic behaviour of concentrated polymer solutions they are not printable by inkjet methods.

4.5 Polymer Solutions in Flow Conditions

When considering polymers in the inkjet printing process the reader must be mindful of the dynamic nature of the inkjet process and the constrained nature of the inkjet head.[55] The orientations a polymer can adopt in solution was discussed in the context of the Mark–Howink–Sakurada equation. However, a polymer chain when exposed to extreme flow conditions can undergo drastic changes in its conformation.[56] The dynamic nature of the polymer chain when exposed to the high stress inkjet head impacts both: the maximum printable concentration of polymers and the molecular weight stability of any polymeric solute. Before concentrating on inkjet specific flow behaviour a brief description of polymer solutions under different rheological conditions will be given.

The behaviour of solutions exposed to a shear flow can be described as Newtonian (no change in viscosity with increasing shear rate), shear thickening (viscosity increase with increasing shear rate) or shear thinning (viscosity drop with increasing shear rate).[57] A shearing flow can be considered a combination of a rotational and elongational component. This results in dynamic processes in the polymer chain with chain extension and relaxation observed along with tumbling of the molecule. Polymer extension and recoil in a shearing flow occurs on a wide variety of timescales.[56] Therefore, a single discrete transition at a critical shear rate from thermodynamically stable coil conformation to chain extended state (the coil–stretch transition) is not observed. Furthermore, the mean fractional extension of polymer chains in shearing flows is not as high as observed in purely extensional flows.[56–58]

Due to the constrained nature of the inkjet head the behaviour of polymer solutions in inkjet printing are best represented by elongation flow experiments; an excellent introduction is given by Barnes.[59] In an elongational flow the rotational component is much reduced compared to a shear flow. There are many parallels between elongational flow experiments and inkjet behaviour of polymer solutions especially with respect to molecular weight stability. The behaviour of high molecular weight polymers at a range of concentrations and polydispersities under elongational flow were investigated by Odell and co-workers.[60–64] The work was carried out using crossed slots

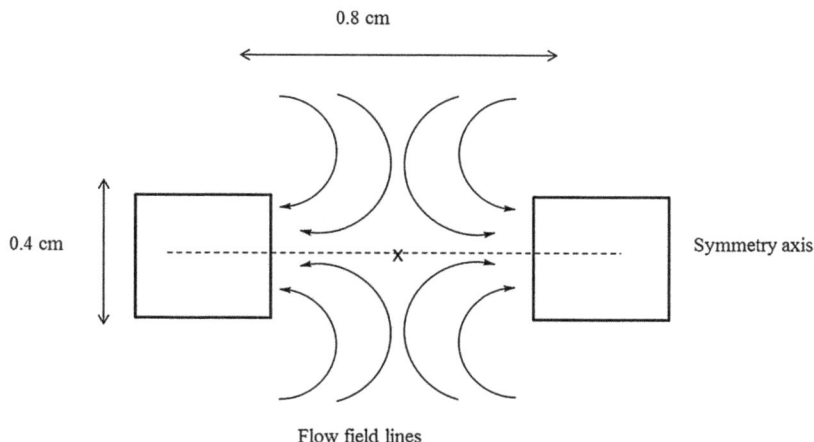

Figure 4.3   A representative schematic of the elongation flow experiments carried out by Odell and co-workers. The flow field is denoted by the curved lines with the stagnation point denoted by x. Polymers caught at the stagnation point chain extend.

apparatus (Figure 4.3) in which two opposing jets of solution with suction apparatus at a right-angle to the jets of polymer solution. This experimental set-up creates a steady Poiselle flow with a stagnation point in the centre of the apparatus which is illuminated by a HeNe laser. When a high molecular weight polymer is caught in the stagnation point the polymer cannot diffuse away. The polymer is extended from its equilibrium Gaussian coil conformation to a chain extended state due to the opposing flows running parallel to the symmetry axis of the experimental set-up; the centre of these opposing flows is the stagnation point.

At equilibrium a polymer solution is isotropic. When a velocity gradient is applied to a polymer solution the statistical distribution of polymer become anisotropic above a critical flow rate. This is due to the polymer molecules lining up with the velocity gradients. This is known as the Maxwell effect, or more commonly birefringence.[65] A birefringent polymer solution has different refractive indices for polarised light parallel or perpendicular to the flow. Birefringence is an appropriate technique to probe polymer solution behaviour as it is innate to the polymer under investigation and is not affected by dust or any (non-polymeric) contaminants. This is in unlike conventional light scattering. Above a critical elongational strain rate the birefringence increases dramatically to a plateau value due to strain induced enhancement of polymer concentration due to molecules caught at the stagnation point.[60–64]

This strain induced birefringence (Figure 4.4(a)) is interpreted as the polymer undergoing the coil stretch transition at the stagnation point; the polymers in solution go from random Gaussian coils to an anisotropic rigid rod conformation. The critical strain rate for the onset of birefringence is found

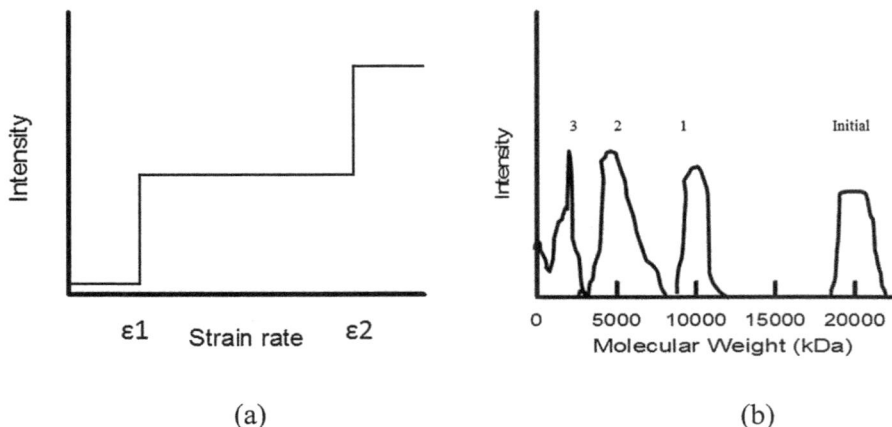

(a)   (b)

Figure 4.4   (a) Idealised birefringence intensity with strain rate in elongational flow experiments. (b) The development of molecular weight of a monodisperse 2000 kDa polymer when exposed to elongational flow. Each peak is labelled to denote the number of degradation events which resulted in polymers with this molecular weight distribution being observed.

to be proportional to $1/M^{1.5}$. This allows birefringence measurements to be used to both measure molecular weight distributions and follow any changes in molecular weight as the polymer is exposed to a high elongational flow (Figure 4.4(b)).[66,67]

The coil stretch process can be quantified by considering the critical Weissenberg number (Wi). Wi is the product of the relaxation time of a process and an applied stress: $Wi = \lambda \gamma_{crit}$. In this case we consider the relaxation time of a polymer from a chain extended state (λ) to its thermodynamically stable Gaussian coil conformation and the elongational strain rate (γ_{crit}). When Wi is greater than 0.5 the highest molecular weight chains in a sample under elongational flow will all be in a chain extended state.[13]

The Zimm relaxation (time is used in the analysis of samples under elongational flow (λ_z) and DOD inkjet printing) as it calculates the longest possible relaxation time of the polymer under consideration:[13]

$$\lambda_z = \frac{\eta_s \cdot [\eta] \cdot M_w}{RT} \qquad (4.3)$$

Where $[\eta]$ is the intrinsic viscosity, η_s is the solvent viscosity, M_w is the weight average molecular weight, R is the Gas constant and T is the absolute temperature. Elongational stress post coil stretch transition will be placed directly onto the polymer backbone. When this reaches a critical value the polymer chain will undergo irreversible molecular weight degradation.[13]

To investigate the molecular weight stability of polymers under elongational flow the molecular weight distribution was determined (both by GPC measurements and birefringence), the strain rate ramped up then reduced and the molecular weight distribution re-measured. It was found that monodisperse polymers exposed to these high strain environments degrade centrosymetrically and by ramping up the strain rate it was possible to observe successive centrosymmetric degradation (Figure 4.4(b)). This approach verified the theoretical conjecture that chain halving would be the predominant scission mechanism. It was found that the fracture strain rate, γ_f, scaled as $1/M^2$. This gives the result of a molecular weight ceiling to coil stretch transitions being possible without an attendant reduction in molecular weight. Coil stretch transitions scale at a strain rate of $1/M^{1.5}$ whereas fracture strain rate γ_f scaled as $1/M^2$. At an M_w value of around 30×10^6 Da the fracture rate is less than that of the coil stretch transition strain rate. This molecular weight to the coil stretch transition has been confirmed experimentally.[60–67]

This behaviour of more concentrated polymer solutions and polydisperse samples under elongational flow is different to that of monodisperse dilute samples; chain overlap effects and a range of relaxation times in solutions result in random degradation of these systems under elongational flow. The forces which cause the polymers to degrade are transmitted through entangled valence bonds rather than the hydrodynamic forces observed in dilute monodisperse polymers. As the constrained polymer attempts to undergo its coil to stretch transition, or relax back to its equilibrium conformation, it can place a strain on overlapping polymers such that they are forced into unfavourable conformations. As there is no chain position dependence on entanglement of polymer chains; the polymer degradation becomes random as the chain is ripped apart at the chain crossover points.

It is also found that polydisperse samples are more susceptible to flow induced degradation than monodisperse samples.[67] Polydisperse samples require a lower elongational strain rate for degradation to be observed. Monodisperse samples have a single relaxation time (in practice a very small range of values due to the small range of differing molecular weight as perfectly monodisperse polymers are not accessible by synthetic methods). Therefore, the polymers will chain extend and relax back to their thermodynamically stable coil state at the same time. It is possible that this process resists the elongational flow or interactions between aligned chains allow it to deform elastically thus conferring extra stability on the monodisperse samples. Monodisperse polymers show greater resistance to flow induced degradation in several high stress environments such as sonication,[68] elongational flow,[67] compressional flow[69] and DOD inkjet printing.[69]

Compressional flows are also of interest and applicable to inkjet printing. Rather than featuring a stagnation point the flow is directed through a narrow channel (Figure 4.5). Polymers in compressional flow also show centrosymmetrical degradation behaviour. This is informative as the polymer chains are not reaching a full coil stretch conformation in solution due to the short residence time in the compressional flow field. This suggests that the polymers are snapping whilst the coil is unravelling; in all probability while adopting a dumbbell configuration as shown in Figure 4.5(b).[69]

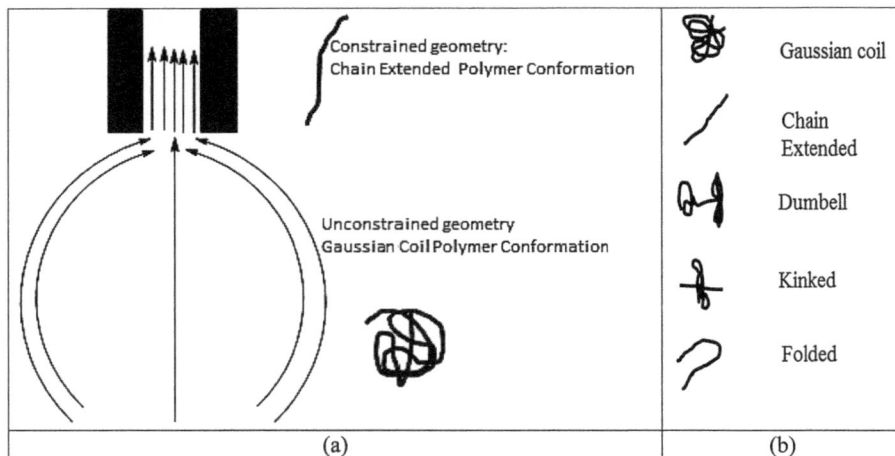

Figure 4.5   (a) An idealised schematic of polymers under flow in constrained and unconstrained geometries. (b) The possible orientations adopted by polymers under flow.

Turbulent flows can also result in polymer degradation.[70] Experiments carried out by Horne and Merril show that both DNA and low dispersity high molecular weight polystyrene degrade centrosymetrically when exposed to a turbulent flow in an apparatus designed to eliminate any macroscale extensional effects. This is informative as DNA is a rigid rod type polymer whereas polystyrene in solution will adopt a coil conformation. Any flow induced strain would obviously be placed at the centre of a DNA molecule, but the similarity in the degradation modes shows the degradation of polystyrene is occurring through the same mechanism; the polymer is adopting a rigid rod conformation and then snapping as the force along the polymer backbone caused by the flow field concentrates on the centre of the macromolecule. The degradation behaviour of dilute high molecular weight monodisperse solutions has been shown to be very similar in both elongational and turbulent flows. This suggests that degradation in turbulent flow is due to areas of high elongational flow in the turbulent systems.

4.6   Inkjet Formulation Rules

Before considering the behaviour of polymer solutes during the inkjet process an introduction will be given into known inkjet formulation rules. The printability of inkjet formulations can be predicted by considering the Reynolds (Re) and Weber (We) numbers. The Reynolds number both relates to the ability of the pressure wave to progress through the ink solution in the print head before drop ejection and the interplay of viscoelasticity in the ligament post drop ejection. The Weber number is important during the early stage of drop ejection where the forward inertia of the fluid needs to be greater than the surface tension of the fluid to force a printing drop out of the printhead.

The Reynolds and Weber numbers are given in eqn (4.4) and (4.5):

$$We = \frac{\rho v^2 L}{\sigma} \tag{4.4}$$

$$Re = \frac{\rho v L}{\eta} \equiv \frac{v l}{v} \quad as \quad v = \frac{\eta}{\rho} \tag{4.5}$$

Where ρ is the density, σ is the surface tension, v is the drop velocity, η is the dynamic viscosity, v is the kinematic viscosity and L is a characteristic length (often drop length).

Simulation work by Fromm calculated different drop morphologies for with $\frac{Re}{We} > 2$.[71] However, this work did not explicitly state printability ranges for ideal inkjet fluids. This approach was refined by Derby and Reis which considers the Ohnesorge number[72,73] (eqn (4.6)):

$$Oh = \frac{\sqrt{We}}{Re} = \frac{\eta}{\rho^{\frac{1}{2}} \sigma^{\frac{1}{2}} L^{\frac{1}{2}}}. \tag{4.6}$$

The advantage of using Oh to describe inkjet phenomena is that it removes the dependence on drop velocity thus making it applicable to many different print configurations and printing equipment. A combination of printing studies and fluid dynamics simulation found the printablity regime for DOD fluids to be $0.1 < Oh < 1$. At low Oh the ligament connecting the printhead to the printing drop shatters forming many satellite drops which are deleterious to print quality; when $Oh > 1$ viscous dissipation of the pressure wave prevents drop formation. Some examples in the literature express the printability in terms of the inverse of Oh (Z). It must be noted that Oh does not predict printability conditions imposed by non-Newtonian behaviour of high molecular weight polymers. However, consideration of Oh is an excellent starting point for the formulation of an inkjet fluid.[72,73]

Before considering how the extensibility of polymeric solutes in inkjet fluids both affects the maximum printable concentration and molecular weight degradation; a qualitative description of the four regimes of inkjet drop formation found in printing polymeric DOD fluids is described in Table 4.1. The relative concentration ranges for each regime for a given ink is determined by the molecular weight of any polymeric solute in the inkjet formulation.[12]

4.7 The Maximum Concentration of Polymers Printable by Inkjet Methods

It is found that the maximum printable concentration of polymeric inks is inversely proportional to molecular weight and can be described using three different regimes.[11,13,74] The transition between the regimes reflects a change in chain dynamics when exposed to the high elongation rate of the inkjet nozzle. Both theoretical studies using finitely extensible non-linear elastic,

Table 4.1   The drop formation behaviour of polymer solutions in the different concentration regimes.

Regime	Drop formation behaviour	Example picture
1	The first regime occurs at either very low concentrations and/or molecular weight, where a long tail is formed that simultaneously disintegrates along its axis to form several satellite droplets. This regime can often be highly chaotic and irreproducible in nature. Characteristic of the dilute polymer regime.	
2	The second regime occurs upon increasing concentration. The ligament breaks into fewer satellites as the ink becomes more viscoelastic and more of the ligament retracts into the printing drop.	
3	Raising concentration or molecular weight yields a single droplet without a tail. Any ligament connecting the printhead to the drop post drop ejection is retracted into the main printing drop. Characteristic behaviour of dilute to semi-dilute solutions. This is ideal printing behaviour.	
4	At high concentration or molecular weight the polymer solution becomes highly viscoelastic. The droplet either is not ejected from the head or the ligament connecting the printing drop to the does not break forming "bungee drops" as the droplet is drawn back into the printhead.	

(FENE), dumbbell simulations and practical studies investigated the effect of inkjet printing polymer solutions at different concentrations were combined to form a model for polymer behaviour in the inkjet head.[11,74] The three observed regimes with the characteristic exponents to the M_w which predict the maximum printable concentration are given in Figure 4.6.

The regime describing the maximum printable concentration of polymer is a function of solvent quality and the critical Weissenberg number. It is possible to calculate the effect of the high extension rate in the inkjet head on

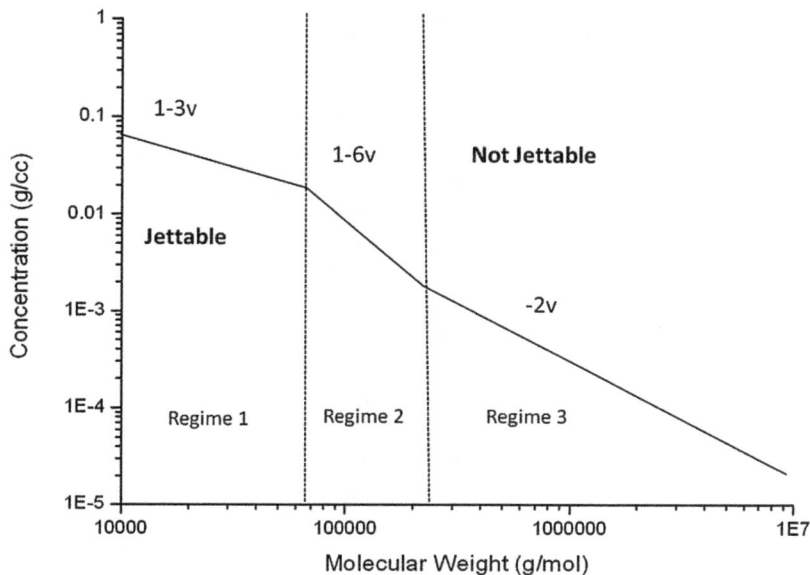

Figure 4.6 The maximum printable concentration of linear polymer solutions by inkjet printing. The representative molecular weight exponent is given for each regime.

polymeric solutes of differing molecular weight. This can be used to explain the maximum printable concentration–molecular weight relationships reported by al-Alamry *et al.*[13] The printable concentration of polymer solution, assuming a constant drop velocity across the measured polymer solutions, scales as an exponent of molecular weight. At low molecular weight the M_w is raised to a factor $(1-3v)$, where v is a solvent quality coefficient, at intermediate molecular weight the exponent is $(1-6v)$ and at high molecular weight the exponent is $-2v$ printing of polymer solutions. In Regime 1 the limiting parameter to the maximum printable concentration is the zero shear viscosity, as the polymer solute is in its Gaussian coil state during the DOD printing process. The transition point between Regime 1 and Regime 2 is a Weissenberg number of 0.5; as previously stated this is when the highest molecular weight fractions of a polymer MWD passes from a random coil to a chain extended state. In this regime the limiting parameter to the maximum printability is the viscoelasticity of the polymer solution. The second transition is where the elongational strain rate is $1/L$, where L is the polymers' finite extensibility limits. The finite extensibility is the ratio of the equilibrium coil to the fully extended length of the polymer. In this regime the limiting parameter is the extensional viscosity as the solution behaves as rigid rods. As rigid rod molecules occupy a considerably larger hydrodynamic volume than coil conformations; this results in a log jam effect blocking the printhead even at low concentrations.[74]

4.8   Degradation of Linear Polymers During DOD Printing

It has previously been described how high molecular weight polymers undergoing a critical elongational strain rate will undergo the coil–stretch transition. This places stresses directly on the polymer backbone; when these stresses rise above a critical value the polymer can undergo irreversible chain scission. It is found that high molecular weight polymers ($M_W > 200$ kDa) can undergo unwanted degradation processes when printed using DOD methods (Figure 4.7).[13]

Below a molecular weight of 200 kDa, in which incidentally the polymer can be printed above its overlap concentration, the polymer does not undergo its coil–stretch transition during the DOD printing process (Regime 1). The molecular weight distributions of these polymers are not affected by the DOD printing process. However, the tacit assumption that all polymeric solutes were not changed by the inkjet printing process was shown to be incorrect by al-Alamry *et al.* in 2011. It was shown that high molecular weight ($M_W =$ 200–750 kDa) monodisperse polymers ($Đ < 1.3$) undergo centrosymmetrical degradation (the polymer snaps at the centre) at a reduced concentration of <0.15. This is strikingly similar to the degradation behaviour observed in elongational flow experiments and is interpreted as the polymer undergoing its coil stretch transition and snapping due to the elongational strain placed on the backbone during the printing process.[13]

It must be noted that ultra-high molecular weight polymers ($Mw > 1000$ kDa) are stable with respect to molecular weight degradation during the DOD printing process. To undergo degradation the polymer must undergo its coil stretch transition and be exposed to a critical elongational flow for a critical residence time. For these ultra-high molecular weight macromolecules the inkjet timescale is not sufficient for these processes to occur. However, this does not offer a great deal of help to formulators due to the extremely limited maximum printable concentrations of these materials and the synthetic

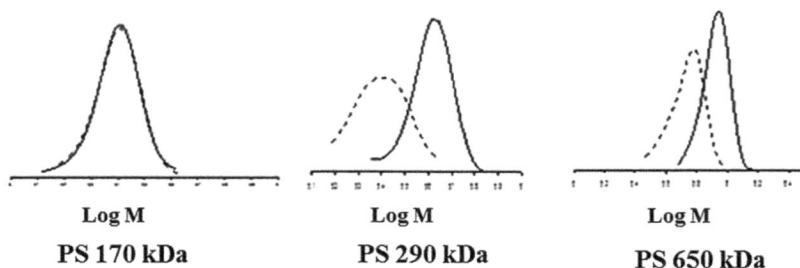

Log M   Log M   Log M

PS 170 kDa   PS 290 kDa   PS 650 kDa

Figure 4.7   The molecular weight distributions for high molecular weight polymeric inks of M_W 170, 290 and 650 kDa pre (solid line) and post (dotted line) inkjet printing. Reproduced from S. D. Hoath, *Fundamentals of Inkjet Printing: The Science of Inkjet and Droplets,* John Wiley and Sons. © 2016, John Wiley and Sons.

difficulty in producing/prohibitive cost of ultra-high molecular weight polymers.[13]

When high molecular weight polydisperse systems are printed by DOD methods random degradation is observed. This suggests that the degradation mechanism in these systems is different to that shown in monodisperse systems. It is postulated that degradation in polydisperse systems is transmitted through bond overlapping as the polymers relax back into their thermodynamically stable Gaussian coil orientation. Due to the range of molecular weights (and thereby relaxation times in the polymers undergoing the inkjet process) random overlapping of polymer chains would occur as the polymers relax back from a chain extended state. This results in potentially strained chain orientations which result in snapping of polymer chains. As these strained conformations could happen at all points along the polymer chain the degradation observed is random. In the case of both monodisperse and polydisperse systems two passes through a DOD printhead is sufficient to degrade a polymer sample molecular weight distribution to one that is stable with respect to further inkjet printing.[13]

Simulations carried out by McIllroy *et al.* rigorously demonstrate that critical elongational strain required for the polymer to undergo the coil stretch transition and the critical fracture strain rate, which results in polymer degradation, can only occur in the printhead.[75] The strain rates in the ligaments rapidly drop below the inverse of the polymer relaxation time post drop ejection and polymeric solute will relax to their thermodynamically stable Gaussian coil orientation. This demonstrates that printhead design is absolutely critical in the stability of high molecular weight polymers and indeed differential stability between monodisperse and polydisperse systems can be shown on the basis of printhead design.[75]

It is shown that polydisperse systems are degraded both using a Dimatix DMP system (average drop speed 6–10 ms^{-1}, nozzle diameter 23 µm, shear rate around 400 kHz) which features a sudden contraction from ink reservoir to nozzle and a Microfab system (average drop speed 1–4 ms^{-1}, nozzle diameter 50 µm, shear rate around 150 000 s^{-1}) which features a cone shaped ink chamber gradually tapering towards the printhead nozzle. Whereas monodisperse systems are only degraded using the Dimatix system which features a higher elongation rate due to the higher velocity of the drops coupled with the almost instantaneous transition of the polymers from a free to a constrained environment due to the printhead design.[13,75]

The elevated stability with respect to high elongation strain rates of monodisperse systems when compared to polydisperse systems is also shown in elongational flow experiments. The enhanced stability is interpreted as elastic effects as all the polymers in a monodisperse sample chain extend and relax in concert which dampens the elongational effects placed on each individual polymer backbone due to light structuring of the fluid. The polydisperse systems as previously mentioned will extend and relax across a range of timescales thus dampening effects are not seen.[13]

4.9 Degradation of Linear Polymers During CIJ Printing

DOD systems only produce printing drops when they are required to be deposited on a substrate and the ink only experiences the drop ejection process once. In contrast in CIJ systems the ink may experience many thousands of drop generation events since it may be recycled many thousands of times through the printhead before being deposited on a substrate. Furthermore, the CIJ process enables for formulations with a higher concentration of polymeric additives such that the inks are in the semi-dilute regime. This presents a challenge to formulators as the inks must be stable with respect to multiple high stress environments over a timescale of thousands of passes. This is in contrast with DOD systems where the ink must be stable with respect to a single pass through a single high stress environment (the printhead itself). It was found by Wheeler *et al.* that not only was there polymer degradation observed for high molecular weight materials during the CIJ printing process but that it was significantly different to that observed during DOD printing. This work was carried out by investigating formulations of PMMA of M_w: 468 kDa, 310 kDa and 90 kDa formulated to a viscosity of 4 mPa.s in methyl ethyl ketone (MEK). The polymers in these inks were significantly above their respective overlap concentrations.[17]

The observed change in molecular weight manifests as an increase in solids content of the inks. As continuous inkjet printers operate with a continuous recycling of ink through the printhead there is some inevitable solvent loss due to evaporation. This solvent loss requires in-built viscosity control systems, which keep the inks in their ideal performance range, and adds extra solvent (make up) as required. The corollary of this is that as the ink thins down due the degradation of high molecular weight polymer chains (which make the greatest contribution to solution viscosity) no extra solvent will be added and an increase in solids content will be observed. A higher concentration of low molecular weight material is required to keep the ink at its ideal printing viscosity.[17]

A striking feature of this increase in solids content is that it occurs across a timescale of many thousands of passes through a CIJ head. This is in contrast to DOD systems where the polymer is fully degraded after two passes through the head. The change in molecular weight distributions with printing time (Figure 4.8(a)) was investigated by GPC; it was found that PMMA 468 kDa and PMMA 310 kDa were degraded during the inkjet printing process. The lowest molecular weight formulation (PMMA 90 kDa) was stable with respect to molecular weight degradation. This is analogous to DOD systems where polymers below a critical molecular weight are not degraded.[17]

The behaviour observed in Figure 4.8(a) is rapid degradation of the highest molecular weight chain; it can be seen that the largest drop in M_w for the highest molecular weight polymer is most significant in the first 20 hours of printing time followed by a gradual decrease in molecular weight. It is also noted that the molecular weight distributions of PMMA 468 kDa and PMMA

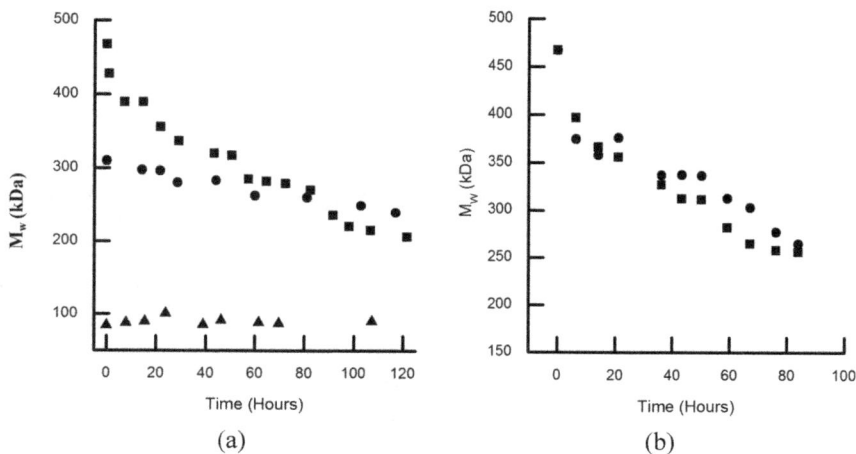

Figure 4.8 (a) The change in molecular weight with respect to printing time for three formulations of PMMA of molecular weight 468 kDa (■), 310 kDa (●) and 90 kDa (▲) (b) The change in molecular weight distribution for formulations of PMMA 468 kDa when printed using a full CIJ system (■) and circulated through the pumps, filters and ink systems of a CIJ printer with the printhead disabled (●). Reproduced under CC-BY 3.0 (https://creativecommons.org/licenses/by/3.0/) from J. S. R Wheeler, S. W. Reynolds, S. Lancaster, V. Sanchez-Romanguera, and S. G. Yeates, *Polym. Degrad. Stab.*, 2014, **105**, 116–121, DOI: 10.1016/j.polymdegrad-stab.2014.04.007. © 2014 The Authors. Published by Elsevier Ltd.

310 kDa are converging after around 80 hours printing time. This degradation is characterised by an increase in dispersity due to the increase in low molecular weight degradation products. This is interpreted as the individual polymer chains being degraded in multiple random degradation events due to the overlapping chains in solution undergoing high stress environments; 120 hours of printing equates to over 1000 statistical passes through the printer.[17]

The design of CIJ systems includes several high stress environments and it is found, counterintuitively, that the locus of polymer degradation in CIJ systems is not the passing of fluid through the printhead but shear induced degradation in the pumps. It can be seen in Figure 4.8(b) that the decrease in molecular weight of two identical solutions of PMMA 468 kDa is identical (not withstanding small experimental variations) whether the CIJ systems investigated featured a printhead or not. Indeed, when the CIJ system was further stripped back and simply consisted of passing a polymer solution through the pump, the same random degradation observed in the full printer was found. Furthermore, it was shown that ink formulation fired through a CIJ printhead with compressed air, at an equivalent drop velocity to a full CIJ system does not result in any change in molecular weight distribution of the polymeric solute. This is despite equivalent shear rates found in Microfab and Dimatix systems; the CIJ printhead nozzle is larger (65 microns) but the drop speed is significantly higher (22 m.s^{-1}). It further demonstrates that simply considering the shear rate at the point of drop ejection is not sufficient to predict degradation behaviour and that the nozzle design is vital when considering polymer stability with respect to the inkjet printing process.[17]

4.10   The Effect of Polymer Branching on Polymer Degradation During Inkjet Printing

It has been described how the challenges with respect to printing high molecular weight polymers such as: the limited maximum printable concentrations and the susceptibility to flow induced degradation are the result of these polymers undergoing the coil stretch transition when under elongational flow in the inkjet head. It is obvious that printing of high molecular weight materials in larger concentrations whilst maintaining the molecular weight of the printed material is desirable both to cut down on printing time (there would be a reduced requirement for complex print-dry-print protocols) and to maintain the physical properties of any deposited functional materials.

It was shown by de Gans *et al.*[55] that inkjet printed high molecular weight star polymers showed significantly lower ligament formation post printing when compared to solutions of high molecular weight linear polymers at equivalent polymer concentrations. The less pronounced ligament formation is due to the star polymer solutions showing lower elasticity during the printing process. This reduced viscoelasticity in the ligament due considerably shorter relaxation times of the star polymers; in the case of branched polymers the molecular weight of the arms rather than the total molecular weight of the polymer determines the relaxation time.[76]

This work was extended by Wheeler *et al.*[18] investigating the synthesis, solution behaviour and inkjet printing performance of a family of hyperbranched polymers of differing molecular weights and levels of branching. It must be noted that although hyperbranched materials do not have the exquisitely defined structures of star polymers (these materials show significant polydispersity), they are considerably simpler to synthesise at high molecular weights and at large scales using the Strathclyde method pioneered by Sherrington and co-workers.[77–79] This synthetic method involves a one-pot reaction of initiator, monomer, difunctional crosslinker and a chain transfer agent to prevent excess crosslinking. Uncontrolled crosslinking of polymer chains results in insoluble gels formation of polymers with molecular weights theoretically tending to infinity. Hyperbranched materials show characteristic multimodal molecular weight distributions. This is due to the difficulty in resolving chromotographically materials with a range of branching and molecular weight.[80]

It is well documented in the polymer literature that branched polymer systems have a lower viscometric contribution to polymer solutions than linear polymers of comparable molecular weight.[81] This is due to the more compact

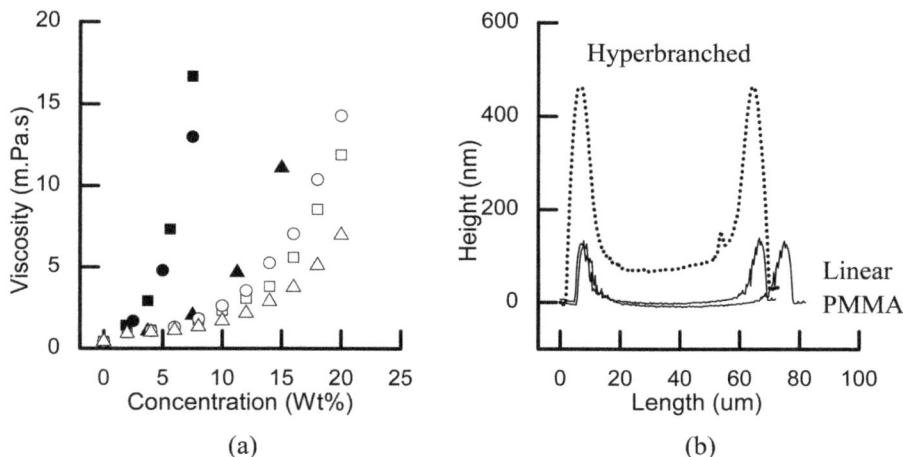

Figure 4.9 (a) Concentration viscosity plots of linear PMMA in MEK with molecular weight (M_W = 468 kDa (■), 310 kDa (●) and 90 kDa (▲)) and hyperbranched materials with molecular weight (609 kDa (□), 360 kDa (○) and 113 kDa (△)). (b) AFM images of individual printed drops of linear PMMA of molecular weight 90 kDa and 468 kDa at its maximum printable concentration (solid line) and a hyperbranched material of molecular weight 609 kDa at its maximum printable concentration (dotted line). Reproduced under CC-BY 4.0 (https://creativecommons.org/licenses/by/4.0/) from J. S. R. Wheeler, A. Longpre, D. Sells, D. McManus, S. Lancaster, S. W. Reynolds and S. G. Yeates, *Polym. Degrad. Stab.*, 2016, **128**, 1–7, DOI: 10.1016/j.polymdegradstab.2016.02.012. © 2016 The Authors. Published by Elsevier Ltd.

orientation of these polymers in solutions which manifests as a lower [η] when compared to equivalent linear polymers. This is demonstrated in Figure 4.9(a) where the viscosity concentration relationships of high molecular weight branched systems is contrasted with linear polymer system.[18]

Furthermore, it is found that it is possible to print high molecular weight branched systems at concentrations above c^*. This results in a much higher maximum printable concentration of branched polymers compared to linear polymers of comparable molecular weight. This is demonstrated in Figure 4.9(b) with the drop profile of a dried branched polymer drop containing significantly more material. Those readers familiar with the drying of inkjet drops will recognise that these drops are showing significant coffee staining; this could be prevented through the use of a secondary high boiling point solvent. This engenders an internal Marangoni flow competing with the flow of material to the pinned edge of the drying drop and results in an even distribution of material across the dried feature.[82,83] An interesting observation arising from the maximum printability of these materials being above c^* reflects that there are no significant extensional effects of these polymers during the inkjet process.

The molecular weight stability of hyperbranched materials with respect to the inkjet printing process was investigated. It was found that branched

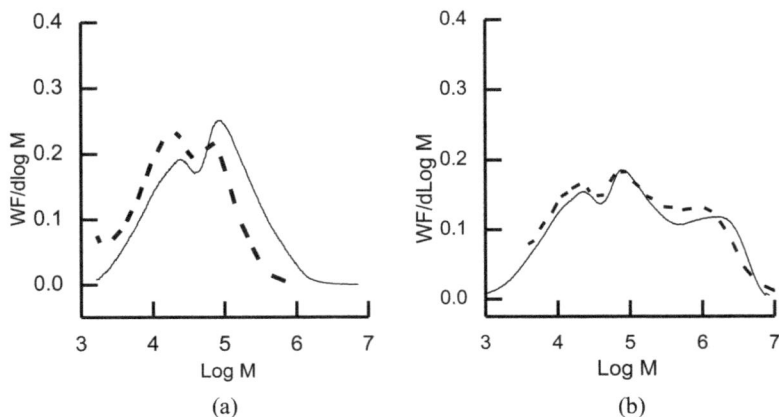

Figure 4.10   (a) The molecular weight distribution of a branched PMMA polymer of M_w of 309 kDa before (solid line) and after inkjet printing (dotted line). (b) The molecular weight distribution of a branched highly PMMA polymer of M_w of 609 kDa before (solid line) and after inkjet printing (dotted line). Reproduced under CC-BY 4.0 (https://creativecommons.org/licenses/by/4.0/) from J. S. R. Wheeler, A. Longpre, D. Sells, D. McManus, S. Lancaster, S. W. Reynolds and S. G. Yeates, *Polym. Degrad. Stab.*, 2016, **128**, 1–7, DOI: 10.1016/j.polymdegradstab.2016.02.012. © 2016 The Authors. Published by Elsevier Ltd.

samples of intermediate molecular weight showed flow induced degradation. It can be seen from Figure 4.10 that the highest molecular weight chains that make up the molecular weight distribution of the branched materials are degraded during the inkjet process. However, when a highly branched (more crosslinker is included in the reaction feed) high molecular weight polymer is investigated it is found there is very little change to the molecular weight distribution. In this case only fractions of the MWD above a molecular weight of 10^6 are affected with a small shaving of the highest molecular weight fractions. This is interpreted as the breaking-off of sections of the highest molecular weight which can chain extend in the inkjet head.[18]

The increased stability of hyperbranched materials with respect to the elongational flow is due to several factors. Although these polymers have high molecular weights, due to the branch points in the structure there are far fewer linear sections above the critical molecular weight which are susceptible to chain extension and degradation in the printhead. Furthermore, due to the enhanced maximum printable concentration of the hyperbranched material, degradation is suppressed by viscous damping. A further factor is analogous to the stability of ultra-high molecular weight linear materials; due to the high molecular weight of these materials they are not able to chain extend and degrade on the inkjet timescale. As shown by the enhanced printability the polymer is being jetted at close to its equilibrium conformation and the elongational flow in the nozzle is not placed directly onto the polymer backbone.

4.11 Mechanistic Insight of Polymer Degradation During Inkjet Printing

Thus far we have presented several molecular weight distributions before and after printing, the results of simulation studies and comparison of inkjet systems to elongational flow experiments to describe the behaviour of polymers during inkjet printing. It will be clear to the reader that it is desirable to carry out measurements directly in the printhead. This is of special interest for studying the degradation of polymeric materials so the degradation mechanism of the inkjet process can be probed.

It is well documented that in well understood systems such as ultrasonication the polymer chain extends and homolytic scission occurs.[84] The free radicals produced during this process can be detected spectroscopically by electron paramagnetic resonance spectroscopy (EPR) Figure 4.11. The EPR experiment is carried out by including a spin trap (such as *N-tert*-butyl-α-phenylnitrone (PBN)) compound in the polymer system under consideration. This species will react with any radicals produced as a result of polymer degradation to form a persistent radical species which is then detected spectroscopically. In the case of PBN the resulting persistent radical is formed on an oxygen section of the spin trap molecule which is not quenched by atmospheric oxygen.

When investigating polymer degradation under ultrasonication as a control experiment it was shown by Wheeler *et al.*[18] that EPR could be used as

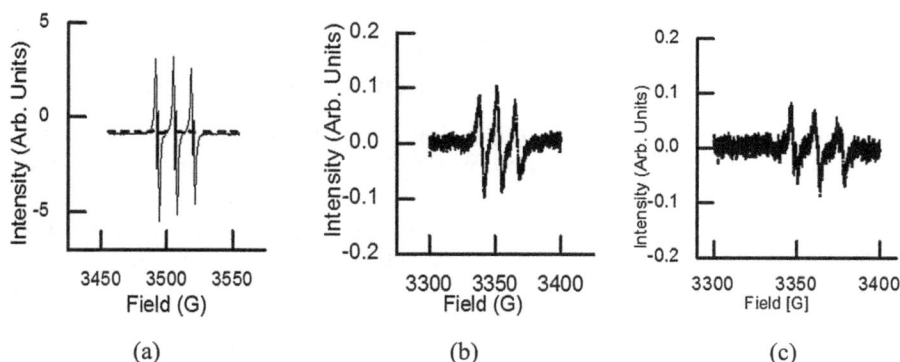

(a) (b) (c)

Figure 4.11 Electron paramagnetic resonance (EPR) spectrum of *N*-tertiary-butyl nitrone (PBN) and polymer, either: 4 wt-% PMMA 468 kDa in toluene or 4 wt-% of a hyperbranched polymer of molecular weight 309 kDa; (a) after sonication of PMMA for 10 mins at a power of 33 W; (b) CIJ printed PMMA after a single pass using a Domino A Series + ink system (c) Hyperbranched polymer at 25 °C single pass through a Dimatix DMP printhead at 40 V. Reproduced under CC-BY 4.0 (https://creativecommons.org/licenses/by/4.0/) from J. S. R. Wheeler, A. Longpre, D. Sells, D. McManus, S. Lancaster, S. W. Reynolds and S. G. Yeates, *Polym. Degrad. Stab.*, 2016, **128**, 1–7, DOI: 10.1016/j.polymdegradstab.2016.02.012. © 2016 The Authors. Published by Elsevier Ltd.

a robust bench top experiment for radical detection without the need to run the experiments in inert atmospheres. Furthermore, EPR can be used to measure the concentration of radicals formed in a system with the signal response being directly proportional to the number of radical species formed.[18]

It was found that when solutions of PMMA of molecular weight 468 kDa in toluene at a concentration of 4 g/dL (it is vital to use a low dielectric solvent when carrying out EPR experiments so any radical signal is not masked) was printed using a CIJ system with the ink doped with PBN there was a measurable increase in radical concentration. This spike in radical concentration shows a degradation mechanism where individual polymer chains are pulled apart resulting in homolytic scission in the backbone.[18]

When this experiment was carried out in a DOD system using a linear high molecular weight polymer system it was not possible to discern any increase in radical concentration. This is because of the vanishingly small concentrations of high molecular weight polymers which are printable by DOD methods. However, when a hyperbranched ink (molecular weight 309 kDa) was printed at a c/c^* of 0.15 an increase in quenched radical signal was detected. These EPR results are interesting for the two reasons: it provides experimental insight into the degradation processes in the printhead and shows it is possible to carry out radical chemical reactions in the jet with degradation products.[18]

4.12 Degradation of Polymers with Tertiary Structure During Drop on Demand Printing

Thus far we have confined the discussion to the behaviour of systems where there are no significant hydrogen bonding interactions between polymers in solution. Systems in which there are substantial intermolecular interactions between polymer chains are described as showing tertiary structure. Polysaccharides are a family of polymers which show significant tertiary structure. There are many polar groups along the polymer backbone for which hydrogen bonding interactions are possible in this class of polymers (Figure 4.12(a)). Due to the extended networks formed in solution by these polymers the maximum printable concentration is limited to below a reduced concentration of 0.25.[85]

It is found that there are significant changes in the molecular weight distribution of the galactomannan polysaccharide pre and post inkjet printing with a Microfab system (Figure 4.12(b)). There is a 30-fold decrease in M_w and a 100-fold decrease in M_N. Although aqueous GPC data is not as accurate as measurements carried out in organic solvents; therefore, these results should be treated as indicative rather than giving absolute molecular weight values; it is clear that there is a significant change in polymer structure due to the printing process. It is unlikely that this difference in degradation behaviour is due to the

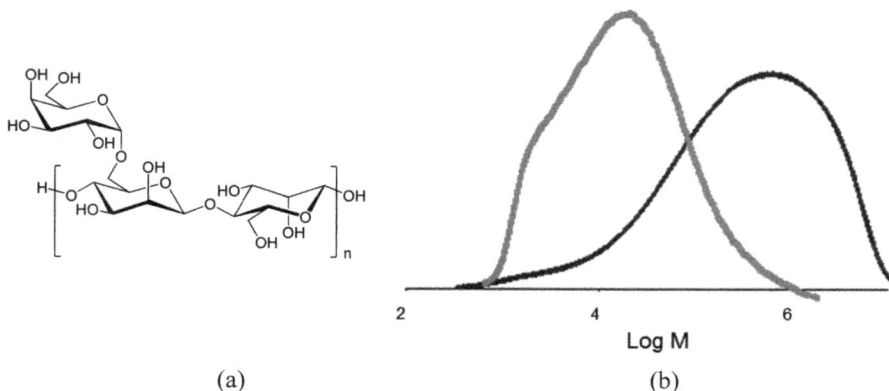

(a)   (b)

Figure 4.12   (a) The molecular structure of the galactomannan polysaccharide. This is a mannose backbone with a branching galactose segment. (b) The molecular weight distributions of galactomannan before inkjet printing (dark curve) and after post inkjet printing (light curve). Reprinted with permission of IS&T: The Society for Imaging Science and Technology, sole copyright owners of *NIP30: International Conference on Digital Printing Technologies and Digital Fabrication 2014*.

relative polymer backbone bond enthalpies in PMMA/PS to polysacharrides. The bond enthalpy of C–C main chain in PS and PMMA, $\Delta Ho \approx 330$ kJmol^{-1} against C–O–C in galactomannan polysaccharides $\Delta Ho \approx 340$ kJmol^{-1} which are the same order of magnitude. Furthermore, the extreme change in the molecular weight distribution of the polymer in a single degradation event (the DOD printing process) is not consistent with that observed in the previously described degradation in DOD systems. A typical O–H .... H hydrogen bond is of the order of 21 kJmol^{-1} and the authors consider that the large decrease in molecular weight of these systems is mainly due to the disruption of the hydrogen bonded tertiary structure of the polymer along with some contribution from main chain scission.[85]

4.13   Conclusions

In conclusion this chapter has introduced the fundamental concepts necessary for the understanding of polymers in inkjet formulation: molecular weight, their solution properties and their behaviour under flow. The formulation challenges of printing high molecular weight polymers arising from these flow behaviours are introduced as are the degradation reactions occurring in both DOD and CIJ systems for synthetic polymers and naturally occurring polymers showing tertiary structure. It is shown how the use of branched polymers in inkjet formulations can be used to address the aforementioned formulation challenges and spectroscopic studies offer the tantalising possibility for radical chemistry to be carried out in the jet stream.

References

1. S. Magdassi, *The Chemistry of Inkjet Inks*, World Scientific Publishing, New York, 2010, ch. 1, pp. 6–15.
2. C. Xue, M. Shi, H. Chen, G. Wu and M. Wang, *Colloids Surf., A*, 2006, **287**, 147–152.
3. S. D. Hoath, *Fundamentals of Inkjet Printing: The Science of Inkjet and Droplets*, Wiley-VCH, Weinheim, 2016, pp. 124–126.
4. Z. Żołek-Tryznowska and J. Izdebska, *Dyes Pigm.*, 2013, **96**, 602–608.
5. Y. Seekaew, S. Lokavee, D. Phokharatkul, A. Wisitsoraat, T. Kerdcharoen and C. Wongchoosuk, *Org. Electron.*, 2014, **15**, 2971–2981.
6. S. H. Eoma, S. Senthilarasu, P. Uthirakumar, S. C. Yoon, J. Lim, C. Lee, H. Seok Lim, J. Lee and S. Lee, *Org. Electron.*, 2009, **10**, 536–542.
7. J. T. Delaney, P. J. Smith and U. S. Schubert, *Soft Matter*, 2009, **5**(24), 4866–4877.
8. G. M. Nishioka, A. A. Markey and C. K. Holloway, *J. Am. Chem. Soc.*, 2004, **126**, 16320–16321.
9. R. E. Saunders, J. E. Gough and B. Derby, *Biomaterials*, 2008, **29**, 193–203.
10. C. Clasen, P. M. Phillips, L. Palangetic and J. Vermant, *AIChE J.*, 2012, **58**(10), 3242–3255.
11. S. D. Hoath, O. G. Harlen and I. M. Hutchings, *J. Rheol.*, 2012, **56**, 1109–1129.
12. D. Xu, V. Sanchez-Romaguera, S. Barbosa, W. Travis, J. de Wit, P. Swan and S. G. Yeates, *J. Mater. Chem.*, 2007, **17**, 4903–4907.
13. K. Al-Alamry, K. Nixon, R. Hind, J. A. Odel and S. G. Yeates, *Macromol. Rapid Commun.*, 2011, **32**(3), 316–320.
14. M. G. Cowie, *Polymers: Chemistry and Physics of Modern Materials*, Blackie Academic and Professional, London, 3rd edn, 2008, pp. 8–12.
15. H. A. Barnes, J. F. Hutton and K. Walters, *An Introduction to Rheology*, Elsevier, Amsterdam, 1989, pp. 97–114.
16. S. D. Hoath, *Fundamentals of Inkjet Printing*, Wiley-VCH, Weinheim, 2016, pp. 339–363.
17. J. S. R. Wheeler, S. W. Reynolds, S. Lancaster, V. Sanchez-Romanguera and S. G. Yeates, *Polym. Degrad. Stab.*, 2014, **105**, 116–121.
18. J. S. R. Wheeler, A. Longpre, D. Sells, D. McManus, S. Lancaster, S. W. Reynolds and S. G. Yeates, *Polym. Degrad. Stab.*, 2016, **128**, 1–7.
19. G. Odian, *Principles of Polymerization*, John Wiley and Sons Inc, New York, 1991, pp. 1–5.
20. P. A. Sørensen, S. Kiil, K. Dam-Johansen and C. E. Weinell, *J. Coat. Technol. Res.*, 2009, **6**(2), 135–176.
21. C. Creton, *MRS Bull.*, 2003, **28**(6), 434–439.
22. G. Odian, *Principles of Polymerization*, John Wiley and Sons Inc, New York, 1991, p. 106.
23. P. W. Wills, S. G. Lopez, J. Burr, P. Taboada and S. G. Yeates, *Langmuir*, 2013, **29**(14), 4434.

24. L. Slade and H. Levine, *Adv. Exp. Med. Biol.*, 1991, **302**, 29–101.
25. G. Odian, *Principles of Polymerization*, John Wiley and Sons Inc, New York, 1991, pp. 109–110.
26. F. W. Billmeyer, *Textbook of Polymer Science*, John Wiley and Sons Inc, New York, 1984, pp. 5–6.
27. F. Rodriguez, C. Cohen, C. Ober and L. A. Archer, *Principles of Polymer Systems*, Taylor & Francis, New York, 2003, p. 9.
28. K. Pang, R. Kotek and A. Tonelli, *Prog. Polym. Sci.*, 2006, **31**, 1009–1037.
29. K. Ohno, Y. Tsujii, T. Miyamot and T. Fukuda, *Macromolecules*, 1998, **31**, 1064–1069.
30. J. V. Crivello, B. Falk and M. R. Zonca, *J. Polym. Sci., Part A: Polym. Chem.*, 2004, **42**, 1630–1646.
31. N. Hadjichristidis, M. Pitsikalis, S. Pispas and H. Iatrou, *Chem. Rev.*, 2001, **101**, 3747–3792.
32. X. Deng, F. Liu, Y. Luo, Y. Chen and D. Jia, *Eur. Polym. J.*, 2007, **43**, 704–714.
33. H. Leea, J. Pietrasik, S. S. Sheikoc and K. Matyjaszewski, *Prog. Polym. Sci.*, 2010, **35**, 24–44.
34. L. Xue, U. S. Agarwal, M. Zhang, B. B. P. Staal, A. H. E. Muller, C. M. E. Bailly and P. J. Lemstra, *Macromolecules*, 2005, **38**, 2093–2100.
35. D. J. Keddie, *Chem. Soc. Rev.*, 2014, **43**, 496.
36. W. A. Braunecker and K. Matyjaszewski, *Prog. Polym. Sci.*, 2007, **32**, 93–146.
37. S. T. Nguyen, L. K. Johnson, R. H. Grubbs and J. W. Ziller, *J. Am. Chem. Soc.*, 1992, **114**(10), 3974–3975.
38. R. F. T. Stepto, R. G. Gilbert, M. Hess, A. D. Jenkins, R. G. Jones and P. Kratochvíl, *Pure Appl. Chem.*, 2009, **81**(2), 351–353.
39. P. J. Flory, *Principles of Polymer Chemistry*, Cornell University Press, Ithaca, 1971, pp. 273–283.
40. P. J. Flory, *Principles of Polymer Chemistry*, Cornell University Press, Ithaca, 1971, pp. 495–505.
41. B. Trathnigg, *Prog. Polym. Sci.*, 1995, **20**, 615–650.
42. L. K. Kostanski, D. M. Keller and A. E. Hamielec, *J. Biochem. Biophys. Methods*, 2004, **58**, 159–186.
43. P. Castignolles, R. Graf, M. Parkinson, M. Wilhelm and M. Gaborieaua, *Polymer*, 2009, **50**, 2373–2383.
44. M. Rubenstein and R. H. Colby, *Polymer Physics*, Oxford University Press, Oxford, 2008, pp. 173–183.
45. C. H. Chan, C. H. Chia and S. Thomas, *Physical Chemistry of Macromolecules*, Apple Academic Press, Toronto, 2014, pp. 26–29.
46. C. H. Chan, C. H. Chia and S. Thomas, *Physical Chemistry of Macromolecules*, Apple Academic Press, Toronto, 2014, pp. 37–46.
47. P. J. Flory, *Principles of Polymer Chemistry*, Cornell University Press, Ithaca, 1971, p. 425.
48. M. Rubenstein and R. H. Colby, *Polymer Physics*, Oxford University Press, Oxford, 2008, p.34.

49. (a) IUPAC, *Compendium of Chemical Terminology, 2nd ed. (The "Gold Book")*, Compiled by A. D. McNaught and A. Wilkinson, Blackwell Scientific Publications, Oxford, UK, 1997; (b) XML on-line corrected version, http://goldbook.iupac.org, 2006, created by M. Nic, J. Jirat and B. Kosata, updates complied by A. D. Jenkins, 882.

50. J. Burke, *AIC Book Paper Group Annual*, 1984, **3**, 13.

51. M. Daoud, J. P. Cotton, B. Farnoux, G. Jannink, G. Sarma, H. Benoit, R. Duplessix, C. Picot and P. G. de Gennes, *Macromolecules*, 1975, **8**(6), 804–818.

52. M. Rubenstein and R. H. Colby, *Polymer Physics*, Oxford University Press, Oxford, 2008, p. 265.

53. M. Doi and S. F. Edwards, *The Theory of Polymer Dynamics*, Clarendon Press, Oxford, 1994, pp. 141–143.

54. P. G. de Gennes, *J. Chem. Phys.*, 1971, **55**(2), 571–572.

55. B. J. de Gans, P. C. Duineveld and U. S. Schubert, *Adv. Mater.*, 2004, **16**, 203–213.

56. T. T. Perkins, D. E. Smith and S. Chu, *Science*, 1997, **276**(5321), 2016–2021.

57. H. A. Barnes, J. F. Hutton and K. Walters, *An Introduction to Rheology*, Elsevier, Amsterdam, 1989, pp. 15–25.

58. M. R. Mackley, *J. Non-Newtonian Fluid Mech.*, 1978, **4**, 111–136.

59. H. A. Barnes, J. F. Hutton and K. Walters, *An Introduction to Rheology*, Elsevier, Amsterdam, 1989, pp. 75–95.

60. A. Keller and J. Odell, *Nature*, 1984, **312**, 98.

61. A. Keller and J. A. Odell, *Colloid Polym. Sci.*, 1985, **263**, 181–201.

62. J. A. Odell and A. Keller, *J. Polym. Sci., Part B: Polym. Phys.*, 1986, **24**, 1889–1916.

63. A. J. Moller, J. A. Odell and A. Keller, *J. Non-Newtonian Fluid Mech.*, 1988, **30**, 99–118.

64. J. A. Odell, A. Keller and M. J. Miles, *Polym. Commun.*, 1983, **24**, 7–10.

65. M. Doi and S. F. Edwards, *The Theory of Polymer Dynamics*, Clarendon Press, Oxford, 1994, pp. 121–125.

66. J. A. Odell, A. J. Mueller, K. A. Narh and A. Keller, *Macromolecules*, 1990, **23**(12), 3092–3103.

67. A. J. Muller, J. A. Odell and S. Carrington, *Polymer*, 1992, **33**, 12.

68. K. S. Suslick and G. J. Price, *Annu. Rev. Mater. Sci.*, 1999, **29**, 295–326.

69. P. A. May and J. S. Moore, *Chem. Soc. Rev.*, 2013, **42**, 7497–7506.

70. A. F. Horne and E. W. Merril, *Nature*, 1984, **312**, 140–141.

71. J. E. Fromm, *IBM J. Res. Dev.*, 1984, **28**, 322–333.

72. B. Derby and N. Reis, *MRS Bull.*, 2003, **28**, 815–820.

73. N. Reis, C. Ainsley and B. Derby, *J. Appl. Phys.*, 2005, **97**, 094903.

74. S. D. Hoath, D. C. Vadillo, O. G. Harlen, C. McIlroy, N. F. Morrison, W. Kai-Hsiao, T. R. Tuladhar, S. Jung, G. D. Martin and I. M. Hutchings, *J. Non-Newtonian Fluid Mech.*, 2014, **205**, 1–10.

75. C. McIlroy, O. G. Harlan and N. F. Morrisson, *J. Non-Newtonian Fluid Mech.*, 2013, **201**, 17–28.

76. B. J. de Gans and L. J. Xue, *Macromol. Rapid Commun.*, 2005, **26**(4), 310–314.
77. F. Isaure, P. A. G. Cormack and D. C. Sherrington, *Macromolecules*, 2004, **37**, 2096–2105.
78. S. Camerlynck, P. A. G. Cormack, D. C. Sherrington and G. Saunders, *J. Macromol. Sci., Part B: Phys.*, 2005, **44**, 881–895.
79. M. Chisholm, N. Hudson, N. Kirtley, F. Vilela and D. C. Sherrington, *Macromolecules*, 2009, **42**, 7745–7752.
80. S. P. Gretton-Watson, E. Alpay, J. H. G. Steinke and J. S. Higgins, *Ind. Eng. Chem. Res.*, 2005, **44**, 8682–8693.
81. B. I. Voit and A. Lederer, *Chem. Rev.*, 2009, **109**, 5924–5973.
82. S. D. Hoath, *Fundamentals of Inkjet Printing*, Wiley-VCH, Weinheim, 2016, pp. 260–263.
83. J. Perelaer, P. J. Smith, E. Van Den Bosch, S. S. C. Van Grootel, P. H. J. M. Ketelaars and U. S. Schubert, *Macromol. Chem. Phys.*, 2009, **210**(6), 495–502.
84. M. M. Caruso, D. A. Davis, Q. Shen, S. A. Odom, N. R. Sottos, S. R. White and J. S. Moore, *Chem. Rev.*, 2009, **109**(11), 5755–5798.
85. J. Wheeler, K. A-Alamry, S. W. Reynolds, S. Lancaster, N. M. P. S. Ricardo and S. G. Yeates, *2014 International Conference on Digital Printing Technologies*, 2014, vol. 4, pp. 335–338.

Reactive Inkjet Printing for Silicon Solar Cell Fabrication

A. J. LENNON

School of Photovoltaic and Renewable Energy Engineering, UNSW, Sydney, NSW 2052, Australia
*E-mail: a.lennon@unsw.edu.au

5.1 Introduction

5.1.1 Crystalline Silicon Photovoltaics

The generation of electricity through the use of a crystalline silicon photovoltaic device was first reported at Bell Labs by Chapin et al.[1] Although Chapin was first discouraged by the fact that, using a one-Watt 14% efficient cell costing US$286, a homeowner would have to pay in excess of US$1M for an array of sufficient size to power the average American house, Raisbeck provided a more positive outlook suggesting that the Bell Labs 'solar battery' would find use where power is needed 'in inaccessible places where no lines go' and 'in doing jobs the need for which we have not yet felt'.[2] Indeed Raisbeck was correct not to be disillusioned; photovoltaics did find its first applications in space and in remote areas, and now today the levelised cost of electricity generated by crystalline silicon photovoltaics is less than that of electricity produced from fossil fuels in many places in the world and silicon photovoltaic modules are priced at less than US0.50 per Watt.[3] Figure 5.1(a) shows an array of crystalline silicon photovoltaic modules in the field and Figure 5.1(b)

Smart Materials No. 32
Reactive Inkjet Printing: A Chemical Synthesis Tool
Edited by Patrick J. Smith and Aoife Morrin
© The Royal Society of Chemistry 2018
Published by the Royal Society of Chemistry, www.rsc.org

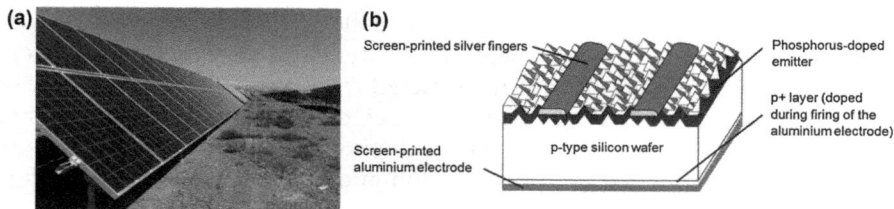

Figure 5.1 (a) Array of crystalline silicon photovoltaic modules; and (b) cross-sectional schematic of an industrial screen-printed solar cell. (b) Reproduced with permission from Patterned masking using polymers: insights and developments from silicon photovoltaics, Z. Li and A. Lennon, *International Materials Review*,[52] copyright © The Institute of Materials, Minerals and Mining, reprinted by permission of Taylor & Francis Ltd, http://www.tandfonline.com on behalf of The Institute of Materials, Minerals and Mining.

is a cross-sectional depiction of an individual industrially-produced silicon solar cell with screen-printed electrodes.

Silicon, being an indirect bandgap semiconductor, is an inefficient absorber and emitter of light. For this reason, when it is used as an absorber material in a photovoltaic device, electrical carriers that are generated by the absorption of light and collected through the use of a p–n junction in the device, need to survive sufficiently long without recombining to enable their collection in the electrodes of the device. This means that the surfaces of the device where the silicon crystal is discontinuous, must be coated in such a way that recombination of electrical carriers is minimised. Although many strategies for this so-called surface passivation (reduction in recombination) have been proposed over the years, the most successful approaches have involved the use of inorganic dielectrics such as thermally-grown silicon dioxide (SiO_2),[4-7] amorphous silicon nitride (SiN_x) deposited using plasma-enhanced chemical vapour deposition (PECVD),[8-11] and amorphous aluminium oxide (Al_2O_3) deposited using either PECVD or atomic layer deposition (ALD).[11-14] These dielectric materials can reduce the density of interface states which represent the sites for carrier recombination (*e.g.*, dangling bonds) and/or through storing charges in the dielectric layer that reduce the concentration of one carrier polarity thereby effectively reducing the recombination rate at the surface. The other key advantage that these inorganic 'electrical passivation' layers share is their stability in an environment where they are exposed to extreme temperature ranges, moisture and UV light for long periods. Crystalline silicon photovoltaic modules are typically sold with warranties of 25+ years operation in the field.

5.1.2 Challenge of Patterning Dielectrics

The inorganic coatings of silicon solar cells also present challenges for device fabrication. In order to form electrical contacts to the solar cell, openings must be made in the coatings. Preferably the openings are as small as possible in order to keep most of the cell's surface well-passivated for achievement of high device voltages, but sufficiently large so as not to incur large resistive losses as current is extracted from the cell. Furthermore, by selectively doping the silicon being contacted, reduced contact resistance can be achieved. The localised heavier doping at the contacts also acts to reduce the minority carrier concentration and hence the recombination at the contact opening.[15]

Historically, two general approaches have been used to form contact openings in the dielectric coatings of silicon solar cells: (i) chemical etching; and (ii) laser ablation. To-date chemical etching has largely been used because it can be difficult to prevent damage to the silicon crystal when using lasers.[16-25] Although chemical etching is typically associated with liquid phase etching, it can also be achieved in solid phase through the use of pastes,[26-29] gas phase through the use of aerosols[30-34] and *via* plasmas as in reactive ion etching.[35-38] The challenge facing chemical etching lies in the toxic and corrosive materials and/or use of high temperatures that are required to etch these stable, resistant dielectrics. In liquid phase etching, SiO_2 and SiN_x are most effectively etched using hydrofluoric acid (HF), which is not only a highly corrosive acid but is also a contact poison due to its ability to penetrate the skin and react with the ubiquitous biologically important ions Ca^{2+} and Mg^{2+}.[39] Although these inorganic dielectric layers can also be effectively etched by screen-printed pastes through the use of glass particles (frits), high 'firing' temperatures are typically required.[27,40-42]

Early laboratory-fabricated cells[43-46] achieved structuring of these layers by using optical lithography to form patterns of openings in an acid-resistant polymer layer and then immersion etching in HF, or buffered variants of this etchant,[47] to etch openings in the inorganic dielectric layers which could either be metallised or used to selectively-dope exposed silicon regions for formation of electrical contacts. Optical lithography is however a very complex and expensive process, and so alternative lower-cost methods have been required for industrial production of solar cells. The current dominant production process for industrial p-type silicon solar cells has evolved to use screen-printed silver pastes to form the metal contacts to cells, with silver pastes being used for the front-surface metal grid and aluminium pastes for the rear contact. The silver pastes contain silver powder, glass frits and metal (typically lead or bismuth) oxides, and when they are rapidly fired at temperatures of 780–850 °C, they 'etch' through the SiN_x layer and silver crystallites form to make electrical contact to the cell. When the screen-printing process was first reported by Spectrolab in the 1970's for the metallisation of silicon solar cells,[48] the width of the metal grid fingers was in the order of 150–200 μm and consequently resulted in significant shading of the cell. However, paste and equipment manufacturers have continued to refine this metallisation process and finger widths are now typically less than 50 μm.[49]

Screen-printing, being a contact printing process, has not always been viewed as the path that will lead to higher industrial solar cell efficiencies and manageable and low manufacturing costs. Contact printing is not easily performed with high yield on very thin cells, and reducing the amount of silicon required per cell is one strategy for reducing cost. Furthermore, the reliance of the front metallisation on silver has in the past (and continues to) raised concerns as the volume of manufacturing increases because manufacturing companies are increasingly exposed to the risk of accretions in the volatile silver price. These concerns have inspired a search for alternative patterning processes that can enable highly efficient devices to be fabricated at high processing throughputs (3000 wafers per hour[50]), however the bar is set very high for silicon photovoltaic manufacturing with regard to low-cost as the industry has progressively reduced costs with a learning rate of ~21% (see Figure 5.2).

5.1.3 Inkjet Printing in Silicon Photovoltaics

Inkjet-mediated metallisation of the front surface of solar cells with metal–organic inks which decompose after printing to form metal (MOD inks) was reported as early as 1987 as a maskless alternative method to conventional optical lithography and screen-printing.[51] Even at this early stage in the evolution of industrial silicon photovoltaics, the advantages of direct writing *via* digital patterns and non-contact silicon wafer processing were proposed

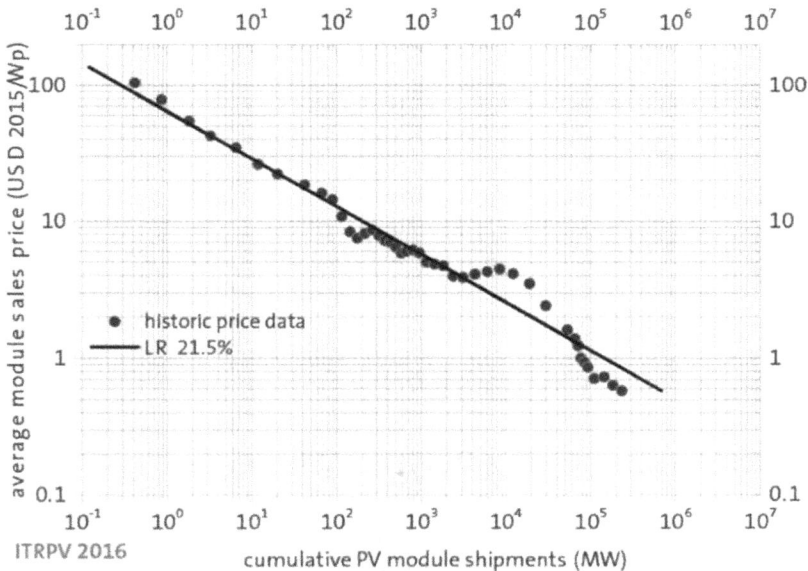

Figure 5.2 Learning rate curve for crystalline silicon module price as a function of cumulative PV shipments. Reproduced with permission from the ITRPV, *International Technology Roadmap for Photovoltaic: 2015 Results*, 2016.[49]

as a pathway to lower cost and enabling use of thinner wafers. Since this initial report by Teng and Vest,[51] many varied inkjet printing applications have been proposed for silicon photovoltaics ranging from masked etching, doping and metallisation[50,52] of silicon to direct (*i.e.*, unmasked) doping,[50,53] etching[54] and metallisation.[55-58] These many diverse applications, which have been recently comprehensively reviewed by Stüwe *et al.*[50] and Li and Lennon,[52] span different manufacturing steps and can be applied to many different solar cell architectures. However, in most of these applications, the inkjet printer was primarily used as a deposition tool with the materials being deposited changing phase at the most (*e.g.*, inkjet-printed phase change waxes for masking[59-62]). This chapter focusses instead on processes that have been developed in which the inkjet-printed material is a component of a chemical reaction on the surface of the wafer, the chemical reaction resulting in material changes that can facilitate a process or step in the fabrication of the solar cell. These examples of reactive inkjet printing have been motivated by the challenges of managing highly corrosive materials, the need to achieve patterning resolutions that are not limited by droplet spreading on a surface and/or the need to reduce the number of steps in a fabrication process.

Although the promise of reactive inkjet printing is to specifically engineer reactions on a surface by separately depositing the reactants,[63] most reactive inkjet printing examples arising from silicon photovoltaics have involved the jetting of a single reactant with the other reactant(s) being either provided as a sacrificial layer on the surface of the cell or by the surface of the device itself. These applications represent, therefore, examples of single reactive inkjet printing.[63] It is not clear whether this focus on single reactive inkjet printing is primarily due to the need for highly reliable processes in silicon photovoltaic manufacturing or to the limited availability of material deposition printers that are capable of jetting multiple fluids in a manufacturing environment. An additional complication presented by silicon solar cells is that the dielectric surfaces that require patterning (*e.g.*, SiO_2 and SiN_x) are hydrophilic and consequently, without surface treatment, deposited fluids are not sufficiently constrained to obtain high resolution patterning. Therefore, the use of a surface layer to constrain the deposited reactants can provide an important function in the patterning process (see *e.g.*, Liu *et al.*[64]).

5.2 Patterning Using Polymer Masks

A logical way in which to approach the simplification of optical lithography for reduced cost is to use direct writing methods such as inkjet printing to create the necessary pattern of openings in acid-resistant polymers. This approach can enable the processes of selective etching, doping and metallization (see Figure 5.3) to be performed substantially as they had been performed using optical lithography for laboratory-fabricated cells. Li and Lennon[52] reviewed the different polymer patterning methods that have been developed inspired by the requirements of patterning for silicon

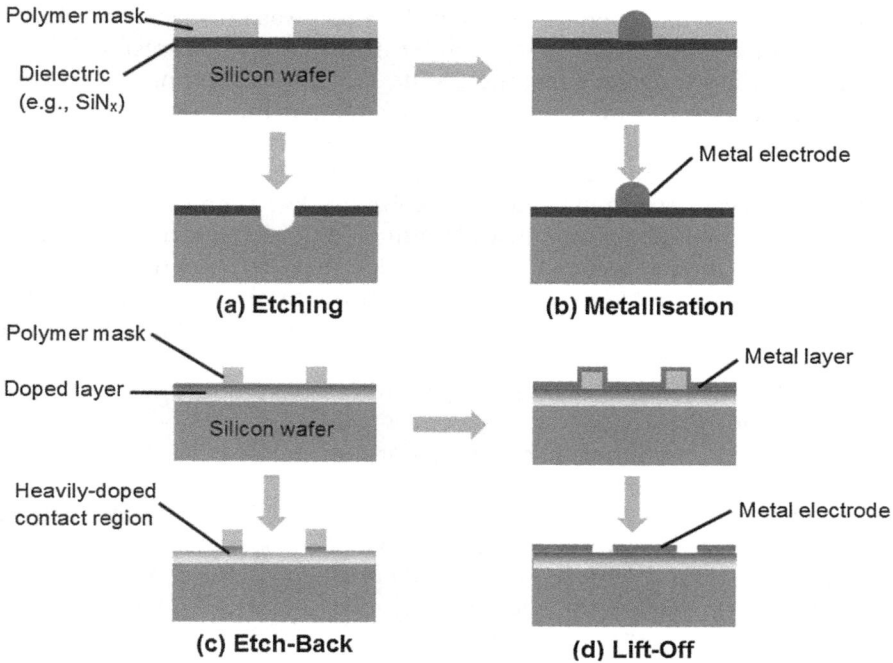

Figure 5.3 Schematic depicting the use of a polymer mask for selective: (a) and (c) etching; and (b) and (d) metallisation. In (a) and (b), the polymer layer covers the majority of the silicon wafer surface, however for the etch-back (c) and lift-off applications (d) the polymer is applied (printed) only in localised regions, typically lines, which cover a small fraction of the wafer surface. Reproduced with permission from Patterned masking using polymers: insights and developments from silicon photovoltaics, Z. Li and A. Lennon, *International Materials Review*,[52] copyright © The Institute of Materials, Minerals and Mining, reprinted by permission of Taylor & Francis Ltd, http://www.tandfonline.com on behalf of The Institute of Materials, Minerals and Mining.

photovoltaics, however, of particular interest to this chapter are the methods in which inkjet printing has been used to deposit reactants for a surface reaction that results in the formation of a patterned polymer mask. The following sections describe three different reactive inkjet printing processes used to form patterned polymer masks. Each of these processes involves the patterning of a routinely-used acid resistant novolac resin (polymer), using different aspects of that polymer's reactivity to achieve the patterning effect.

5.2.1 Novolac Resin Patterning by Dissolution

Polymer patterning can be achieved through the inkjet deposition of organic solvents for the polymer. Kawase *et al.*[65–67] demonstrated the inkjet deposition of ethanol to make circular openings with a diameter of 30–40 µm in a 250 nm thick polyvinylphenol (PVP) films, the opening diameter increasing to ~80 μm when the polymer thickness was increased to 750 nm. The openings had a characteristic 'crater-like' shape which resulted from the dissolved polymer redepositing on the side of the openings as the solvent evaporated.[68-70] However this polymer re-distribution process, which has been proposed as a means of fabricating microlense arrays,[70,71] requires a relatively large amount of solvent to create complete openings in a polymer layer. Kawase *et al.* reported the need to use five aligned layers of 5 pL droplets to make openings in just 250 nm of PVP. In their analysis of this solvent-driven process, Zhang *et al.* concluded that there is a finite penetration limit for each polymer/solvent combination, beyond which polymer is no longer removed from the opening due to the high polymer concentration in a small space.[72] The limiting penetration depth is therefore expected to reduce with smaller openings thereby limiting the application of this process to thin polymer layers.

Polymer dissolution mediated by inkjet printing has also been used to pattern novolac resins for selective-etching of buried silicon layers and electrical contacting of thin film crystalline silicon solar cells.[73] However, instead of using an organic solvent, Young and Lasswell inkjet-printed a solution comprising 15% (*w/v*) KOH to form circular openings in the polymer (cresol-novolak in propylene glycol monomethyl ether acetate (PGMEA) without photo-active components sold by AZ Eelectronic Materials as AZ(R) P150 protective coating). Novolac resins are typically 'developed' in alkaline solutions when used for optical lithography, and so Young and Lasswell sought to form openings in the insulation layer (which was retained as part of the final thin film silicon module) by locally developing the resin.

The diameter of the openings created in 4–5 μm thick resin films was ~100 μm in earlier reports[73] but was refined to be ~80 μm.[74] As with polymer dissolution by printing of solvent, a relatively large volume of alkaline solution was required to form openings in the resin, with Young reporting the need for between 20 to 240 pL of 15% (*w/v*) KOH to form openings.[75] Although it was not clear why such a large volume range was needed, a larger volume would be expected in order to dissolve the thicker polymer layers, which were necessary in this application as the patterned resin was retained as an insulation layer in the crystalline silicon module. Figure 5.4 shows a cross-sectional schematic of the thin crystalline silicon device,[76] showing how the formed holes enabled electrical contact to both the n and p-type polarity of the device.

Alkaline dissolution is limited by its relatively slow kinetics. The mechanism by which novolac resins dissolve in alkaline solutions differs from that which occurs with dissolution in organic solvents. The hydroxyl ions directly react with the phenol moieties of the resin, deprotonating them and enabling them to be solvated by water.[77,78] Once a sufficient number of phenol groups are ionized, the polymer chains detach from the matrix and transfer into solution. Dissolution proceeds *via* water and hydroxyl ions simultaneously entering the polymer matrix at surface hydrophilic sites, where they begin to form phenolate groups. The cations from the solution follow to maintain electroneutrality and a more fluid phase of the polymer, called the penetration zone,

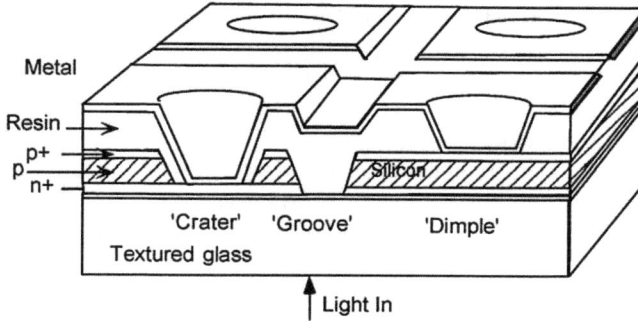

Figure 5.4   Cross sectional schematic showing the arrangement of openings to the n-type and p-type silicon regions through a novolac resin layer of 4–5 μm thickness of the CSG Solar thin silicon device. Reprinted from *Solar Energy*, 77(6), M. A. Green, P. A. Basore, N. Chang, D. Clugston, R. Egan, R. Evans, D. Hogg, S. Jarnason, M. Keevers, P. Lasswell, J. O'Sullivan, U. Schubert, A. Turner, S. R. Wenham and T. Young, Crystalline silicon on glass (CSG) thin-film solar cell modules, 857–863, Copyright (2004) with permission from Elsevier.[76]

forms at the surface. Once a critical conversion of phenol to phenolate has occurred, the penetration zone collapses and the polymer dissolves into the solution.[77,78]

A second key difference between this dissolution reaction and polymer dissolution in thermodynamically good solvents is the way the alkaline solution diffuses into the solid polymer matrix. Unlike the uniform diffusion front that occurs with an organic solvent, the charged hydroxyl groups percolate through the matrix being attracted to the hydrophilic phenol/phenolate groups whilst being repelled by the hydrophobic polymer chains.[77,79] Due to the relatively slow dissolution process, openings formed in the resin ~1 min after printing. This relatively slow reaction time presented a limit to the patterning resolution as the alkaline solution also diffuses laterally, extending the size of the opening. Although lateral diffusion also occurs with organic solvents, the solvent is being constantly removed by evaporation which places a limit on the extent of lateral diffusion. De Gans *et al.* demonstrated that the size of the openings for solvent dissolution was limited by the solvent spread on the surface of the polymer (*i.e.*, the wettability).[70]

Novolac patterning by dissolution in printed KOH was demonstrated in commercial production by CSG Solar AG, however inkjet printing alignment and reliability of the printheads were challenging. Use of KOH as the jetting fluid precluded the use of silicon MEMS printheads which offered a path to lower drop volumes and increased drop placement resolution and accuracy.[80,81] Indeed the need to use corrosive alkaline fluids for photovoltaic patterning applications to either directly etch silicon or dissolve resins, as described above, has motivated the development of inkjet printheads that are compatible with alkaline solutions.[82]

Polymer dissolution is perhaps a very simple example of single reactive ink-jet printing; however, it does highlight the demands placed on an inkjet printing approach when a large volume of solvent is required in order to dissolve the polymer. One strategy to increase throughput is to use a printhead able to jet drops of larger volumes, however this will result in a larger wetted area and hence a larger volume of polymer to be dissolved. Thicker polymer layers therefore present challenges, however in many cases a sufficiently thick mask may be required in order cover all regions of a pyramidally-textured silicon surface.[83]

5.2.2 Resin Plasticising for Repeated Patterning Steps

With optical lithography, it is generally assumed that a polymer layer can only be used for a single patterning step. The polymer is then removed by chemical dissolution or by using a plasma process, and then further polymer application and patterning are required for each patterning step. This can add cost in terms of both material and processing time when multiple steps are required. For back-contacted solar cells, such as the CSG Solar device structure described in Section 5.2.1, patterning for the contacting to both semiconductor polarities is required. Utama[84–86] and Lennon[87] reported the inkjet printing of a plasticiser for novolac resin (in PGMEA) to enable the etching of underlying dielectric and silicon layers. Plasticisers weaken intramolecular bonds and push the polymer chains apart thereby making the polymer more plastic.[88] This causes some polymer redistribution as described in Section 5.2.1, although in this case a poor solvent, diethylene glycol (DEG), for the polymer was printed and while the polymer was partially redistributed in the plasticised regions, openings were not formed (see Figure 5.5(a)).

Acidic etchants for silicon-containing dielectrics and for silicon itself can however permeate the plasticised regions without removal of the polymer from the surface. Lennon *et al.* showed that once a first set of openings in a dielectric layer had been formed, the resin's permeability to the acidic etchants could be reversed by saturating the polymer in PGMEA for 15 min, and printing a second pattern of DEG on the polymer to plasticise regions for the other polarity contacts.[87] In this process, which is depicted in Figure 5.5(b), the resin was sacrificial and therefore removed after the two patterning steps were complete.

A key advantage of this processing sequence from a perspective of increasing processing throughput, was that only 4 pL of DEG was required per point opening in the dielectric.[87] This is considerably less than would be required for complete dissolution of the polymer through printing of either a solvent or alkaline solution. A further advantage of the process over the alkaline dissolution reported by Young and Lasswell,[73] was the ability to reliably jet the DEG using a widely-used silicon MEMS printhead (Dimatix Material Deposition Printer; DMP-2831) using the small droplet volume of 1 pL. This reduced the 'wetted' area enabled point openings of diameter 35–40 μm to be formed,[87] which were significantly smaller than the openings reported by

(a) **(b)**

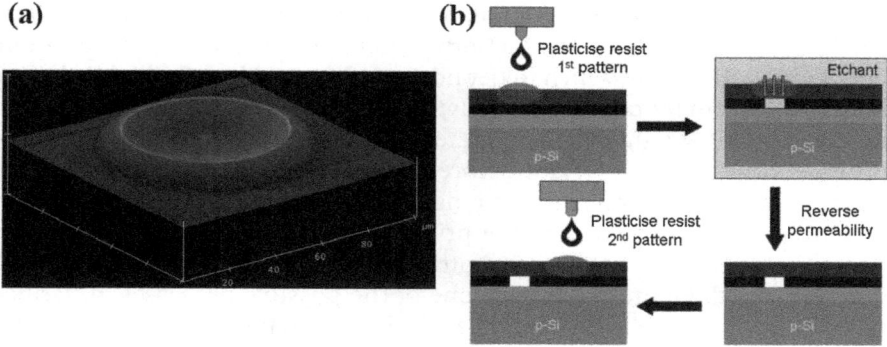

Figure 5.5   (a) An atomic force microscope image of a novolac resin surface after inkjet deposition of an 8 pL droplet of 100% DEG; and (b) schematic flowchart showing how the resin plasticising was used to achieve two sets of contact regions for a back-contact silicon solar cell. (b) Reprinted from *Solar Energy Materials and Solar Cells*, **92**(11), A. J. Lennon, R. Y. Utama, M. A. T. Lenio, A. W. Y. Ho-Baillie, N. B. Kuepper and S. R. Wenham, Forming openings to semiconductor layers of silicon solar cells by inkjet printing, 1410–1415, Copyright (2008) with permission from Elsevier.[87]

Young and Lasswell for novolac resin removal through the deposition of an alkaline solution.

5.2.3   Inkjet Lithography

A logical way in which to eliminate the need for pre-purposed screens or lithographic plates is to inkjet print a chemical that directly initiates, or alternatively inhibits, a resin hardening (*i.e.*, crosslinking) reaction thereby permitting the non-hardened polymer to be dissolved and removed using alkaline development as for photoresists. Etching or metallisation can then be achieved through this patterned mask as for optical lithography, however with the benefit of having the additional flexibility to design new patterns on the fly using digital printing. Since this reactive inkjet printing process, which does not necessarily need to use UV illumination to initiate polymer hardening, closely resembles optical lithography, it is subsequently referred to in this chapter as 'inkjet lithography'.

The potential of inkjet lithography was first recognized in the British patent GB1431462 assigned to Agfa Gevaert Aktiengessellshaft and published in 1976. This publication discloses the idea of inkjet printing a substance that can alter the solubility of a polymer with respect to a given solvent so that when the polymer is exposed to the solvent, the solubility of the polymer is changed in the 'printed' areas and a relief image is obtained. This patent presented a number of different examples of this process, including examples in which the polymer's solubility was changed by altering its acid value and examples where the polymer did not need to be exposed to light for pattern/relief formation. Hallman then extended this concept in a later patent assigned to Sun Chemical Corporation, in which he described a number of different ways in which inkjet lithography could be achieved including inkjet printing of monomers and polymerization catalysts, initiators or intermediates and UV-light blocking materials.[89] He introduced the possibility of inkjet-printing the reactant before or after the untreated polymer layer was formed and presented a wide range of different patterning mechanisms including mechanisms where the printed regions were not hardened and hence developed and where the printed regions were the only regions which were hardened. Figure 5.6 depicts one of the possible processes, described by Hallman, in which the printed material was a polymerization initiator that caused polymer hardening, leaving the non-printed polymer area able to be developed (*i.e.*, removed).

More recently, an inkjet lithography process was reported by Li *et al.* and used to form a novolac resin mask for the selective metallisation of silicon solar cells.[52,83,90,91] This inkjet lithography variant exploited the chemical amplification technique which had been developed to improve the throughput of optical lithography.[92,93] With many chemically-amplified resists, exposure to UV light results in the formation of a photo-acid that, on subsequent heating, catalyses cross-linking causing resin hardening.[94] Rather than selectively generate the acid by optical exposure through a mask, Li *et al.* reported the inkjet-printing of a acidity modifier onto the polymer surface that could change the acid value of the polymer and thereby induce an

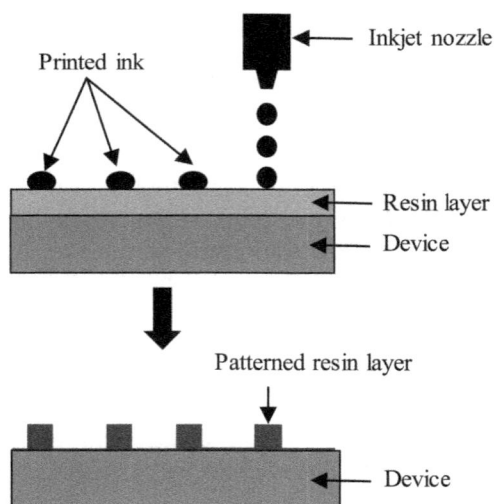

Figure 5.6   Printing of an ink reactant that causes the resin to harden and not be developed in the printed regions. Adapted from N. Platzer, *Journal of Polymer Science A: Polymer Chemistry*[88] with permission from John Wiley and Sons, Copyright © 1982 WILEY-VCH Verlag GmbH & Co. KGaA, Weinheim.

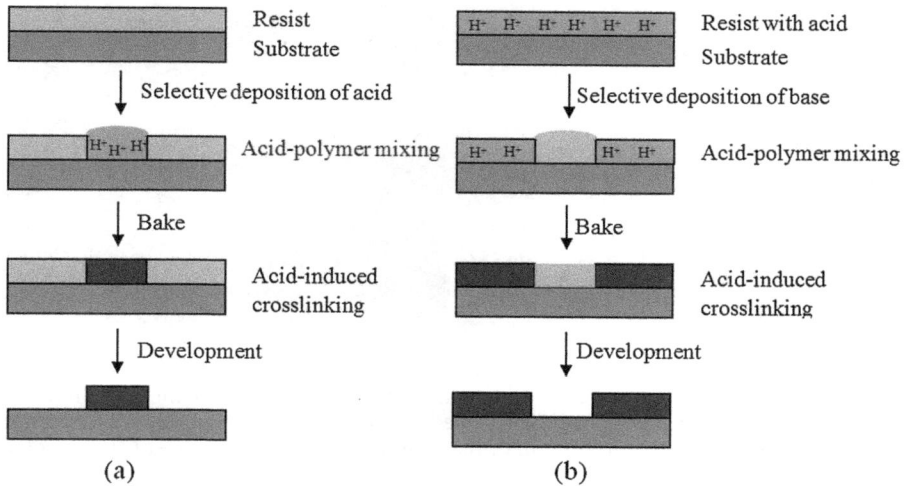

Figure 5.7   Inkjet lithography approaches involving acid value modification showing in: (a) the selective deposition of an acidic fluid resulting in polymer cross-linking in the printed area; and in (b) selective deposition of
an alkaline fluid neutralising the acid pre-added to the polymer and
resulting in cross-linking only in the non-printed regions. Reproduced
with permission from Patterned masking using polymers: insights
and developments from silicon photovoltaics, Z. Li and A. Lennon,
International Materials Review,[52] copyright © The Institute of Materials,
Minerals and Mining, reprinted by permission of Taylor & Francis Ltd,
http://www.tandfonline.com on behalf of The Institute of Materials,
Minerals and Mining.

amplified chemical reaction which could result in either the printed region
being cross-linked (see Figure 5.7(a)) or the non-printed area being cross-
linked (see Figure 5.7(b)) depending on the components in the polymer and
reactive ink.[52]

One of the key advantages of this inkjet lithography process is that only
small amounts of the printed reactant was required to control the solubility
of the chemically-amplified polymer. Li *et al.* reported the formation of point
openings in a 3–4 μm thick novolac resin layer by depositing a single 1 pL
droplet of the reactant.[52] Due to the small volume of reactant required to be
deposited, the dimensions of the opening were reduced significantly from
those reported for methods involving polymer dissolution (Section 4.2.1) or
plasticisation (Section 4.2.2) with holes having diameters as small as 20 μm
and lines being 20 μm wide as shown in Figure 5.8.

Although use of a polymer mask for the metallisation of industrial p-type
cells may not provide any cost advantages compared to the incumbent
screen-printed metallisation, it may provide a cost-effective way in which to
fabricate plated copper electrodes for the higher efficiency silicon heterojunction cells where a transparent conducting oxide (TCO) is used on the
cell surfaces for increased lateral conductivity.[52,95] With this cell structure,

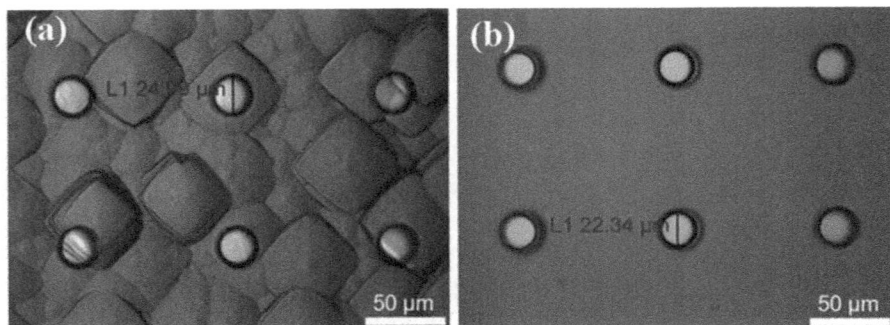

Figure 5.8   Arrays of point openings etched in a SiN$_x$ layer on: (a) an alkaline-etched silicon surface; and (b) a polished silicon surface, using the chemical lithography method described by Li *et al*. Each opening resulted from the deposition of 1 pL of reactive ink that acted to prevent the novolac resin hardening during a subsequent thermal treatment. Reproduced with permission from Patterned masking using polymers: insights and developments from silicon photovoltaics, Z. Li and A. Lennon, *International Materials Review*,[52] copyright © The Institute of Materials, Minerals and Mining, reprinted by permission of Taylor & Francis Ltd, http://www.tandfonline.com on behalf of The Institute of Materials, Minerals and Mining.

plating of metal grids requires masking of some form in order to prevent plating to the entire TCO surface. Li *et al*. demonstrated the potential to form plated metal grids for these cells using their inkjet lithography process to form the mask. Figure 5.9(a) shows a ~20 μm wide copper finger which had been plated on the TCO of a silicon heterojunction cell through a novolac resin mask patterned using their inkjet lithography process. The width of plated fingers can be effectively constrained by plating through the formed resin openings which have straighter slopes than can be typically achieved when the masking material is directly printed.[61,96] Figure 5.9(c) depicts two single 156 mm cell mini-modules patterned and plated in this way and interconnected using SmartWire Interconnection Technology[97,98] (see Figure 5.9(c)).

Etching to form point metal contacts for back contact cells can also be performed using masks patterned using inkjet lithography. Currently the highest solar cell efficiencies have been recorded with back contact solar cells,[99-101] however in order to achieve these high energy conversion efficiencies it is necessary to maximise the current extracted from the cells whilst minimising recombination (in the cell and at the surfaces). Strategies that can be adopted include forming small-area contacts, minimising electrical shunting between the n and p polarities and ensuring that light is not parasitically absorbed in the rear metal structures. Li *et al*. explored the potential of using a novolac resin mask, patterned using their inkjet lithography process, to first form the openings in the rear dielectric layer and then retain the polymer mask as an insulating layer that can minimise

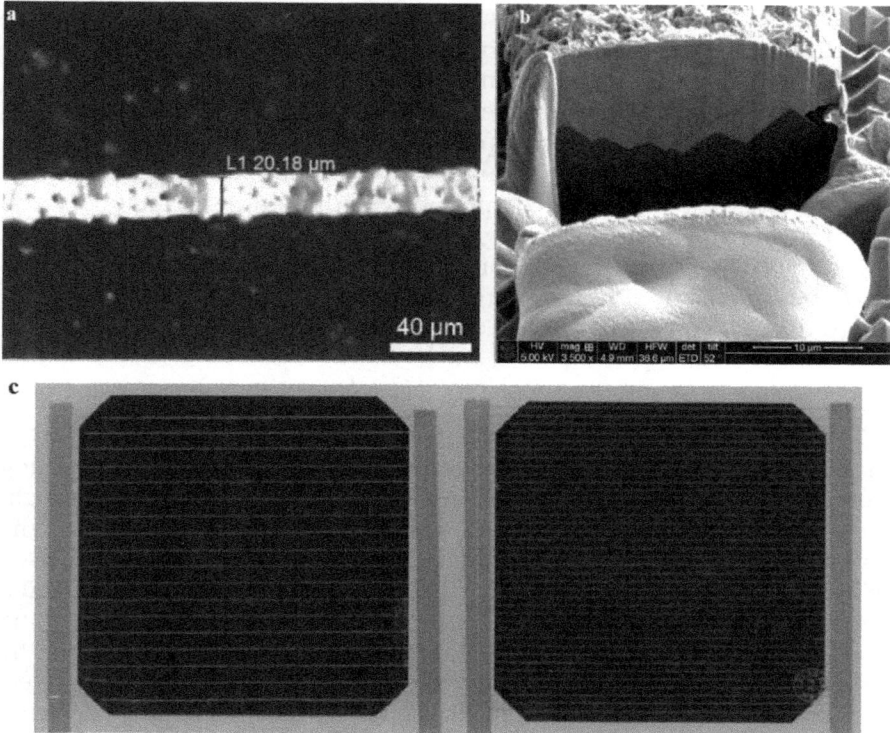

Figure 5.9 (a) A copper grid finger plated directly on the TCO of a heterojunction cell through a novolac mask that was patterned using inkjet lithography; (b) cross-section focused-ion-beam cut section through one of the plated copper fingers; and (c) photos of 156 mm 1-cell modules formed using SmartWire Interconnection Technology. Reprinted from *Energy Procedia*, **67**, Z. Li, P.-C. Hsiao, W. Zhang, R. Chen, Y. Yao, P. Papet, and A. Lennon, Patterning for Plated Heterojunction Cells, 76–83, Copyright (2015) with permission from Elsevier.[83]

electrical shunting and also displace the rear metallisation a sufficient distance from the cell so that parasitic absorption can be minimised.[102,103] The retention of the polymer as part of the device can not only potentially enhance device performance, but it also eliminates the cost that is associated with polymer removal and managing the organic waste generated through the removal.[52]

In addition to the applications described above, inkjet lithography may find applications in fields other than photovoltaics where masked patterning is required. A further advantageous feature of this general approach is that there is potentially a wide range of polymer modifiers that can be readily printed using industrial silicon MEMS printheads and can impart a solubility change in a polymer. Solubility modifiers can be readily incorporated into solvent-based inks which have a vapour pressure, surface tension and viscosity in ranges that are suitable for either piezo and thermal ink-jet printheads. This ink compatibility makes it possible for processes to use the latest developments in printhead technology. For example, reduced drop volumes can result in higher printing resolution, improved nozzle architectures can increase jetting (drop placement) accuracy and increased jetting frequencies and wider printheads can result in faster deposition.[104] Furthermore, improved ink compatibility with printhead materials and improved management of printing errors[105,106] can result in low equipment downtimes in a production environment and improved yield. All these factors can lead to a lower barrier to transitioning a process into a production environment.

5.3   Direct Etching of Inorganic Dielectrics

The use of resist material and steps to pattern the resist add cost to a patterning process. As discussed in Section 5.1.2, low-cost and high processing through-put are critical for viable silicon solar cell manufacturing, consequently the ability to directly pattern the silicon-containing dielectrics commonly-used for cells promises significant cost reductions. Lennon *et al.*[107,108] described a 'direct etching' reactive inkjet printing process in which a fluoride-containing fluid was jetted onto a benign spin-coated water-soluble polymer to form the corrosive etchant HF *in situ* on the surface of a wafer from where it can selectively etch an underlying SiO_2 or SiN_x layer. In this process, which is depicted in Figure 5.10, the deposited fluoride ions abstract protons from the polymer's carboxylic acid groups to form small quantities of HF locally in an aqueous environment according to:

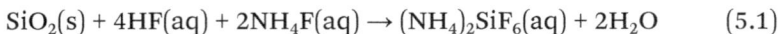

$$SiO_2(s) + 4HF(aq) + 2NH_4F(aq) \rightarrow (NH_4)_2SiF_6(aq) + 2H_2O \qquad (5.1)$$

The etching product, $(NH_4)_2SiF_6$, is highly soluble in NH_4F solutions which means that it can be readily rinsed in water without forming precipitates.[107,109]

Point openings of ~45 µm were achieved in 270 nm thick thermally-grown silicon dioxide layers (see Figure 5.11), and linear openings were demonstrated in oxides of up to 400 nm in thickness.[107] Unlike the polymer dissolution methods discussed previously, there is no re-distribution of material to the edges of the openings (see Figure 5.11(d)) as the solid inorganic dielectric is solubilised in an aqueous polymer region which presumably forms on deposition of the NH_4F solution. The diameter of the formed openings was largely determined by the initial wetted area highlighting the role of the polymer layer in constraining the spread of the aqueous solution on impacting the surface.

The etching reaction is fast (in comparison to the polymer dissolution processes described in Section 5.2.1), and once pattern printing is complete, the polymer and reaction products can be removed by rinsing in water. Not only

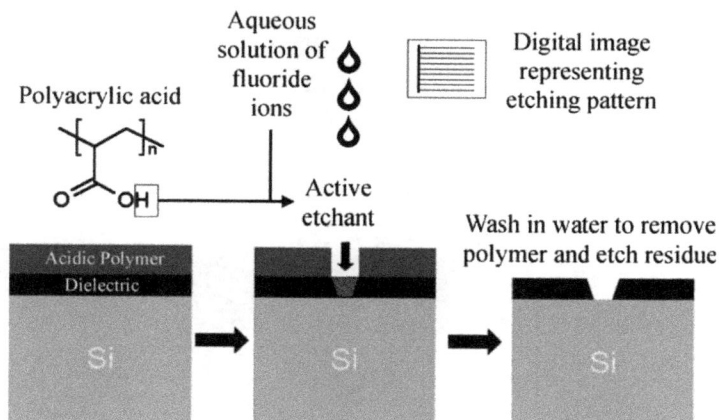

Figure 5.10   Schematic depicting the direct etching of a dielectric layer by inkjet printing a source of fluoride onto an acidic water soluble polymer for localised generation of the etchant, HF. Reprinted from *Solar Energy Materials and Solar Cells*, 93, A. J. Lennon, A. W. Y. Ho-Baillie, and S. R. Wenham, Direct patterned etching of silicon dioxide and silicon nitride dielectric layers by inkjet printing, 1865–1874, Copyright (2009) with permission from Elsevier.[54]

is this process safer for processors than using immersion etching in HF solutions, it also reduces the amount of fluoride waste generated as the HF is only formed in the localised regions where etching is required. This can enable the fluoride waste levels to be reduced to <5 ppm which can eliminate the need to treat the aqueous polymer waste. However, the process requires a relatively large volume of the NH_4F solution to be printed. To ensure small openings, a 1 pL DMP 2831 printhead was used to deposit the NH_4F solution, however as many as 50 printing layers were required to completely etch point openings in a 270 nm thick SiO_2 dielectric (see Figure 5.11). Although, the number of printing layers was reduced to 30 when 75 nm SiN_x layers were etched to form metal grid openings for plated silicon solar cells,[108] the large number of layers required for etching made the process difficult to consider for pilot production.

The etching efficiency of this single reactive inkjet process was calculated to be at best ~30%,[107] highlighting a number of inefficiencies inherent in the process. The need to first solvate the polymer represents an overhead that may be addressed by two component reactive inkjet etching in which both the acid and the fluoride reactants are printed. However, thought would need to be given to reducing the spread of the deposited reactants on the hydrophilic dielectric surface if high resolution patterning is to be achieved. In the above mentioned direct etching process, the acidic polymer layer also acts to constrain the spread of the printed fluid and this property of PAA surface layers was exploited Liu *et al.* in the development of a geometrically confined etching process.[64]

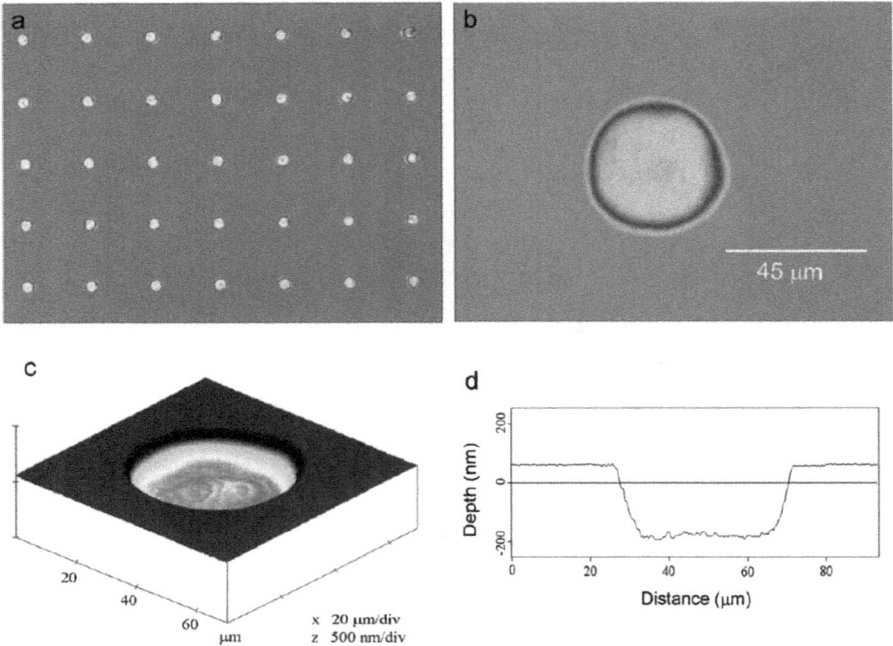

Figure 5.11   (a) An array of hole openings etched in a 270 nm thick thermally-grown silicon dioxide layer on a polished silicon wafer. The holes were etched by depositing 50 layers of a solution containing 12.8% (w/v) NH_4F, 20% (v/v) polyethylene glycol 400, adjusted to pH of 9. The polyacrylic acid layer was 2.3 μm thick and the platen temperature was maintained at 50 °C during the etching process. (b) A magnified view of a single opening from (a), (c) An AFM profile; and (d) an AFM cross-section of the hole depicted in (b). Reprinted from *Solar Energy Materials and Solar Cells*, **93**, A. J. Lennon, A. W. Y. Ho-Baillie, and S. R. Wenham, Direct patterned etching of silicon dioxide and silicon nitride dielectric layers by inkjet printing, 1865–1874, Copyright (2009) with permission from Elsevier.[54]

5.4   Direct Metallisation

Direct metallisation refers to the process of directly depositing metal on the solar cell surface without the need to separately etch openings or form a mask through which metal can be deposited. Since the surfaces of solar cells need to be electrically-passivated to reduce recombination, direct metallisation requires that the metal can in some way, during either the deposition process or a subsequent heating step, penetrate through the ARC to make electrical contact with the underlying silicon device. The requirement to etch through the ARC distinguishes this process as a 'reactive' processing step rather than a simple metal deposition step. Direct metallisation is achieved for the dominant screen-printed silver metallisation process used for industrial p-type solar cells by rapidly heating, or 'firing' the silver pastes at temperatures of 780 to 830 °C so that they can etch the SiN_x ARC and make electrical contact to the underlying silicon (see Section 5.1.1).

5.4.1 Metallisation through a SiN_x Antireflection Coating

An alternative to the masked metallisation process described in Section 5.2, is to directly inkjet print the metal on the surface of the SiN_x ARC and 'mimic' the screen printing 'fire-through' process so that the printed metal can directly contact the silicon. However, inkjet printing requires that the printed material is considerably less viscous than the pastes used by screen-printers. This limits the solid content of the deposited fluid, making it very challenging to print the entire metal contact grid without using many printing passes.

Ebong *et al.*[110] reported the inkjet-printing of a silver seed layer for front surface metal grids on silicon solar cells using an ink that was essentially a diluted form of the silver paste used for screen printing. This work extended earlier work by Horteis and Glunz where a similar ink was aerosol-printed directly on the SiN_x and fired to etch the dielectric and form a seed metal layer for the front surface electrical contact of the solar cell.[111,112] The printed silver ink comprised a silver powder, glass frits (ground down from the larger particles used for the paste), a metal oxide (M_xO_y) dispersed in a glycol ether and *N*-methyl pyrrolidone (NMP). In order to be jetted, the size of the particles was reduced to 5–50 nm and the ink's viscosity and surface tension were adjusted to be 8–20 cP and 28–32 mN/m, respectively.[55] The ink was printed using an Xjet Solar printer fitted with a 2.5–4 pL printhead having nozzles with a 20–30 μm opening.[55]

After printing of a seed layer of silver (40 μm wide and 3 μm high), cells were fired to enable the silver to penetrate through the SiN_x ARC to make electrical contact to the underlying silicon. The surface reaction that occurs for the inkjet-printed ink is substantially as described by Fields *et al.* for screen-printed pastes and can be summarised as comprising the following temperature dependent steps.[27]

1. As the temperature increases above 500 °C, the frits melt and wet the Ag/SiN_x interface;
2. Between 500 and 650 °C, the M_xO_y (*e.g.*, PbO or Bi_2O_3) in the frits reacts with and penetrates the SiN_x ARC (see Figure 5.12(a)) according to:

$$2M_xO_y + ySiN_x \rightarrow 2xM + ySiO_2 + \frac{xy}{2}N_2 \qquad (5.2)$$

3. The metal (*e.g.*, lead or bismuth) that is formed then alloys with and assists in liquid-phase mediated sintering of the silver (see Figure 5.12(b));

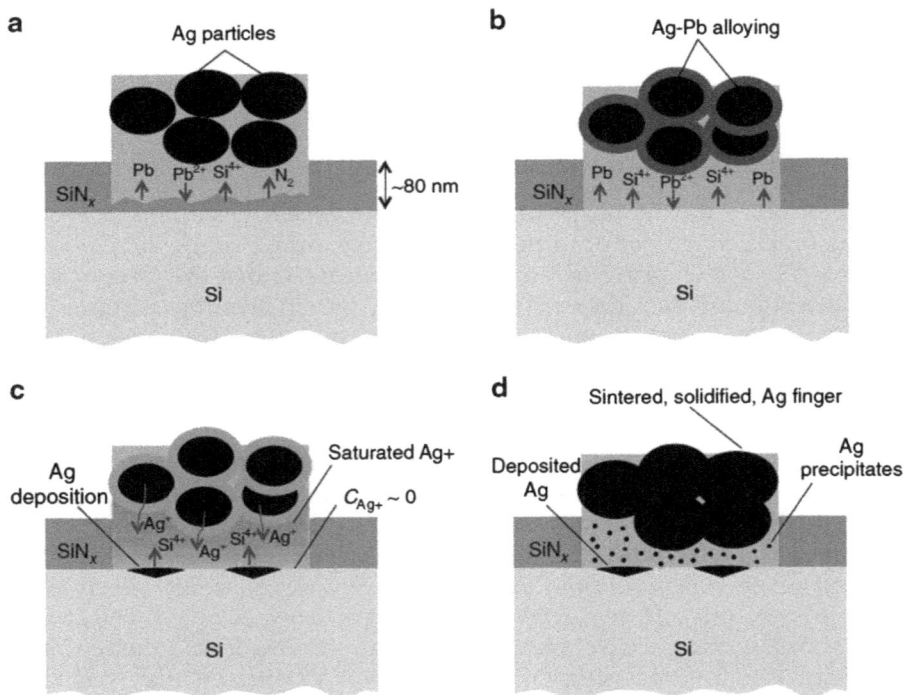

Figure 5.12 Schematic illustrations showing stages of electrical contact formation for screen-printed pastes during firing: (a) SiN_x etching by PbO in the frit; (b) Ag–Pb alloying; (c) silver transport through molten frit and deposition at the silicon surface; and (d) resulting fired-contact morphology, with inclusion of small silver precipitates within the glass intermediate layer. Reproduced under CC-BY 4.0 (https://creativecommons.org/licenses/by/4.0/) from J. D. Fields, M. I. Ahmad, V. L. Pool, J. Yu, D. G. Van Campen, P. A. Parilla, M. F. Toney and M. F. A. M. van Hest, *Nature Communications*, 7, 11143, DOI: 10.1038/ncomms11143.[27] Copyright © 2016, Nature Publishing Group, a division of Macmillan Publishers Ltd.

4. Above 650 °C, the silver dissolves in the glass frit to form Ag_2O which diffuses towards the silicon surface where it reacts with the silicon form SiO_2 and metallic silver (see Figure 5.12(c)) according to:

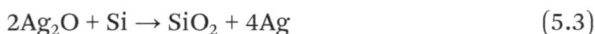

$$2Ag_2O + Si \rightarrow SiO_2 + 4Ag \tag{5.3}$$

5. On cooling, the solubility of silver in the melt decreases and nanocrystals precipitate from within the glass matrix (see Figure 5.12(d)).

The seed metal layer was then either electroplated to form silver fingers that were 65 μm wide and 25 μm high.[110] The large-area commercial-sized mono-crystalline solar cells metallised by Ebong *et al.* with the metal grid thickened by silver plating had efficiencies of up to 18.7% which were comparable or slightly higher than values routinely achieved using screen-printing at the time. The 'plating up' process offered a slight reduction in contact resistance which was proposed to be due to further reduction in the metal oxide at the silicon interface; however, the introduction of the electroplating step added further cost without a sufficiently large increment in efficiency to offset the increased cost.

Although inkjet printing of the entire grid was also demonstrated by Ebong *et al.* with metal fingers having a width and height each of ~45 μm,[110] the cell efficiency was less than that achieved by inkjet-printing a seed layer, firing that layer and then electroplating to thicken the metal contacts, due to a non-optimal firing process. In later reports,[55,56] this problem appears to have been rectified and the fabrication of 156 mm cells with fully inkjet-printed silver grids having slightly improved performance over the screen-printed controls was demonstrated.[56] It was estimated that inkjet printing the entire front metal grid could reduce the amount of silver required per cell because less silver could be printed at the busbars where sufficient conductivity can be achieved through the subsequently soldered interconnection tabs. Achieving different busbar and finger heights with screen printing requires two printing steps, one for the entire grid (*i.e.*, fingers and busbars) and then a second printing pass just to thicken the fingers. This screen-printing process is referred to a 'double print'. Assuming 2012 costs of silver, the cost of silver per Watt of power produced was reduced from US$0.17 (275 g of Ag) to US$0.043 (100 g of Ag) when comparing double-screen printing and all-inkjet printing using the XJet process.

A limitation of the seed layer process used by Ebong *et al.*[55,56,113] was that the smaller glass frits, which were required in order to inkjet print the diluted screen printing silver paste, were not as effective at etching through the SiN_x as the larger frits in the screen-printing pastes.[114] This has motivated interest in developing fire-through inks specifically for inkjet printing that use metal oxides, either formed during firing from soluble metal nitrates in the ink[115] or though the direct addition of metal oxide nanoparticles to the inkjet-printed ink,[116] to enable etching of the SiN_x ARC. Although these studies are assisting in the elucidation of the specific reactions that occur during the rapid high temperature etching of SiN_x, to date the contact resistivity of contacts formed through the inkjet printing of these new chemistries remain higher than the value of ~1 mΩ cm^2 achieved using routinely used screen-printed silver pastes.[116,117] Although further ink improvements are likely, a key challenge for inkjet-printed front surface metallization is the reduction in printing passes required to form a metal grid that is sufficiently thick to minimise resistive losses.[118]

In addition, the screen-printing process has continued to improve with new developments in pastes and screens leading to the ability to print much thinner metal features on cells which can reduce cell shading.

In 2016 the average silver finger width for front surface metallisation was ~45 µm with further reductions in feature size expected in future leading to screen-printed fingers ~25 µm in width by 2026.[49] Although challenges remain for screen-printed silver metallisation of p-type solar cells with respect to the high and volatile cost of silver and the expense of screens that can reliably enable high resolution printing, it is difficult to see how reactive inkjet printing can be competitive unless new reactive inks utilising lower cost metals such as copper can be developed. Screen-printing of copper is difficult to achieve due to the oxidation of the copper particles in the pastes, however if reactive copper inks could be developed (see Section 5.4.3) which could react on deposition to both etch through the SiN_x and reduce deposited copper ions, then cost reductions may be possible, especially as the required height of the printed grid is reduced due to the use of more busbars.[49]

5.4.2 Direct Metallisation of Silicon Heterojunction Cells

Reactive inkjet printing may also find applications in the direct metallisation of silicon heterojunction cells where the surfaces are coated with TCOs to increase their lateral conductivity. For these cells, silver screen-printing requires special (more expensive) pastes that can be fired at lower temperatures so as to not damage the cell's electronic properties. For this reason, masking and copper plating (see Section 4.2) and inkjet printing of silver nanoparticle inks[61,119,120] or reactive inkjet printing of silver salts[121,122] have been proposed as cost-effective alternatives to silver screen-printing. An arguably more cost-effective alternative may be to use reactive inkjet printing and print a nickel or copper salt or complex, followed by a reducing agent to form metal tracks (see Figure 5.13),[123–126] or alternatively print a mixture of the nickel or copper salt/complex and reducing agent with a subsequent treatment to reduce the metal ions.[127,128]

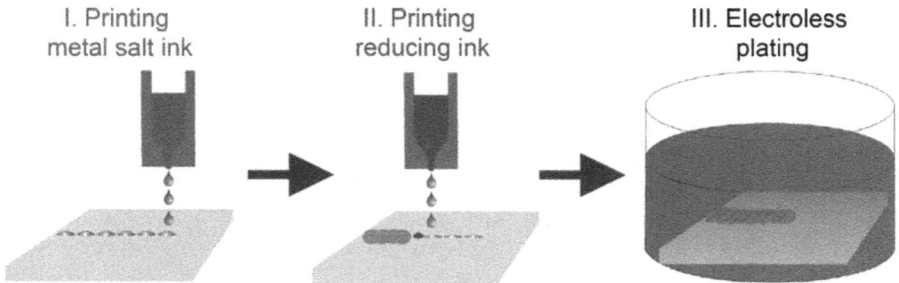

Figure 5.13   Schematic flow for reactive inkjet printing used for direct metallization. Reprinted from *Materials Letters*, **132**, D. I. Petukhov, M. N. Kirikova, A. A. Bessonov, M. J. A. Bailey, Nickel and copper conductive patterns fabricated by reactive inkjet printing combined with electroless plating, 302–306, Copyright (2014) with permission from Elsevier.[126]

The use of reactive inkjet printing for the formation of printed metal grids from metal salt inks has a number of advantages over nanoparticle inks, especially with respect to copper. First, because the copper is already in its oxidised state, inks comprising metal salts are stable and can be readily stored. Second, unlike the nanoparticle inks, inks comprising metal salts do not result in ink nozzle clogging due to particle aggregation[129] which is a considerable advantage for photovoltaic applications where reliable patterning over 156 mm wafers is required for 24/7 manufacturing. Finally, metal salt inks can use higher vapour pressure solvents (*e.g.*, glycols) which can result in more stable printing.

The approach of printing the metal salt with a reducing agent and then using low temperature heating to generate the reducing environment is attractive from the perspective of process simplicity. Reactive inkjet printing of a copper formate $(Cu(HCOO)_2)$ ink, in which the $Cu(HCOO)_2$ was complexed with 2-amino-2-methyl-1-propanol (CuF–AMP), has been used to generate metallic copper patterns *via* pyrolysis at low temperatures.[127,128] The use of formate as a reducing agent allows for a high copper load and decomposition products that are volatile, thereby leaving no undesired residues in the resulting metal that can impact the long 25 year lifetime of a photovoltaic module. Furthermore, the low decomposition temperature enables printing at low temperatures, making the process suitable for use with silicon heterojunction cells. Finally, the self-reduction reaction by formate (see eqn (5.4)) is accompanied by the release of hydrogen gas, which can further contribute to the reducing atmosphere, thus assisting in preventing the oxidation of the formed copper.

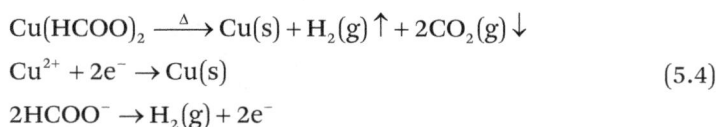

$$Cu(HCOO)_2 \xrightarrow{\Delta} Cu(s) + H_2(g) \uparrow + 2CO_2(g) \downarrow$$

$$Cu^{2+} + 2e^- \rightarrow Cu(s) \tag{5.4}$$

$$2HCOO^- \rightarrow H_2(g) + 2e^-$$

Although reactive inkjet printing of copper salts in a reducing environment seems to be a promising route to direct metallisation of silicon heterojunction cells, to-date only direct silver metallisation has been reported for these cells.[119,121,122] Given that the TCOs commonly-used for silicon heterojunction cells have been shown to act as effective barriers for copper diffusion and can prevent the ingress of printed copper into the cell where it can impact the carrier lifetimes and hence the cell performance,[130] there are no real concerns regarding reliability of direct printing of copper rather than silver.

However, one general challenge facing printing of low viscosity metal salt inks on hydrophilic TCO surfaces is spreading of the deposited fluid on the surface on impact. Jeffries *et al.* demonstrated metallization of silicon heterojunction cells using silver reactive inkjet printing however, due to ink spreading, the width of the printed lines was 100–200 μm,[121] considerably wider than the lines of ~35 μm that were reported by Hermans *et al.* using a silver nanoparticle ink on TCO surfaces,[61,119] and the 25 μm wide lines achieved by Li *et al.* using copper plating through a polymer mask.

5.5 Concluding Remarks

Reactive inkjet printing has been explored on many occasions as an approach to achieve patterned etching and/or metallisation for a range of silicon solar cell structures. In most cases, the value presented by inkjet printing was non-contact deposition, a deposition process that promises compatibility with the thinner silicon wafers that are increasingly being adopted for commercial production. Separating supply of reactants to a cell surface using reactive inkjet printing can also minimise generation or use of corrosive and toxic etching products and, in doing so, reduce the environmental impact of a manufacturing process. However, the key challenges faced by inkjet printing processes in general for silicon photovoltaic applications are high processing throughput, high yield and low cost.

Numerous innovative reactive inkjet printing processes have been developed over the relatively short silicon photovoltaic industry's history to pattern polymers to form masks for selective etching, doping and metallisation and to directly pattern inorganic dielectrics. This research has been motivated by the challenge of developing cell fabrication processes which can achieve higher energy conversion efficiencies and in doing so make photovoltaic electricity generation more affordable. Although the requirement for high processing throughput and reliability has limited industrial use of many of the developed processes, the processes may also find applications in other fields, or inspire the development of further processes. This silicon photovoltaic reactive inkjet printing research has introduced the concepts of using plasticized polymers for multiple step patterning, localised generation of corrosive etchants such as HF directly on a surface and exploiting the properties of chemically-amplified resists for high resolution, high throughput patterning of novolac resins.

Reactive inkjet printing has also been used to address the challenge of direct metallization of solar cells with *in situ* etching of an inorganic ARC. A 156 mm pseudosquare silicon solar cell can generate 8–9 A at its operating voltage and consequently any direct metallisation process must be capable of depositing sufficient metal to ensure that the solar cell's current can be extracted with low resistance. This typically requires multiple printing passes for low viscosity reactive inks which is a challenge for high throughput processing. However, this challenge has inspired some inkjet integrators to develop high throughput wafer-based inkjet printing systems with inbuilt redundancy to address reliability concerns. Furthermore, inkjet-printed MOD inks may provide a pathway by which the industry can transition to lower cost nickel/copper metallisation and a more flexible medium through which an increased understanding of the ARC glass-based etching/contacting process may be achieved.

References

1. D. M. Chapin, C. S. Fuller and G. L. Pearson, *J. Appl. Phys.*, 1954, **25**, 676–677.
2. G. Raisbeck, *Sci. Am.*, 1955, **193**, 102–110.
3. PVinsights.com, http://pvinsights.com/, accessed 6 February 2017.
4. A. G. Aberle, *Prog. Photovoltaics: Res. Appl.*, 2000, **8**, 473–487.
5. A. G. Aberle, *Crystalline Silicon Solar Cells: Advanced Surface Passivation and Analysis*, UNSW, 1999.
6. A. G. Aberle, S. W. Glunz, A. W. Stephens and M. A. Green, *Prog. Photovoltaics: Res. Appl.*, 1994, **2**, 265–273.
7. S. Glunz, A. Sproul, W. Warta and W. Wettling, *J. Appl. Phys.*, 1994, **75**, 1611–1615.
8. A. G. Aberle, *Sol. Energy Mater. Sol. Cells*, 2001, **65**, 239–248.
9. M. Kerr, J. Schmidt and A. Cuevas, *Prog. Photovoltaics: Res. Appl.*, 2000, **8**, 529–536.
10. J. Schmidt, M. Kerr and A. Cuevas, *Semicond. Sci. Technol.*, 2001, **16**, 164.
11. R. Hezel and K. Jaeger, *J. Electrochem. Soc.*, 1989, **136**, 518–523.
12. B. Hoex, J. Schmidt, R. Bock, P. Altermatt, M. C. M. van de Sanden and W. M. M. Kessels, *Appl. Phys. Lett.*, 2007, **91**, 112107–112109.
13. J. Schmidt, A. Merkle, R. Brendel, B. Hoex, M. C. M. v. de Sanden and W. M. M. Kessels, *Prog. Photovoltaics: Res. Appl.*, 2008, **16**, 461–466.
14. G. Agostinelli, A. Delabie, P. Vitanov, Z. Alexieva, H. F. W. Dekkers, S. De Wolf and G. Beaucarne, *Sol. Energy Mater. Sol. Cells*, 2006, **90**, 3438–3443.
15. A. Cuevas and Y. Di, *IEEE J. Photovoltaics*, 2013, **3**, 916–923.
16. M. Abbott, P. Cousins, F. Chen and J. Cotter, *Proc. of the 31st IEEE Photovoltaics Specialist Conference*, Colorado Springs Resort, Lake Buena Vista, Florida, USA, 2005.
17. S. R. Wenham, *Laser Grooved Silicon Solar Cells*, PhD, UNSW, 1986.
18. Y. Hayafuji, T. Yanada and Y. Aoki, *J. Electrochem. Soc.*, 1981, **128**, 1975–1980.
19. X. Wang, Z. H. Shen, J. Lu and X. W. Ni, *J. Appl. Phys.*, 2010, **108**, 033103.
20. H. M. van Driel, *Phys. Rev. B*, 1987, **35**, 8166–8176.
21. A. A. Brand, F. Meyer, J.-F. Nekarda and R. Preu, *Appl. Phys. A*, 2014, **117**, 237–241.
22. S. Hermann, T. Dezhdar, N.-P. Harder, R. Brendel, M. Seibt and S. Stroj, *J. Appl. Phys.*, 2010, **108**, 114514.
23. S. Rapp, G. Heinrich, M. Wollgarten, H. P. Huber and M. Schmidt, *J. Appl. Phys.*, 2015, **117**, 105304.
24. G. Heinrich and A. Lawerenz, *Sol. Energy Mater. Sol. Cells*, 2014, **120**, 317–322.
25. G. Heinrich, M. Wollgarten, M. Bähr and A. Lawerenz, *Appl. Surf. Sci.*, 2013, **278**, 265–267.
26. M. Nejati, W. Zhang and L. Huang, *IEEE J. Photovoltaics*, 2013, **3**, 669–673.
27. J. D. Fields, M. I. Ahmad, V. L. Pool, J. Yu, D. G. Van Campen, P. A. Parilla, M. F. Toney and M. F. A. M. van Hest, *Nat. Commun.*, 2016, **7**, 11143.

28. G. C. Cheek, R. P. Mertens, R. V. Overstraeten and L. Frisson, *IEEE Trans. Electron Devices*, 1984, **31**, 602–609.

29. C. Ballif, D. M. Huljić, G. Willeke and A. Hessler-Wyser, *Appl. Phys. Lett.*, 2003, **82**, 1878–1880.

30. Y. L. Chen, J. R. Brock and I. Trachtenberg, *Aerosol Sci. Technol.*, 1990, **12**, 842–855.

31. Y. L. Chen, J. R. Brock and I. Trachtenberg, *Appl. Phys. Lett.*, 1987, **51**, 2203–2205.

32. S.-M. Lee, B.-D. Chon and S.-G. Kim, *Korean J. Chem. Eng.*, 1991, **8**, 220–226.

33. A. Lennon, M. Renn, B. King and S. R. Wenham, *Proc. of the 24th European Photovoltaic Solar Energy Conference*, Hamburg, Germany, 2009.

34. J. Rodriguez, A. Lennon, H. Mei, C. Chan, P. H. Lu, Y. Yao and S. R. Wenham, *Proc. of the Digital Fabrication Conference*, Minneapolis, USA, 2011.

35. S. Schaefer and R. Lüdemann, *J. Vac. Sci. Technol., A*, 1999, **17**, 749–754.

36. P. N. K. Deenapanray, C. S. Athukorala, D. Macdonald, W. E. Jellett, E. Franklin, V. E. Everett, K. J. Weber and A. W. Blakers, *Prog. Photovoltaics: Res. Appl.*, 2006, **14**, 603–614.

37. H. Jansen, H. Gardeniers, M. Boer, M. Elwenspoek and J. Fluitman, *J. Micromech. Microeng.*, 1996, **6**, 14.

38. G. S. Oehrlein, *J. Mater. Sci. Eng. B*, 1989, **4**, 441–450.

39. R. D. Cox and K. A. Osgood, *J. Toxicol., Clin. Toxicol.*, 1994, **32**, 123–136.

40. M. M. Hilali, A. Rohatgi and B. To, *Proc. of the 14th Workshop on Crystalline Silicon Solar Cells and Modules*, Winter Park, CO (US), 2004.

41. E. Cabrera, S. Olibet, J. Glatz-Reichenbach, R. Kopecek, D. Reinke and G. Schubert, *J. Appl. Phys.*, 2011, **110**, 114511.

42. G. Schubert, F. Huster and P. Fath, *Sol. Energy Mater. Sol. Cells*, 2006, **90**, 3399–3406.

43. R. M. Swanson, S. K. Beckwith, R. A. Crane, W. D. Eades, K. Young Hoon, R. A. Sinton and S. E. Swirhun, *IEEE Trans. Electron Devices*, 1984, **31**, 661–664.

44. J. Zhao, A. Wang, P. P. Altermatt, S. R. Wenham and M. A. Green, *Sol. Energy Mater. Sol. Cells*, 1996, **41–42**, 87–99.

45. J. Zhao, A. Wang and M. A. Green, *Prog. Photovoltaics: Res. Appl.*, 1999, **7**, 471–474.

46. A. Blakers, A. Wang, A. Milne, J. Zhao and M. A. Green, *Appl. Phys. Lett.*, 1989, **55**, 1363–1365.

47. K. R. Williams, K. Gupta and M. Waslik, *J. Microelectromech. Syst.*, 2003, **12**, 761–778.

48. E. L. Ralph, *Proc. of the 11th IEEE Photovoltaics Specialists Conference*, Scottsdale, Arizona, USA, 1975.

49. ITRPV, *International Technology Roadmap for Photovoltaic: 2015 Results*, 2016.

50. D. Stüwe, D. Mager, D. Biro and J. G. Korvink, *Adv. Mater.*, 2015, **27**, 599–626.

51. K. F. Teng and R. W. Vest, *Proc. of the 19th IEEE Photovoltaic Specialists Conference*, New Orleans, LA, 1987.
52. Z. Li and A. Lennon, *Int. Mater. Rev.*, 2016, **61**, 416–435.
53. D. Stüwe, R. Keding, A. Salim, M. Jahn, R. Efinger, F. Clement, B. Thaidigsmann, J. G. Korvink, C. Tüshaus, S. Barth, O. Doll and D. Biro, *Proc. of the 29th European Photovoltaic Solar Energy Conference*, Amsterdam, The Netherlands, 2014.
54. A. Lennon, A. Ho-Baillie and S. R. Wenham, *Sol. Energy Mater. Sol. Cells*, 2009, **93**, 1865–1874.
55. A. Ebong, I. B. Cooper, B. Rounsaville, A. Rohatgi, M. Dovrat, E. Kritchman, D. Brusilovsky and A. Benichou, *IEEE Electron Device Lett.*, 2012, **33**, 637–639.
56. A. Ebong, *Jpn. J. Appl. Phys.*, 2015, **54**, 08KD24.
57. A. Ebong, I. B. Cooper, B. Rounsaville, A. Rohatgi, M. Dovrat, E. Kritchman, D. Brusilovsky and A. Benichou, *Proc. of the 26th European Photovoltaic Solar Energy Conference and Exhibition*, Hamburg, Germany, 2011.
58. S. Glunz, A. Aberle, R. Brendel, A. Cuevas, G. Hahn, J. Poortmans, R. Sinton, A. Weeber, A. Kalio, A. Richter, M. Hörteis and S. W. Glunz, *Energy Procedia*, 2011, **8**, 571–576.
59. T. Lauermann, A. Dastgheib-Shirazi, F. Book, B. Raabe, G. Hahn, H. Haverkamp, D. Habermann, G. Demberger and C. Schmid, *Proc. of the 24th European Photovoltaic Solar Energy Conference*, Hamburg, Germany, 2009.
60. P. Papet, J. Hermans, T. Söderström, M. Cucinelli, L. Andreetta, D. Bätzner, W. Frammelsberger, D. Lachenal, J. Meixenberger, B. Legradic, B. Strahm, G. Wahli, W. Brok, J. Geissbühler, A. Tomasi, C. Ballif, E. Vetter and S. Leu, *Proc. of the 28th European Photovoltaic Conference and Exhibition*, Paris, France, 2013.
61. J. Hermans, P. Papet, K. Pacheco, W. J. M. Brok, B. Strahm, J. Rochat, T. Söderström and Y. Yao, *Proc. of the 29th European Photovoltaic Solar Energy Conference and Exhibition*, Amsterdam, The Netherlands, 2014.
62. J. Hermans, R. van Knippenberg, E. Kamp, W. Brok, P. Papet, B. Legradic and B. Strahm, *Proc. of the 28th European Photovoltaic Solar Energy Conference and Exhibition*, Paris, France, 2013.
63. P. J. Smith and A. Morrin, *J. Mater. Chem.*, 2012, **22**, 10965–10970.
64. L. Liu, X. Wang, A. Lennon and B. Hoex, *J. Mater. Sci.*, 2014, **49**, 4363–4370.
65. T. Kawase, S. Moriya, C. J. Newsome and T. Shimoda, *Jpn. J. Appl. Phys.*, 2005, **44**, 3649–3658.
66. T. Kawase, T. Shimoda, C. Newsome, H. Sirringhaus and R. H. Friend, *Thin Solid Films*, 2003, **438–439**, 279–287.
67. T. Kawase, H. Sirringhaus, R. H. Friend and T. Shimoda, *Adv. Mater.*, 2001, **13**, 1601–1605.
68. B.-J. de Gans, P. C. Duineveld and U. S. Schubert, *Adv. Mater.*, 2004, **16**, 203–213.
69. R. D. Deegan, O. Bakajin, T. F. DuPont, G. Huber, S. R. Nagel and T. A. Witten, *Nature*, 1997, **389**, 827–829.

70. B.-J. de Gans, S. Hoeppener and U. S. Schubert, *J. Mater. Chem.*, 2007, **17**, 3045–3050.

71. I. A. Grimaldi, A. De Girolamo Del Mauro, G. Nenna, F. Loffredo, C. Minarini and F. Villani, *J. Appl. Polym. Sci.*, 2011, **122**, 3637–3643.

72. Y. Zhang, C. Liu and D. C. Whalley, *J. Phys. D: Appl. Phys.*, 2015, **48**, 455501.

73. T. L. Young and P. Lasswell, *Method of Forming Openings in an Organic Resin Material*, US7585781, 2009.

74. P. A. Basore, *Proc. of the 31st IEEE Photovoltaics Specialists Conference*, Lake Buena Vista, Florida, 2005.

75. T. Young, *Improved Method of Etching Silicon*, EP1665353, 2006.

76. M. A. Green, P. A. Basore, N. Chang, D. Clugston, R. Egan, R. Evans, D. Hogg, S. Jarnason, M. Keevers, P. Lasswell, J. O. O'Sullivan, U. Schubert, A. Turn, S. R. Wenham and T. Young, *Sol. Energy*, 2004, **77**, 857–863.

77. J. P. Huang, T. K. Kwei and A. Reiser, *Macromolecules*, 1989, **22**, 4106–4112.

78. A. Reiser, Z. Yan, Y.-K. Han and M. Soo Kim, *J. Vac. Sci. Technol., B*, 2000, **18**, 1288–1293.

79. T. F. Yeh, H. Y. Shih and A. Reiser, *Macromolecules*, 1992, **25**, 5345–5352.

80. T. Ryhanen and H. Pohjonen, in *Handbook of Silicon Based MEMS: Materials and Technologies*, ed. T. M. M. Tilli, V.-M. Airaksinen, S. Franssila, M. Paulasto-Krockel and V. Lindroos, Elsevier, 2nd edn, 2015, p. 826.

81. A. L. Brady, M. M. McDonald, S. N. Theriault and B. Smith, *Proc. of the NIP & Digital Fabrication Conference*, 2005.

82. T. Chen, *NIP & Digital Fabrication Conference, 2009*, 2009, pp. 635–638.

83. Z. Li, P.-C. Hsiao, W. Zhang, R. Chen, Y. Yao, P. Papet and A. Lennon, *Energy Procedia*, 2015, **67**, 76–83.

84. R. Utama, *Inkjet Printing for Commercial High Efficiency Solar Cells*, PhD, UNSW, 2009.

85. R. Utama, A. Lennon, A. Ho-Baillie, M. Lenio, N. Borojevic and S. R. Wenham, *Proc. of the 17th International Photovoltaic Science and Engineering Conference*, Fukuoka, Japan, 2007.

86. R. Utama, A. Lennon, M. Lenio, N. Borojevic, A. Karpour, A. Ho-Baillie and S. R. Wenham, *Proc. of the 23rd European Photovoltaic Solar Energy Conference*, Valencia, Spain, 2008.

87. A. J. Lennon, R. Y. Utama, M. A. T. Lenio, A. W. Y. Ho-Baillie, N. B. Kuepper and S. R. Wenham, *Sol. Energy Mater. Sol. Cells*, 2008, **92**, 1410–1415.

88. N. Platzer, *J. Polym. Sci., Polym. Lett. Ed.*, 1982, **20**, 459.

89. R. W. Hallman, K. I. Shimazu and H. Zhu, *Process for the Production of Lithographic Printing Plates*, EP0776763, 1997.

90. Z. Li, *Chemical Lithography and its Applications for Commercial High Efficiency Silicon Solar Cells*, PhD, UNSW, 2016.

91. Z. Li, R. Chen, Y. Yao and A. Lennon, *Proc. of the 40th IEEE Photovoltaics Specialist Conference*, Denver, Colorado, USA, 2014.

92. H. J. Levinson, *Principles of Lithography*, SPIE, Bellingham, WA, 2nd edn, 2005.

93. L. F. Thompson, C. Grant Willson and M. J. Bowden, *Introduction to Microlithography*, American Chemical Society, Washington, DC, 1994.
94. W. E. Feely, J. C. Imhof and C. M. Stein, *Polym. Eng. Sci.*, 1986, **26**, 1101–1104.
95. S. De Wolf, A. Descoeudres, C. Holman Zachary and C. Ballif, *Green*, 2012, **2**, 7.
96. P. Papet, T. Söderström, J. Ufheil, S. Beyer, J. Hausmann, J. Meixenberger, B. Legradic, W. Frammelsberger, D. Bätzner, D. Lachenal, G. Wahli, B. Strahm, J. Zhao, T. Hoes, M. Blanchet, E. Vetter, A. Richter and S. Leu, *Proc. of the Fourth Workshop on Metallization for Crystalline Silicon Solar Cells*, Constance, Germany, 2013.
97. A. Schneider, L. Rubin and G. Rubin, *Proc. of the 2006 IEEE 4th World Conference on Photovoltaic Energy Conference*, 2006.
98. T. Söderström, P. Papet and J. Ufheil, *Proc. of the 28th European Photovoltaic Solar Energy Conference and Exhibition*, Paris, France, 2013.
99. J. Nakamura, N. Asano, T. Hieda, C. Okamoto, H. Katayama and K. Nakamura, *IEEE J. Photovoltaics*, 2014, **4**, 1491–1495.
100. D. Smith, G. Reich, M. H. Baldrias, G. Harley, P. Loscutoff, M. Reich, N. Boitnott and G. Bunea, *Proc. of the 43rd IEEE Photovoltaics Specialists Conference*, Portland, Oregon, US, 2016.
101. K. Masuko, M. Shigematsu, T. Hashiguchi, D. Fujishima, M. Kai, N. Yoshimura, T. Yamaguchi, Y. Ichihashi, T. Mishima, N. Matsubara, T. Yamanishi, T. Takahama, M. Taguchi, E. Maruyama and S. Okamoto, *IEEE J. Photovoltaics*, 2014, **4**, 1433–1435.
102. Y. Li, Z. Li, Z. Lu, J. Cui, Z. Ouyang and A. Lennon, *Proc. of the 40th IEEE Photovoltaic Specialist Conference*, Denver, CO (USA), 2014.
103. U. Römer, N. Song, Z. Li, Y. Li and A. Lennon, *Proc. of the Silicon PV 2017*, Freiburg, Germany, 2017.
104. S. D. Hoath, *Fundamentals of Inkjet Printing: The Science of Inkjet and Droplets*, Wiley, 2016.
105. M. R. Rossell and S. P. Aramendia, *Method of Controlling a Printer and Printer Having at Least One Print Bar*, US20150197082, 2015.
106. H. Gothait, R. A. Peleg, O. Baharav and M. Dovrat, *Method and System for Nozzle Compensation in Non-contact Material Deposition*, WO2009153795, 2009.
107. A. J. Lennon, A. W. Y. Ho-Baillie and S. R. Wenham, *Sol. Energy Mater. Sol. Cells*, 2009, **93**, 1865–1874.
108. A. Lennon, V. Allen, A. Ho-Baillie and S. R. Wenham, *Proc. of the 24th European Photovoltaic Solar Energy Conference*, Hamburg, Germany, 2009.
109. H. Kikyuama, N. Miki, K. Saka, J. Takano and I. Kawanabe, *IEEE Trans. Semicond. Manuf.*, 1991, **4**, 26–35.
110. A. Ebong, B. Rounsaville, B. Cooper, K. Tate, A. Rohatgi, S. Glunz and M. Horteis, *Proc. of the 35th IEEE Photovoltaics Specialists Conference*, Hawaii, USA, 2010.

111. M. Hörteis, J. Bartsch, V. Radtke, A. Filipovic and S. Glunz, *Proc. of the 24th European Photovoltaic Solar Energy Conference*, Hamburg, Germany, 2009.

112. M. Hörteis and S. Glunz, *Prog. Photovoltaics: Res. Appl.*, 2008, **16**, 555–560.

113. A. Ebong, I. B. Cooper, B. Rounsaville, K. Tate, A. Rohatgi, B. Bunkenburg, J. Cathey, S. Kim and D. Ruf, *Prog. Photovoltaics: Res. Appl.*, 2010, **18**, 590–595.

114. J. Robert, F. Marco, E. Markus, S. Jochen, U. Florian, W. Andreas and M. Alexander, *J. Micromech. Microeng.*, 2015, **25**, 125021.

115. H.-G. Kim, S.-B. Cho, B.-M. Chung, J.-Y. Huh and S. S. Yoon, *J. Nanosci. Nanotechnol.*, 2012, **12**, 3620–3623.

116. R. Jurk, M. Fritsch, M. Eberstein, J. Schilm, F. Uhlig, A. Waltinger and A. Michaelis, *J. Micromech. Microeng.*, 2015, **25**, 125021.

117. H. Yang, X. Lei, H. Wang and M. Wang, *Clean Technol. Environ. Policy*, 2013, **15**, 1049–1053.

118. A. Ebong, X. Solar and T. Zhou, *Proc. of the High Capacity Optical Networks and Emerging/Enabling Technologies*, 2012.

119. J. Hermans, P. Papet, K. Pacheco, Y. Yao, W. J. W. Brok and B. Strahm, *Proc. of the 30th European Photovoltaic Conference*, 2015.

120. A. Gautrein, C. Schmiga, M. Glatthaar, W. Trmel and S. Glunz, *Proc. of the 29th European Photovoltaic and Solar Energy Conference and Exhibition*, Amsterdam, The Netherland, 2014.

121. A. Jeffries, A. Mamidanna, J. Clenney, L. Ding, O. Hildreth and M. Bertoni, *Proc. of the 42nd IEEE Photovoltaic Specialist Conference*, 2015.

122. A. M. Jeffries, A. Mamidanna, L. Ding, O. Hildreth and M. Bertoni, *Proc. of the 43rd IEEE Photovoltaic Specialists Conference*, Portland, OR, USA, 2016.

123. P. Bentley, J. Fox, A. Hudd and M. Robinson, *Method of Forming a Conductive Metal Region on a Substrate*, EP 1590500, 2005.

124. D. Li, D. Sutton, A. Burgess, D. Graham and P. D. Calvert, *J. Mater. Chem.*, 2009, **19**, 3719–3724.

125. K. Kim, S. I. Ahn and K. C. Choi, *Curr. Appl. Phys.*, 2013, **13**, 1870–1873.

126. D. I. Petukhov, M. N. Kirikova, A. A. Bessonov and M. J. A. Bailey, *Mater. Lett.*, 2014, **132**, 302–306.

127. Y. Farraj, M. Grouchko and S. Magdassi, *Chem. Commun.*, 2015, **51**, 1587–1590.

128. D.-H. Shin, S. Woo, H. Yem, M. Cha, S. Cho, M. Kang, S. Jeong, Y. Kim, K. Kang and Y. Piao, *ACS Appl. Mater. Interfaces*, 2014, **6**, 3312–3319.

129. J. Perelaer, P. J. Smith, D. Mager, D. Soltman, S. K. Volkman, V. Subramanian, J. G. Korvink and U. S. Schubert, *J. Mater. Chem.*, 2010, **20**, 8446–8453.

130. J. Yu, J. Bian, W. Duan, Y. Liu, J. Shi, F. Meng and Z. Liu, *Sol. Energy Mater. Sol. Cells*, 2016, **144**, 359–363.

CHAPTER 6

Reactive Inkjet Printing: From Oxidation of Conducting Polymers to Quantum Dots Synthesis

GHASSAN JABBOUR*, HYUNG WOO CHOI, MUTALIFU ABULIKAMU, YUKA YOSHIOKA, BASMA EL ZEIN AND HANNA HAVERINEN

Electrical Engineering, College of Engineering, University of Ottawa, Ottawa, ON, Canada
*E-mail: gjabbour@uottawa.ca

6.1 Introduction to Reactive Inkjet Printing

In addition to their traditional usage in the production of information displays including newspapers and books, printing technologies have been employed in, and/or enabled, the production steps of other applications in several industries including ceramic/glass tiles and windows, auto and aerospace, circuit boards, RFID labels, and much more.

Recently, various printing approaches have been investigated in research fields such as optics and photonics, electronics, biology and medicine, smart packaging, to mention a few. Printing techniques such as gravure, offset, screen, contact, and inkjet printing have been used in many areas including

Smart Materials No. 32
Reactive Inkjet Printing: A Chemical Synthesis Tool
Edited by Patrick J. Smith and Aoife Morrin
© The Royal Society of Chemistry 2018
Published by the Royal Society of Chemistry, www.rsc.org

the fabrication of organic solar cells,[1-4] silicon solar cells,[5,6] light emission,[7,8] batteries,[9] capacitors,[10,11] sensors,[12-14] bioapplications,[15-17] and others.[18]

The majority of reported scholarly articles on printed materials and devices, in various areas of research and technology, utilize the printer as a vehicle to place (deposit) a certain amount of given material, at certain locations on the substrate surface, without any chemical reactions taking place. On the other hand, a growing branch of research involving the use of reactive chemical components (inks) in the printing process is finding its way into several areas. This approach, often called "reactive printing", can be used in any printing technique, as long as certain measures are taken to avoid the reactivity between the ink and any of the printer's hardware (*e.g.*, corrosion of metallic parts of the printer due to a given oxidizing ink). In what follows, we will focus our discussion on reactive inkjet printing (RIJ).[19] Either thermal or piezoelectric disposal of printed droplets can be used in RIJ. The power of inkjet to accurately control the disposal of inks to extremely small volumes (*e.g.*, picoliters) endows on such approach an attractive feature regarding combinatorial discovery of materials, and optimization of devices and properties. Due to its ability to dispense ultra-small ink volumes, RIJ can then be used as a tool to optimize a certain process. In this case, the knowledge gained and discoveries made can be transferred, and process variables adjusted accordingly, to initiate a roll-to-roll, or any high throughput reactive printing process, thus saving the production process much in materials and development costs, and shortening the time to market.

6.2   Grey Scale Sheet Resistivity *via* RIJ

Perhaps the first report, to the best of our knowledge, on the use of RIJ in device setting was the demonstration of controlled modification of the sheet resistivity of conducting polymer (PEDOT:PSS).[20,21] In this work, a continuous, and/or digitized, change of sheet resistivity can be obtained through the use of an oxidizing agent as ink, and a color control program (*e.g.* Power Point) to drive the ink disposal onto the substrate. The approach is capable of generating on-demand sheet resistivity values in a thin PEDOT:PSS film. The first demonstration of RIJ sheet resistivity modification was carried guided by grey scale color function to control the placement of the oxidizing agent ink on top of PEDOT:PSS, and thus the term "grey scale resistivity".[20]

To formulate the required ink in our case, an aqueous solution containing an oxidizing agent of sodium hypochlorite (2 wt%), DI water, and surfactant was prepared and optimized to yield the required surface tension and viscosity for best controlled wetting over the surface of the conducting polymer film. The ink was injected into the precleaned black cartridge of a commercial desktop multicolor HP printer. The HSL color function was used in this case. Only the luminosity (L) was varied to obtain the required darkness of a given image, keeping the saturation (S) at zero, and holding the hue (H) at a constant value. In other words, the amount of the printed oxidizing ink is

color	No.	luminosity (L)	darkness (%)
	# 1	255	0
	# 2	230	10
	# 3	215	16
	# 4	200	22
	# 5	170	33
	# 6	140	45
	# 7	125	51
	# 8	110	57
	# 9	50	80
	# 10	0	100

Figure 6.1   Values of L used to control ink loading on top of conducting polymer layer. Here $S = 0$, and H was picked arbitrarily. Reproduced from Y. Yoshioka, P. Calvert, and G. E. Jabbour, *Macromolecular Rapid Communications*, John Wiley and Sons, © 2005 WILEY-VCH Verlag GmbH & Co. KGaA, Weinheim.[20]

controlled *via L* only. A Power Point program was used to generate various shades on the grey scale, as shown in Figure 6.1,[20] which in turn was used to control the amount of oxidizing ink printed per given area.

The oxidizing ink was printed on top of commercially available PEDOT:PSS thin layers precoated on flexible substrates of polyethylene terephthalate (OrgaconTM EL-350- courtesy of Agfa). Upon printing, the ink chemically interacts with the conducting polymer surface. The reaction penetration depth can be controlled by several factors including the number of printed drops (average ink volume per area as controlled by the darkness of the color chosen), the reaction time, temperature, and the chemical concentration of the oxidizing ink. After printing, the substrates were kept at around 40–60 °C, to allow for better and more uniform oxidant-PEDOT:PSS reaction. Subsequently, the substrates were thoroughly rinsed in DI water then dried in an oven.

To determine the end of the oxidation reaction at room temperature, the sheet resistivity was measured over a period of nearly 3 hours, immediately after printing ended. A four-point probe was used to measure the sheet resistivity. Figure 6.2 indicates that a plateau is reached in the measured sheet resistivity of PEDOT:PSS that received oxidizing ink on top of it. This indicates that the oxidation reaction has ended after around 70 minutes at room temperature. In this case, an increase of two orders of magnitude in sheet resistivity (around 7×10^4 Ohm sq^{-1}) was observed, compared to that of pristine PEDOT:PSS layer (around 370 Ohm sq^{-1}). This increase seems to be most rapid in the first few minutes (around 10 min). As time progresses, the remaining oxidant from the printed ink slowly diffuses deeper into the conducting polymer surface oxidizing it even further, which in turn results in less and less effective conducting polymer thickness, thus leading to further increase in sheet resistivity (Figure 6.2).

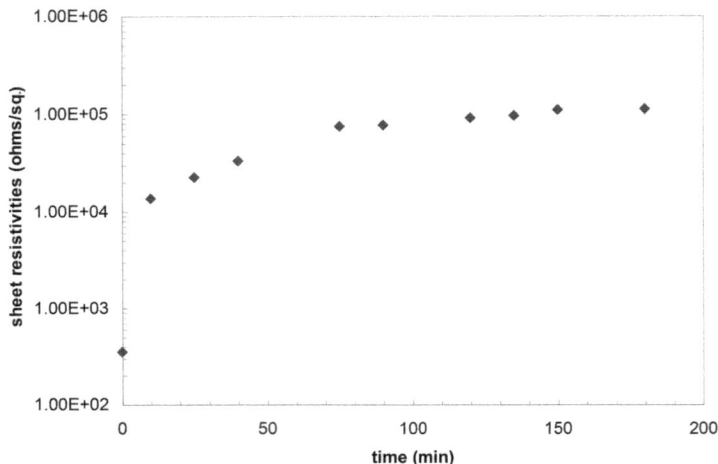

Figure 6.2 Room temperature, four-point probe measurements of sheet resistivity as a function of time for printed areas having $L = 0$ and darkness values shown in Figure 6.1. Reproduced from Y. Yoshioka, P. Calvert, and G. E. Jabbour, *Macromolecular Rapid Communications*, John Wiley and Sons, © 2005 WILEY-VCH Verlag GmbH & Co. KGaA, Weinheim.[20]

A judicial increase in temperature of the substrate during (or after) the printing processes is expected to lead to an accelerated oxidation process reaching to more depth within the conductive polymer layer, thus resulting in further increase in sheet resistivity of the PEDOT:PSS film.

The different ink loadings according to the values of L shown in Figure 6.1 yield different values of the sheet resistivity. In this case, an increase in sheet resistivity can be seen with increasing darkness values (ink drops per area). The values reported in Figure 6.3 were measured 1 hour after printing of the oxidizing ink.

One benefit of color inkjet printing technique (*i.e.*, multi-cartridge or multi-ink compartments) is the facile and highly controlled combinatorial experiments it allows for given inks. To demonstrate the power of such an approach, a simple experiment was carried by placing in the pre-cleaned yellow (Y) cartridge an ink made of a mixture of deionized (DI) water and 0.05 wt% surfactant, and an ink of DI water and 5.6 wt-% sodium hypochlorite mixture in the pre-cleaned cyan (C) cartridge.

The dispensing of respective ink droplets was controlled using the Cyan, Magenta, Yellow, and Black (CMYK) color model. In our case, only the two colors Y and C are mixed as indicated above. For each color, a value between 0 and 100 was chosen. These values can be obtained using PPT software. The experiments were carried out with the DI water/surfactant printed first (Y color cartridge), followed by printing of the oxidizing agent ink (C color cartridge). Printing C ink first results in rapid oxidation of the conducting polymer surface, and hence limits the effects of the second ink.

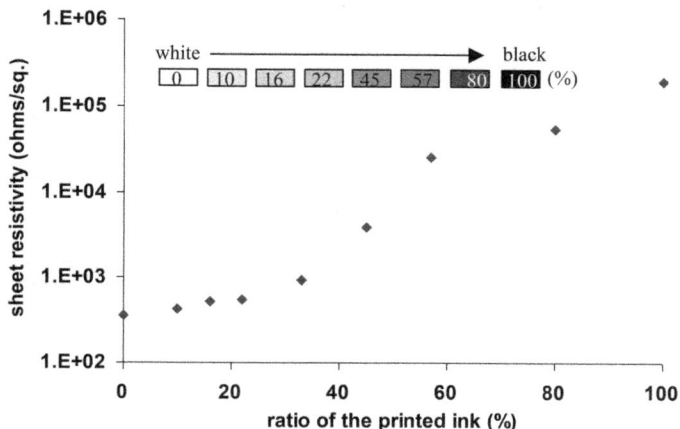

Figure 6.3   Sheet resistivity of PEDOT:PSS at 8 printed locations where $L = 255$ is for white (pristine surface), and $L = 0$ is for black (maximum ink per printed area, resulting in a fully oxidized surface). The four-point probe measurements were done in air, around 1 hour after the printing. Reproduced from Y. Yoshioka, P. Calvert, and G. E. Jabbour, *Macromolecular Rapid Communications*, John Wiley and Sons, © 2005 WILEY-VCH Verlag GmbH & Co. KGaA, Weinheim.[20]

By adjusting the experimental conditions, it is possible to obtain uniform mixing of the two inks as dictated by the C–Y combined color given by the PPT program. A representative set of rectangular areas each having a defined value of C and Y is shown in Figure 6.4. The figure also shows the respective color resulting from mixing of the various values of C and Y. The printed library of Figure 6.4 was dried in air for 3 hours, at a temperature of around 55 °C, followed by a DI water rinse to remove any residual oxidant material.

After processing and drying, the sheet resistivity of each of the oxidized PEDOT:PSS areas corresponding to samples 1–18 was measured, and the resulting values are shown in Figure 6.5. Very high sheet resistivity ($\gg 10^7$ ohm sq^{-1}, not shown) was measured for samples 1, 2 and 3, which have a common C value of 100%. This indicates a complete passivation of the thin layer of PEDOT:PSS. A constant Y value of 100, *i.e.* full DI water/surfactant coverage of first printed layer, results in a sharp rise in sheet resistivity for printed samples having C value between 22 and 29.

Figure 6.5 shows that sheet resistivities of 10^3 ohm sq^{-1} and 10^5 ohm sq^{-1} are obtained for samples having ($Y100$, $C9$) and ($Y100$, $C25$–$C29$), respectively. For the printing process using a single pre-mixed ink of DI water/surfactant/2 wt% oxidizing agents, similar sheet resistivity values are obtained for samples having ($L170$, darkness 33) and ($L50$, darkness 80), respectively, as shown in Figure 6.3. The agreement in the measured values indicates successful mixing of the two inks over the PEDOT:PSS surface during printing. The attraction of using the CMYK color model (with two inks in our case) allows us to access a myriad of sheet resistivity values not possible with single ink component printing approach using the grey-scale model.

a)

Y   color   C

concentration
profile

H₂O and surfactant	oxidizing agent	
1) printed yellow [H₂O + surfactant]	2) printed cyan [oxidizing agent]	3) H₂O and oxidizing agent were mixed on the surface and reacted with the PEDOT-PSS.

b)

sample number	1	2	3	4	5	6	7	8	9
yellow function	0	50	100	100	100	100	100	100	100
cyan function	100	100	100	67	47	39	32	29	25
resulted color									

sample number	10	11	12	13	14	15	16	17	18
yellow function	100	100	100	100	100	100	100	100	100
cyan function	22	20	19	18	17	16	15	9	0
resulted color									

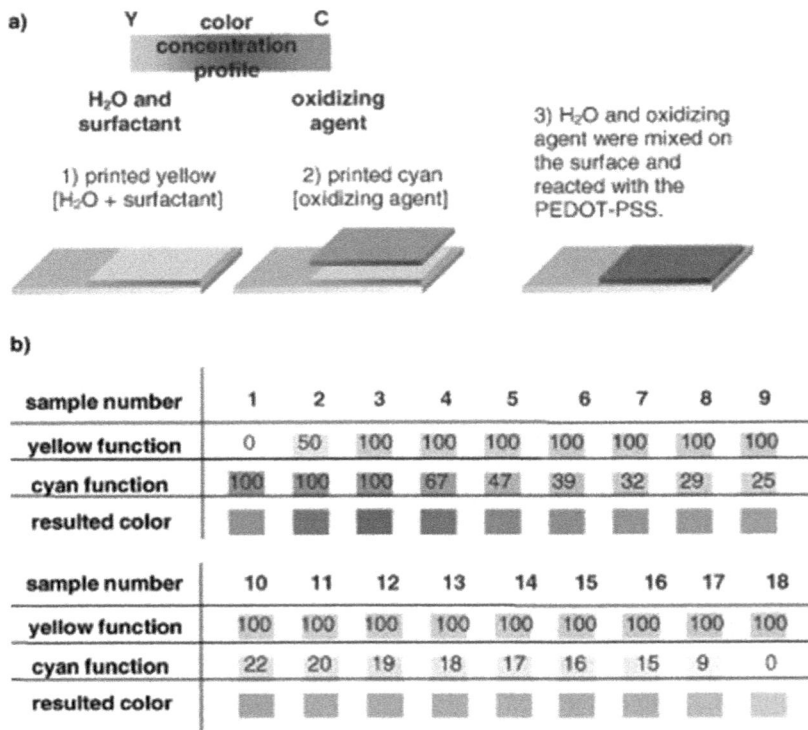

Figure 6.4   (a) A representative color concentration profile obtained by mixing C and Y colors upon printing on the substrate. Y was printed first, followed by C to yield the mixed layer. (b) A library of C and Y values used to produce various sheet resistivities over the surface of the PEDOT:PSS. In all of these experiments, *M* was set to 0. Reproduced from Y. Yoshioka, P. Calvert, and G. E. Jabbour, *Macromolecular Rapid Communications*, John Wiley and Sons, © 2005 WILEY-VCH Verlag GmbH & Co. KGaA, Weinheim.[20]

6.3   Transitioning to Tool Friendly Oxidizing Agent

The significant corrosive ability of sodium hypochlorite, when used at high concentrations, has detrimental effects on printer metallic parts and cartridge electronics (*e.g.* corrosion), leading to frequent costly maintenance and replacement of parts. Although the diluted oxidizing ink (*e.g.* 2 wt%) resulted in an on-demand control of the sheet resistivity of conducting polymer layer, above its pristine value, several experimental aspects might not be conducive to full industrial scale operation. These include the corrosion factor mentioned above, the relatively long processing time (up to 3 hours) needed for reaction completion between oxidizing ink and the polymer, and the extra DI water rinse step needed to remove residual crystals of sodium hypochlorite left behind on the polymer surface. In this regard, we turned our attention to a search for milder oxidizing agents. While several candidates

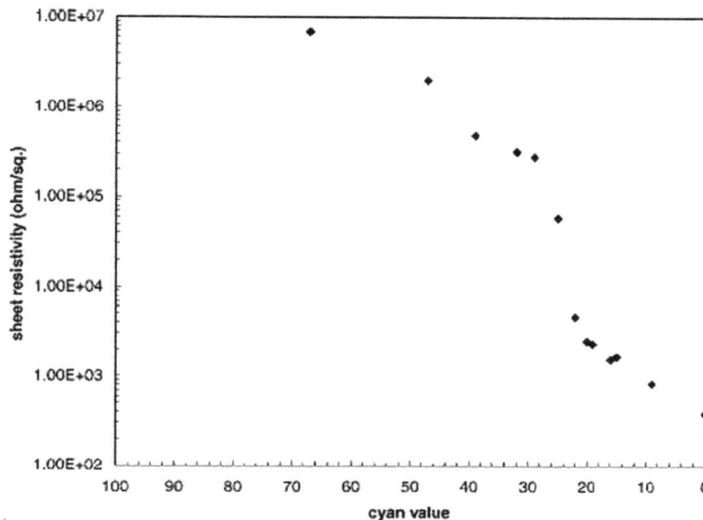

Figure 6.5 Sheet resistivity values for printed color combinations of Figure 6.4. Note the sharp increase in sheet resistivity appears around C values of 22–29. Reproduced from Y. Yoshioka, P. Calvert, and G. E. Jabbour, *Macromolecular Rapid Communications*, John Wiley and Sons, © 2005 WILEY-VCH Verlag GmbH & Co. KGaA, Weinheim.[20]

were experimented with, we will present here results pertaining to the use of hydrogen peroxide as an oxidizing agent.

In addition to being more environmentally friendly than sodium hypochlorite, an aqueous solution of hydrogen peroxide evaporates, depending on the concentration, completely at temperatures as low as 115 °C, without leaving any chemical powder residues. This is an attractive feature as it eliminates the DI water rinse step needed in the case of sodium hypochlorite ink. Furthermore, it is also possible to increase the concentration of the hydrogen peroxide ink without causing damage to the cartridge or the printer parts. In fact, an ink of 50 wt% hydrogen peroxide did not cause any observable damage to the inkjet cartridge or printer parts during repeated printing experiments, thus leading to less frequent need of replacement parts, and reducing the down time of the process.

To demonstrate the feasibility of using high concentration of hydrogen peroxide as an ink, we present the results of a similar inkjet printing experiment using the HSL model as described earlier, but with an aqueous ink 50 wt% hydrogen peroxide loaded in the cartridge. By setting the value of S to 0 and holding H constant, we varied L over 9 different values (255, 230, 215, 200, 170, 140, 110, 50, and 0), as shown in Figure 6.6, each used in printing of a rectangular feature over the PEDOT:PSS layer (Figure 6.6 insert).

Although the boiling temperature of the ink used is around 126 °C, heating the sample up from room temperature to 50 °C is found to decrease the sheet resistivity. However, the change in sheet resistivity with decreasing L

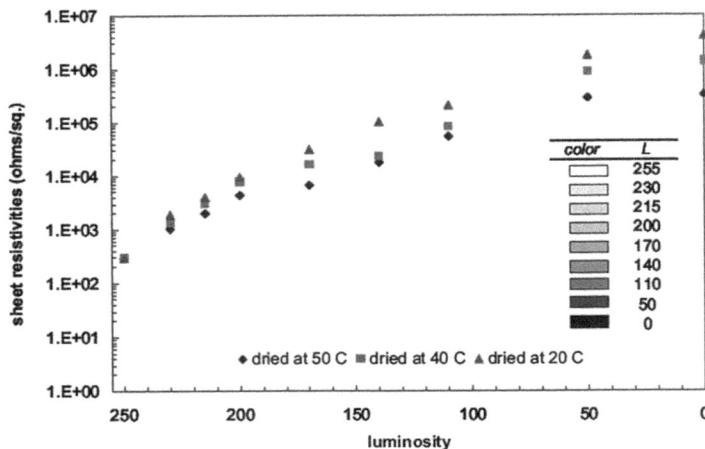

Figure 6.6   Values of sheet resistivity for various L obtained from four-point probe measurements at 3 different temperatures of 20 °C (triangle), 40 °C (square), and 50 °C (diamond), respectively. Reproduced from Y. Yoshioka and G. E. Jabbour, *Advanced Materials*, John Wiley and Sons, © 2006 WILEY-VCH Verlag GmbH & Co. KGaA, Weinheim.[21]

values is highest at 20 °C, followed by 40 °C and 50 °C, respectively. At 20 °C less ink is leaving the substrate in the form of vapor and thus more oxidation can occur than is the case at 50 °C, resulting in increased sheet resistivity (Figure 6.6). A gradual increase in sheet resistivity with decreasing L values is observed for each temperature.

We also generated a rectangular strip of continuous color change (grey gradient with white and black) using the "Fill Effect" capability of the software. Over this strip, the L values varied gradually from black ($L = 0$) to white ($L = 255$). The values of sheet resistivity measured at various locations on the strip are depicted in Figure 6.7. In similar fashion to the discrete prints above, the sheet resistivity increases with decreasing L values, reaching a maximum around 10^6 ohm sq^{-1}. at $L = 0$.

A current was passed through a 4 inch long oxidized strip of PEDOT:PSS, shown in the inset of Figure 6.7, and the temperature was measured at several locations on the strip, as shown in Figure 6.8. As anticipated, the differing sheet resistivity values for the various oxidized regions gave rise to increasing temperature with increasing sheet resistivity, Figure 6.8 (samples dried at 50 °C). A gradual increase in temperature can be seen along the length of the strip starting at 21.5 °C for the non-oxidized PEDOT:PSS surface, and ending with 25.5 °C for the mostly oxidized location of the conducting polymer layer surface (*i.e.*, at $L = 0$). Such a moderate increase in temperature along the length of the strip can find applications in several areas including biology, body heaters, window heaters, defense, IR maps and embedded images, to mention a few.

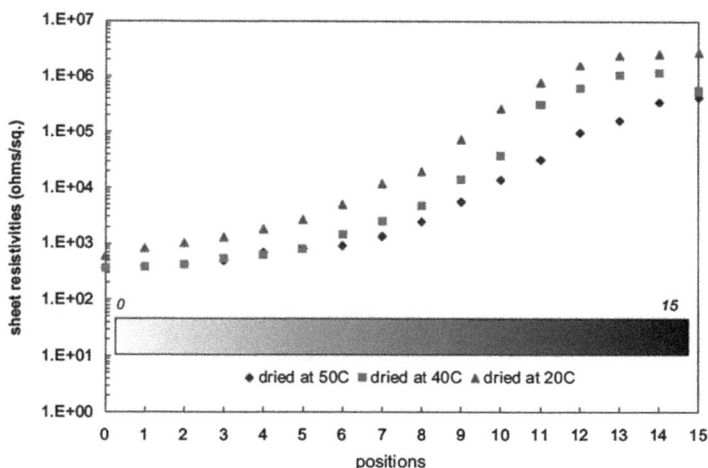

Figure 6.7   Sheet resistivity measurements for gradually changing *L* values (insert), over a 4 inch long strip, taken after reaction between hydrogen peroxide ink and PEDOT:PSS at different temperatures of 20 °C (triangle), 40 °C (square), and 50 °C (diamond), respectively. Reproduced from Y. Yoshioka and G. E. Jabbour, *Advanced Materials*, John Wiley and Sons, © 2006 WILEY-VCH Verlag GmbH & Co. KGaA, Weinheim.[21]

6.4   Reaction Mechanism and Characterization of Oxidation Process

6.4.1   Sodium Hypochlorite Case

A possible reaction route between PEDOT:PSS and sodium hypochlorite is shown in Figure 6.9. The oxidizing agent acts on converting the thiophene (part (A) of Figure 6.9) group of the PEDOT to thiophene-1-oxide (B). Thiophene-1-oxide (B) is effortlessly transformed to thiophene-1,1-dioxide (C). Thiophene-1-oxide (B) is recognized previously as an intermediate prior to the formation of thiophene-1,1-dioxide (C).[22,23] The loss of SO_2 occurs as a result of further oxidation of thiopene-1,1-dioxide (C) accompanied nucleophilic attack that results in the linkage of hydroxyl groups (due to the presence of water), thus producing the structure shown in (D).

FTIR spectroscopy was performed on an oxidized PEDOT:PSS sample and pristine one (Figure 6.10). The measurements depict a slight difference in relative band intensities. The vibrational mode due to asymmetric and symmetric stretching of the S–O sulfonic group is shown at wavenumber of 1343 cm^{-1} and 1132 cm^{-1}, respectively.[23–25] The absorbance resulting from the symmetrical stretching of sodium sulphonate group (PSS-Na) is depicted at 1044 cm^{-1}.[24,25] To take a closer look at the conclusion of the oxidation reaction, we synthesized structure (D) from a dried sample (powder) of diluted PEDOT-PSS dispersion (Baytron P, Bayer) overloaded with NaOCl aqueous solution. A FTIR study was carried on this powder sample along with a separate powder sample

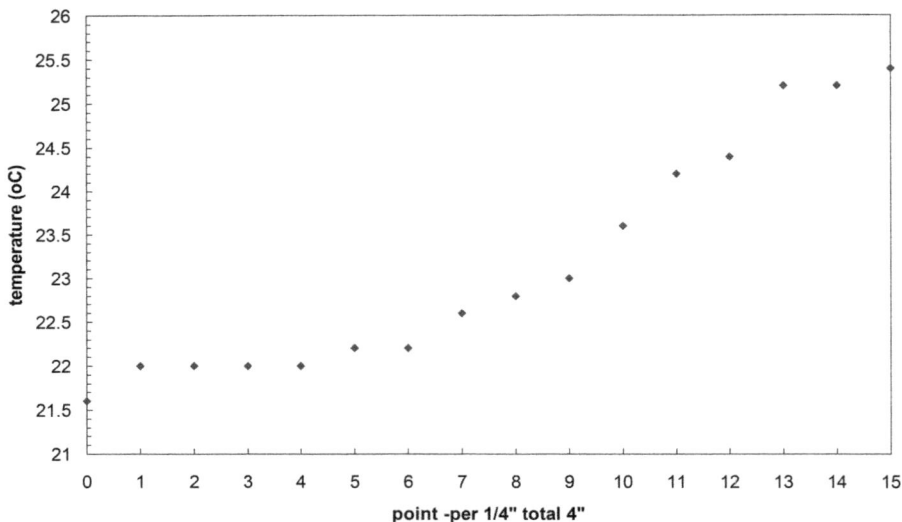

Figure 6.8 Gradual temperature increase across the continuously changing L values. Data shown for PEDOT:PSS treated at 50 °C after printing of hydrogen peroxide ink. Measurements were taken at 1/4 inch intervals over the 4 inch long strip, with $L = 0$ being the highest sheet resistivity (pt 15 on scale).

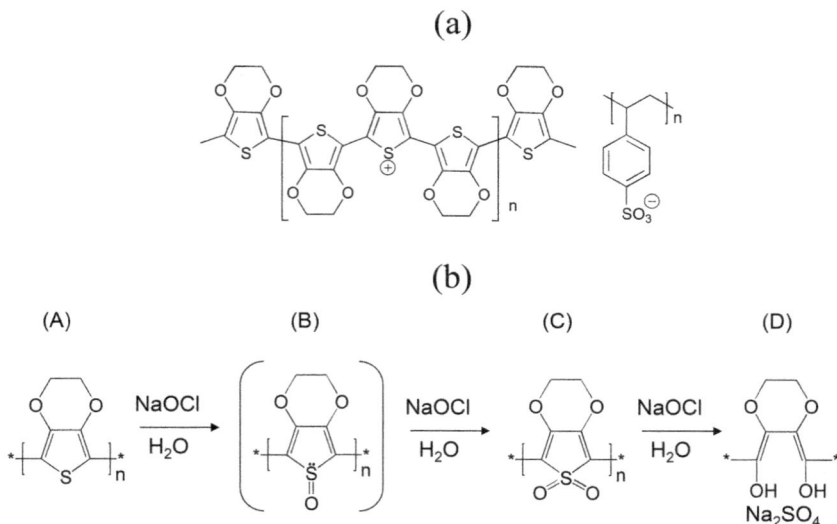

(a)

(b)

Figure 6.9 (a) Chemical structure of PEDOT:PSS, and (b) candidate oxidation route of PEDOT:PSS reaction with sodium hypochlorite ink. Reproduced from Y. Yoshioka, P. Calvert, and G. E. Jabbour, *Macromolecular Rapid Communications*, John Wiley and Sons, © 2005 WILEY-VCH Verlag GmbH & Co. KGaA, Weinheim.[20]

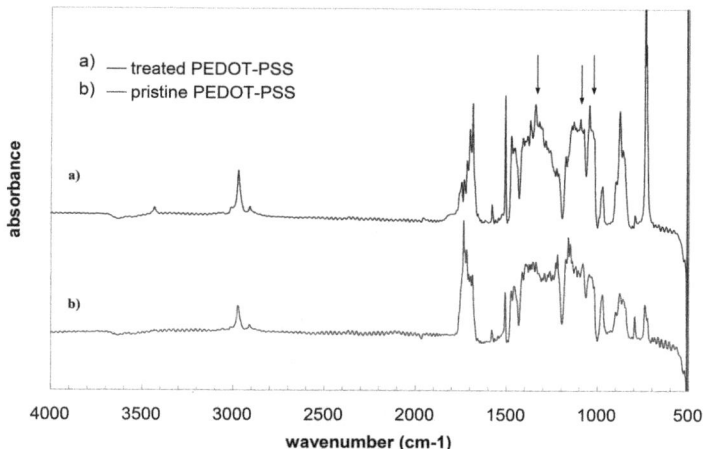

Figure 6.10   FTIR spectra for oxidized (treated-a) and prisitine (b) PEDOT:PSS layer. Reproduced from Y. Yoshioka, P. Calvert, and G. E. Jabbour, *Macromolecular Rapid Communications*, John Wiley and Sons, © 2005 WILEY-VCH Verlag GmbH & Co. KGaA, Weinheim.[20]

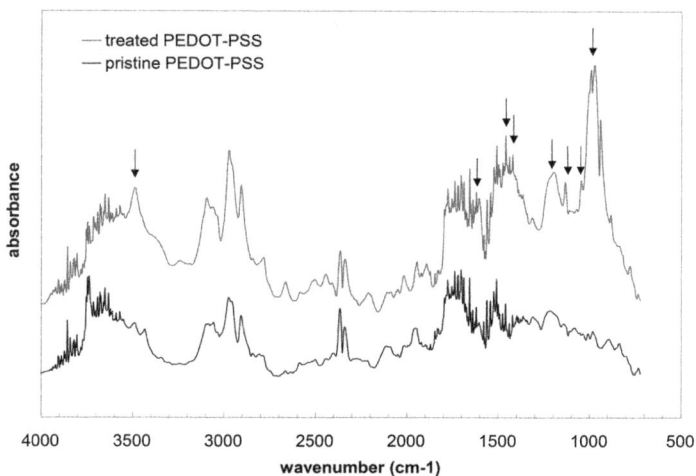

Figure 6.11   FTIR spectra of a dry sample of fully oxidized PEDOT-PSS (a), and a dry sample of pristine PEDOT-PSS (b), mixed with KBr and pressed into pellets. Reproduced from Y. Yoshioka, P. Calvert, and G. E. Jabbour, *Macromolecular Rapid Communications*, John Wiley and Sons, © 2005 WILEY-VCH Verlag GmbH & Co. KGaA, Weinheim.[20]

of pristine PEDOT:PSS. Each of the powders was mixed with KBr and pressed into pellets. In both cases, resolutions of 4 cm^{-1}and 128 scans were used.

The results are shown in Figure 6.11. As shown in the figure a noticeable difference in band intensity can be seen. The stretching and bending of O–H, and stretching of C–O bonds are clearly visible at 3493 cm^{-1}, 1410 cm^{-1}, and 1201 cm^{-1}, respectively. The stretching of C–C bonds can be seen at 1616 cm^{-1}. SO$_4$

stretching mode gives rise to the absorbance at 1134 cm^{-1}.[25] An O–Cl vibrational band is seen at 974 cm^{-1}, which arises from non-reacted sodium hypochlorite.[26]

6.4.2   Hydrogen Peroxide Case

A possible chemical reaction between hydrogen peroxide and the conducting polymer PEDOT:PSS is shown in Figure 6.12. The displayed reaction involves the conversion of thiophene into its oxide version, thiophene dioxide,[27] Figure 6.12(a). Hydrogen peroxide can also impact the integrity of oxyethylene rings to result in esters and sulfoxide.[28] However, FTIR studies were not sensitive enough to clearly detect this change. This is mainly due to the relatively small number of oxidant molecules compared to that of EDOT monomers in the PEDOT:PSS thin film. Thus, it is not possible to have significant amount of EDOT molecules reacting to yield the needed FTIR signal. To circumvent this, we prepared pristine PEDOT:PSS films, from commercially available Baytron P, on silicon substrates. Some of the samples were fully oxidized by applying excess of hydrogen peroxide. The FTIR of the reacted samples and the pristine ones are shown in Figure 6.12(b). A clear difference between the FTIR spectra of both cases can be clearly seen at 1309 cm^{-1}, 1309 cm^{-1}, and 931 cm^{-1}, respectively. This can be attributed to the partial dissociation and/or protonation of the sulfonate group (in PSS) upon oxidation with the hydrogen peroxide. The dominance of PSS in PEDOT:PSS film[29] might overwhelm the spectra, thus making the FTIR data representing mostly that of PSS. The vibrational aspects of the sulfone group are observed at 1360 cm^{-1} and 1182 cm^{-1}, while the bands observed at 1751 cm^{-1}, 1678 cm^{-1}, and 1416 cm^{-1} are those of ester group, which is a product of the oxidation reaction.

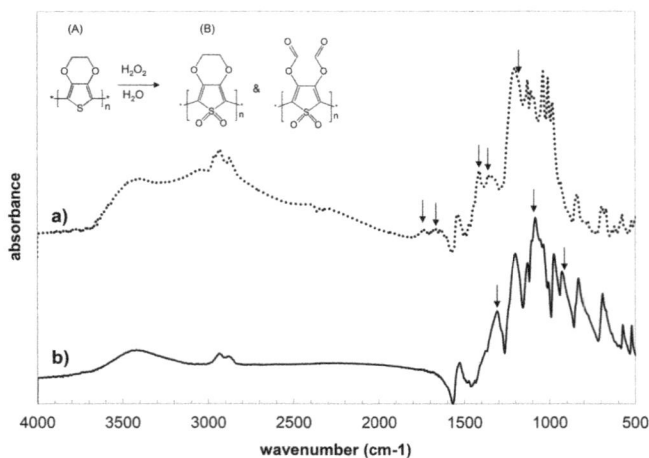

Figure 6.12   A and B represent a possible chemical reaction route between hydrogen peroxide and PEDOT:PSS and FTIR spectra of oxidized PEDOT:PSS using hydrogen peroxide (a), and (b) unoxidized pure PEDOT:PSS. Reproduced from Y. Yoshioka and G. E. Jabbour, *Advanced Materials*, John Wiley and Sons, © 2006 WILEY-VCH Verlag GmbH & Co. KGaA, Weinheim.[21]

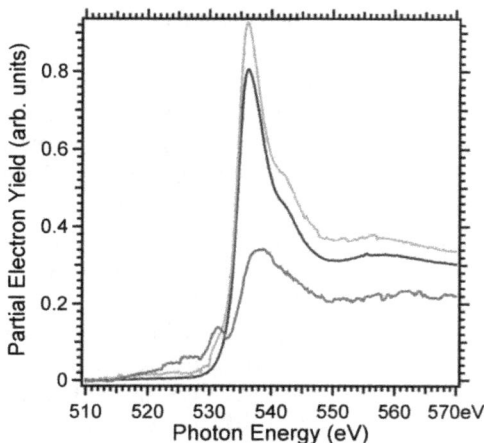

Figure 6.13   NEXAFS spectra of PEDOT:PSS pristine film (middle curve), treated PEDOT:PSS film (top curve), and PET substrate (bottom curve).

Near edge X-ray absorption fine structure (NEXAFS) spectroscopy was performed on pristine PEDOT:PSS film, oxidized PEDOT:PSS, and the bare plastic substrate (PET) on which PEDOT:PSS film was deposited. The results shown in Figure 6.13 demonstrate increased oxygen intensity in the oxidized conducting polymer sample, (treated sample) due to the formation of more C–O, thus leading to more insulating behavior of the polymer film at the locations where it interacted with the printed oxidizing ink, thus reducing drastically its electrical properties.

Further characterization of the oxidized and pristine PEDO:PSS films were carried using XPS and UPS. The results support our conclusion in regard to PEDOT:PSS oxidation and the respective formation of insulting species (disruption of PEDOT chemical integrity), as indicated in our discussion of the FTIR data given above.

6.5   RIJ Printing: From Surface Modification to Nanomaterials and Quantum Dots Assembly

In what follows we will present examples demonstrating the promising potential of RIJ in the synthesis of self-assembled nanoparticles of single metal element, alloy, compound, and quantum dots.

Interest in Au nanoparticles has grown in recent years due to their potential application in biology and medicine,[30–32] energy and power generation,[33–39] forensics,[40] sensing,[41] and photonics,[42,43] to mention a few.[44] The non-toxic nature, chemical stability, good heat and electrical conductivity,[45] quick response to magnetic and electric fields stimuli, and relative ease of synthesis, favour the use of Au NPs over many of its metallic counterparts, such as Ag, Cu, Al, and others.

Many methods have been reported concerning the synthesis of Au NPs including approaches such as lithography,[46-48] high vacuum deposition,[49,50] wet chemistry,[51-53] and biological,[54] to mention a few. Nevertheless, the dominant approaches to gold synthesis remain the liquid based methods. Regardless of the liquid approach used, the preparation of colloidal Au NPs, purification, and their stabilization with surface moieties involves several steps, thus requiring from a few hours to more than a day (depending on the approach adopted) to achieve the final product. In addition to imparting better NPs dispersion in a given liquid, the surface moieties can also have electrical, optical or other desired functions.

Inkjet and other printing techniques have been used to deposit and pattern liquid suspension of Au NPs.[54-56] However, inkjet printing has several features that make it more attractive, in many settings, than other printing approaches including the need for no stencils, ability to jet ultrafine droplet volumes with high reproducibility, on-demand printing of chosen materials anywhere on the substrate surface, real time patterning capability and the ability to change the pattern on the fly, and near 100% utility of formulated inks.

Our aim is to move away from the passive approach of inkjet printing of Au NPs, to an active role where the inkjet facilitates control of chemical reactions and materials self-assembly over a given surface.[57,58] Indeed, we will show here that RIJ can be used to synthesis self-assembled Au NPs, NPs alloy, and non-metallic compound NPs. Furthermore, it is possible to use RIJ approach to synthesize quantum dots (QDs).

Our experiments reported in the rest of this chapter rely on the use of Dimatix Materials Printer DMP-2800. In all cases, we used a 10 pL cartridge having a line of 16 nozzles with inter-distance (spacing) of 254 μm. Each nozzle has a diameter of 22 μm. Silicon substrates were cleaned using successive ultrasonic baths of acetone, ethanol, and isopropanol, respectively, each bath lasted 10 min. High purity nitrogen (99.999) was used to dry the samples after each bath. Unless otherwise specified, all experiments were carried in ambient environment at room temperature. Immediately prior to printing, we used plasma ashing to remove any organic residues from the Si surface, and placed the cleaned substrates at a 0.25 mm distance under the printing nozzle. The viscosity values of the various inks used in this discussion were roughly in the range between 1 cP to 5 cP.

6.6   Self-assembled Au NPs

In these experiments, we prepared an ink mixture of 10 ml and 1 ml 1,2-dichlorozene (dispersion solvent) and oleylamine (reducing and capping agent), respectively. The ink was placed in a cartridge (ink A). Another ink (ink B) composed of a gold(III) chloride hydrate $HAuCl_4 \cdot 3H_2O$ at 0.12 mmol, and 10 ml of dimethyl sulfoxide (DMSO). The contact angle on silicon surface and surface tension of ink A were 28.60° and 29.68 mN m^{-1}, respectively. The contact angle and surface tension of ink B were 35.75°, and 43.88 mN m^{-1}, respectively.[57] A Kruss EasyDrop was used to measure the contact angle at room

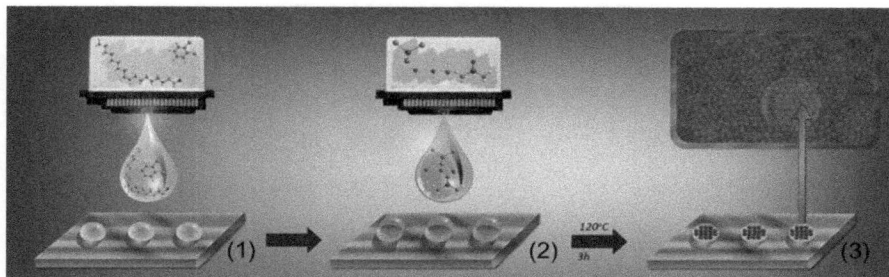

Figure 6.14   Self-assembled Au NPs synthesis using RIJ. (1) Printing of ink A, (2) printing of ink B on top of ink A, and (3) self-assembled Au NPs grown on substrate after heat treatment of printed inks at 120 °C for 3 hours. Reproduced from M. Abulikemu, E. Daas, H. Haverinen, and G. E. Jabbour, *Angew. Chem. Int. Ed.*, John Wiley and Sons, © 2014 Wiley-VCH Verlag GmbH & Co. KGaA, Weinheim.[57]

temperature (around 22–23 °C), and a Kruss Tensiometer (K100MK2/SF/C) was used (at 22.6 °C) to measure the surface tension. The two inks were loaded in separate cartridges. The printing process was carried out by printing ink A first, followed by ink B on top of ink A. A depiction of the experimental process is shown in Figure 6.14.

Ink A and ink B were printed in sequence, with ink B placed exactly on top of ink A printed locations on the silicon substrate. The inks were printed in the form of a square array having 30 μm spacing between adjacent drops. A post printing heat treatment was carried out for 3 hours and at 120 °C.

The substrates were then transported in air and placed in a high-resolution SEM (HRSEM-Nova NANO 600) for imaging and analysis *via* energy dispersive X-ray spectroscopy (EDS). For randomly selected elements of the printed array, the HRSEM inspection revealed a self-assembled layer of NPs having an average diameter of around 8 ± 2 nm. A typical image is shown in Figure 6.15, which indicates a densely packed layer of NPs with relatively uniform spacing between them. Figure 6.16 shows the EDS analysis of the NPs, which reveals the presence of gold and carbon along with a sizable silicon signal originating from the substrate. The uniform distance between adjacent NPs is attributed to the capping ligands of oleylamine attached to the surface of the NPs, which are the origin of the carbon signal in the EDS data. It is worth mentioning that the Au NPs are synthesized and capped with oleylamine in one single step, as opposed to the many steps followed in traditional synthesis approaches! The average length of a free standing oleylamine ligand is around 2 nm.[59] However, when oleylamine ligands from different NPs are brought into close proximity to each other, an interdigitated structure between the ligands from one NP is formed with those of the adjacent NP, which results in an overall NP–NP distance that is shorter than the length of two olyelamine ligands touching each other in an end-to-end manner. In our case, the average distance between adjacent NPs is around 2.6 nm, as obtained from Figure 6.15.

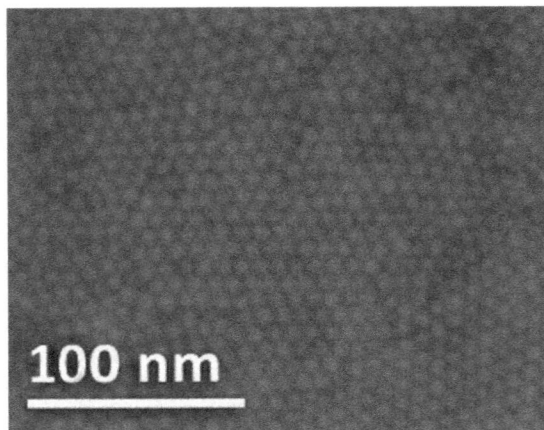

Figure 6.15   HRSEM of heat-treated samples obtained by printing ink A followed
by ink B on top. Relatively uniform self-assembled NPs of diame-
ter around 8 ± 2 nm are obtained. Reproduced from M. Abulikemu,
E. Daas, H. Haverinen, and G. E. Jabbour, *Angew. Chem. Int. Ed.*,
John Wiley and Sons, © 2014 Wiley-VCH Verlag GmbH & Co. KGaA,
Weinheim.[57]

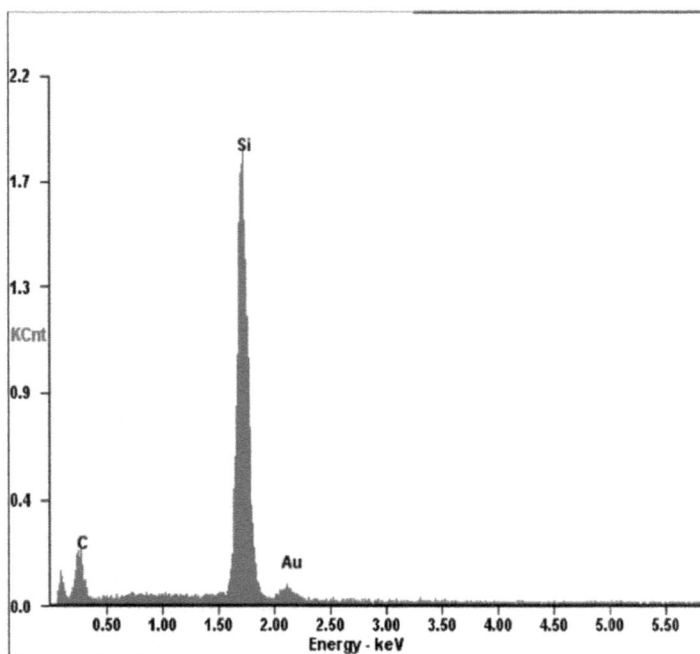

Figure 6.16   EDS analysis of NPs after heat treatment of substrate as described in
the text. The metal Au signal in the data originates from the NPs, the
Si signal is due to the substrate, and the carbon signal is from the
organic ligands surrounding each NP.

To compliment the above studies and characterize the samples further, a high-resolution X-ray diffraction (XRD) was carried out to reveal the crystal structure of the Au NPs, using a Bruker D8 Discover. Further study using a Titan ST 300 kV (FEI) high-resolution transmission electron microscope (HRTEM) was also performed on the samples. In this case, a focused ion beam (Helios 400s by FEI) with lift-out method was used in the cross sectional TEM analysis of the samples, where a Ga ion beam (30 kV, 0.28 nA) was used to thin down the lamellar. All samples were cleaned at 2 kV and 47 pA. Figure 6.17 shows the results of XRD and HRTEM characterization, where the crystalline structure of the NPs is confirmed as that of face centered cubic crystal structure of Au, with 2.35 Å and 2.03 Å lattice spacing for the planes (111) and (002).[60,61]

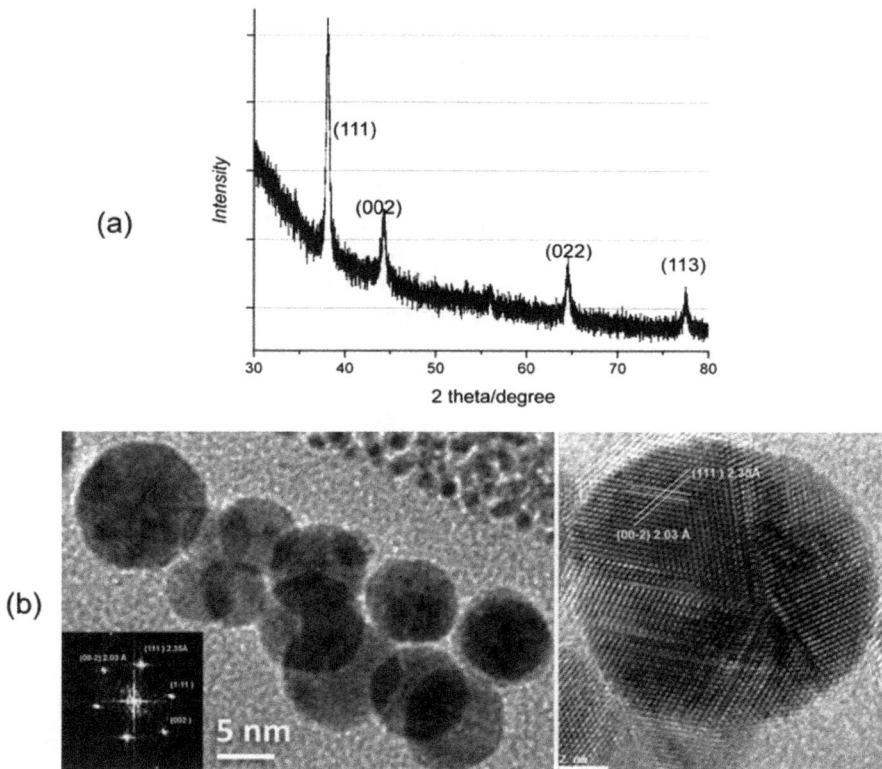

Figure 6.17 (a) XRD data indicating face centered cubic structure, which is also confirmed by HRTEM analysis in (b). The data is consistent with that of crystal structure of pristine Au. Reproduced from M. Abulikemu, E. Daas, H. Haverinen, and G. E. Jabbour, *Angew. Chem. Int. Ed.*, John Wiley and Sons, © 2014 Wiley-VCH Verlag GmbH & Co. KGaA, Weinheim.[57]

Several factors that affect the growth of the NPs in the traditional liquid phase approach are also critical when using the RIJ printing method. These include the nature of the precursor and its polarity, the solvent chosen, the concentration of chemical components involved, temperature and time. Additional factors were found to affect the growth of NPs in RIJ printing involve the sequence of printing steps, wetting of various inks on top of substrate and on top of each other, and others. The details of some of the above-mentioned parameters and their impact on Au NPs growth will be discussed elsewhere.

6.7   Single Ink Approach to Au NPs Self-assembly

It is possible to accelerate the overall growth process of Au NPs by resorting to the use of a single ink. To prepare the ink, 0.18 mmol of gold precursor ($HAuCl_4 \cdot 3H_2O$) and 10 ml of oleylamine were mixed together at room temperature. The mixing was stopped when a clear solution (ink) was obtained. To clean the ink from any particulates that might be detrimental to the processing, a 0.45 μm filter was used prior to the ink injection into the cartridge. The ink was then printed, at a substrate temperature of 60 °C. After the printing step, a 2 h interval of heat treatment at 120 °C was performed, followed by SEM imaging to assess the resulting morphology of dried film. Figure 6.18 shows a SEM image of the printed surface after heat treatment. The image reveals the formation of uniform size Au NPs that self-assemble on top of the substrate surface. The average diameter of the NPs is around 22 nm.

It is important to note that once the ink is prepared, it must be used within several hours. A prolonged shelf storage will allow the growth of Au NPs

Figure 6.18   SEM image of the printed single ink after heat treatment indicating the formation of relatively uniform Au NPs. Scale bar is 400 nm.

in solution, rendering the ink unsuitable for future printing experiments. However, this should not be a detriment for an industrial adoption of such an approach as the components of the ink can be mixed and fed into the cartridge during the production process, thus eliminating the negative effects of single ink short shelf lifetime.

6.8   Increasing NPs Coverage and Reducing the Self-assembly Time to 1 Minute

In growing the Au NPs with RIJ, it is possible to reduce significantly the post printing processing time. A noticeable reduction in growth time is translated into reduction of the cost of Au NPs. In fact, research efforts in this area have been reported for the traditional liquid chemical approach. For example, a close to 2 minute Au NPs synthesis approach using CO gas as reducing agent,[62] and PVP reduction *via* NaOH[63] has been reported. In our case, a rapid photonic heating process was used immediately after the printing process.

The experiment proceeded by first preparing the two inks. We used for ink A, a solution of 1 ml oleylamine in 10 ml chloroform, and for ink B a solution of 0.1 mM $HAuCl_4 \cdot 3H_2O$ in 10 ml of DMSO. To optimize the surface coverage of Au NPs, we studied the impact of ink B on the growth of Au NPs. Three different cases were characterized where ink B droplets printed on top of ink A droplets were in the ratio of 1:1 (ink A:ink B), 1:2, and 1:3, respectively. As the volume of ink B printed on top of ink A increases, the surface area increases accordingly, with 1:3 ratio representing the largest printed wet area (around 50 μm in diameter). A rapid heat treatment step was carried out immediately after the printing of the two. A high power pulsed light source was used to heat the sample to 170 °C (±10 °C) for 1 minute. The morphology of the heat-treated samples was characterized with an SEM. Figure 6.19 indicates the formation of Au NPs, and that a droplet ratio of 1:3 resulted in best coverage of NPs growth over the dried printed area (note that the wet printed area is larger than the final dried region).

6.9   Au:Ag Nanoparticle Alloys *via* RIJ

We have extended the use of RIJ to fabricate binary alloys of several elements. Here, we will present a brief summary of the results pertaining to Au:Ag alloy, only. In this case, we prepared three inks: Au based, Ag based, and reducing agent-based. The Au ink consisted of a 0.1 mmol $HAuCl_4 \cdot 3H_2O$ in DMSO. The mixture was vigorously stirred for 2 hours prior to its introduction into the inkjet cartridge. A similar concentration to that of ink A was used to prepare a silver perchlorate hydrate ($AgClO_4 \cdot xH_2O$) in DMSO. This is done so we can match accurately the mixing ratio of both metal sources in order to have better control of alloy formation. This ink was then injected into a separate inkjet cartridge to the Au based ink. The reducing agent ink was prepared by stirring, for one hour, a mixture of 1 ml oleylamine and 10 ml chloroform. The ink was introduced into a separate inkjet cartridge.

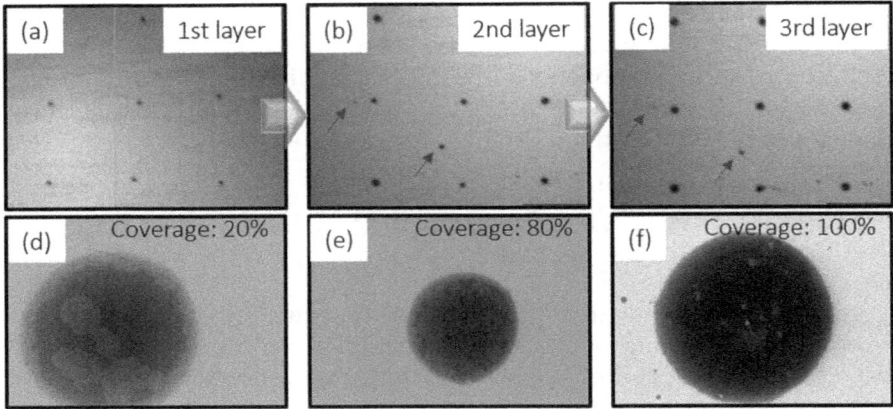

Figure 6.19   Printed dot array before heat treatment [(a), (b), (c)], SEM images of
printed drops upon photonic heat treatment [(d), (e), (f)]. An 80% surface coverage of Au NPs is readily achieved with 1 droplet of ink A and
2 droplets of ink B. While, nearly full surface coverage can be seen for
a printed region having 1 printed droplet of ink A and 3 droplets of
ink B (f).

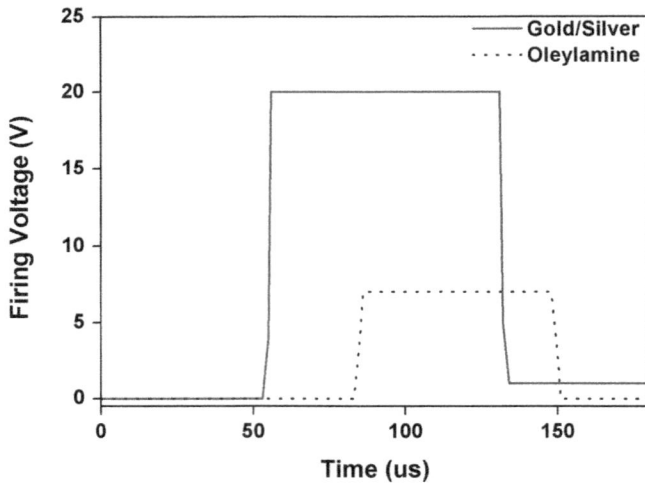

Figure 6.20   Firing voltage used for the gold ink and silver ink (full curve), and for
the reducing agent ink of oleylamine (dotted curve).

The printing conditions were optimized to yield the best printing quality.
Among such conditions is the waveform of the firing voltage that controls
the cartridge ability to dispense various liquid droplets, thus has a noticeable
impact on our experiment. In this case, an optimum firing voltage waveform
is shown in Figure 6.20. As the figure indicates, a 20 V step with a duration
of around 90 µs was used for the firing voltage during the printing of Au and
Ag inks. A voltage step of 7 V and around 70 µs was used to jet the ink of the

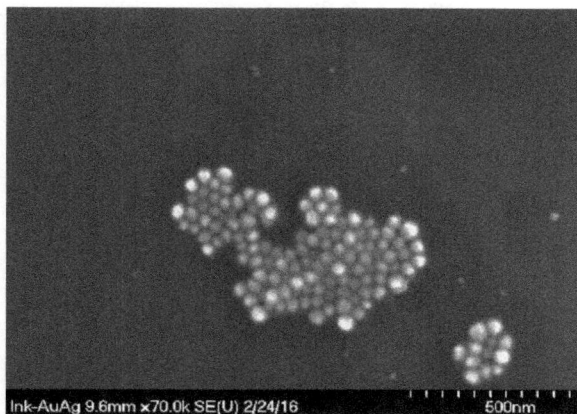

Figure 6.21 SEM image of nanoparticles formed upon printing of the three inks in a sequence of Au ink, Ag ink, and oleylamine ink, respectively, and precisely on top of each other. Scale bar is 500 nm.

reducing agent. Statistically, it is possible for some Au or Au atoms to nucleate quickly at room temperature when reacting with the reducing agent. To prevent this from happening, the printing sequence proceeded by jetting first the Au-based ink, then Ag-based ink, followed by printing the reducing agent ink. Great care was taken to ensure that accurate droplet placement is maintained during printing of the three different inks on top of each other, at the same location on the substrate. Upon printing the metal-based inks on top of each other, a 1 minute residence time was used to allow proper mixing of the printed droplets, over the substrate surface, prior to printing the reducing agent. Upon printing of the three inks, the substrates were heated to 120 °C for one hour. The morphology was investigated using an SEM, as shown in Figure 6.21. The SEM image reveals the growth of nanoparticles of an average size around 50 nm.

Figure 6.22 shows the optical absorption due to the surface plasmon resonance (SPR) of the RIJ synthesized nanoparticles. As expected, the peak SPR absorption shifts to the red, from that of pure Ag nanoparticle to that of pure gold nanoparticle, as the content of Au is increased in the NP. Moreover, the absence of a dual peak in the SPR absorption data indicates the formation of Au : Ag alloyed nanoparticles. This is supported by previously published data on the formation of NP alloys of Au : Ag, when their ionic forms from starting materials (inks) were reduced simultaneously by sodium citrate.[64] Further characterization of such NPs and methods to optimize their size uniformity are currently under investigation.

6.10 RIJ Growth of ZnO Nanostructures

In this section, we will discuss the use of RIJ printing in the growth of ZnO nanostructures. Although the results are preliminary, they attest to the versatility of RIJ in various fields of materials synthesis. ZnO nanostructures can

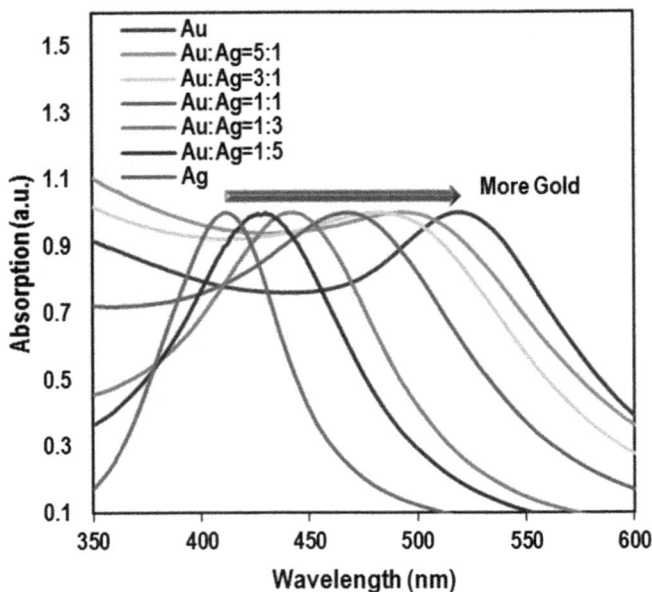

Figure 6.22 SPR absorption of various NPs of Au:Ag with varying ratio from Au = 0% (pure silver NPs) to Au = 100% (pure gold NPs).

be grown *via* various approaches.[65–67] However, to the best of our knowledge, and at the time of writing this chapter, there has been no published report on the growth of ZnO nanostructures using RIJ, and with the combination of inks described below.

A mixture of 1 mM of Zinc Nitrate Hydrate $(Zn(NO_3)_2 \cdot xH_2O)$ in water was prepared. The mixture was stirred and the resulting ink was injected into the printer cartridge. The reducing agent ink consisted of 1 ml of oleylamine in 10 ml chloroform (CF). Prior to printing the inks, a seed layer was prepared. The layer was formed using zinc acetate $(Zn(CH_3COO)_2 \cdot 2H_2O)$ dissolved in ethanol, which was spin-coated on a silicon substrate. A dense seed layer was achieved after 5 spin coating rounds were carried. The resulting coating was heat treated at 400 °C for 3 hours in air, resulting in the formation of a smooth ZnO layer. Upon cooling to room temperature, the substrates were placed on the printer substrate holder and heated to 60 °C. At this substrate temperature, the printing was performed for 2 hours. ZnO source ink was printed first followed by the printing of reducing agent ink on top. An SEM scan of the printed substrate surface reveals the growth of nanostructures of ZnO having size range on the order of 50 nm to 400 nm, as shown in Figure 6.23.

Further SEM investigation, of a given NP of Figure 6.23, at higher magnification reveals a nano-leaflet (nanowalls) growth structure, with a thickness on the order of 10 nm, as shown in Figure 6.24.

Current efforts focus on studying the printed structures using various characterization techniques, and fine-tuning the experimental details to

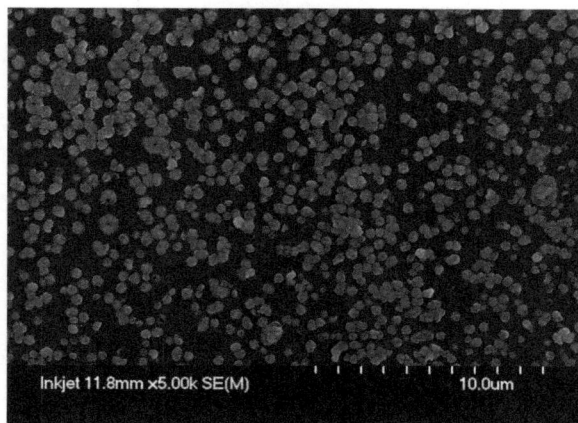

Figure 6.23   SEM image of nanostructures of ZnO obtained *via* RIJ. Scale bar is 10 μm.

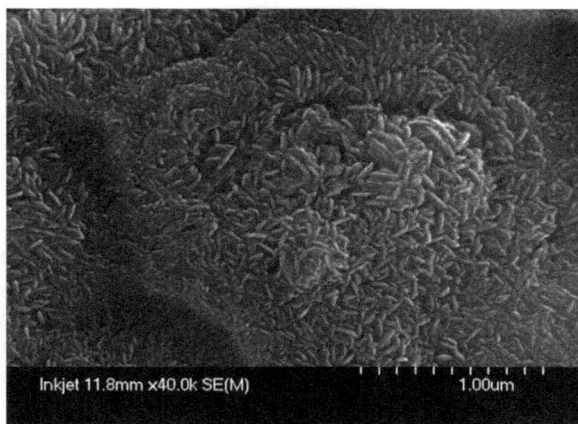

Figure 6.24   Nanowall structures of a given ZnO NP of Figure 6.23 as observed at higher SEM magnification. Scale bar is 1 μm.

accurately and reproducibly grow ZnO nanostructures, with narrower size distribution.

In order to demonstrate an all RIJ printed ZnO nanostructure growth, we resorted to printing an Au seed layer in the array-format discussed in prior sections of this chapter. Once the Au dotted array seed layer is obtained, the ZnO starting, and reducing, inks were printed as described above. Figure 6.25 is an SEM scan of the surface of the Au printed spots which reveals the formation of ZnO nanorods on top of the Au seed layer. Although the growth is not in a controlled and vertically oriented manner, the results demonstrate the potential of RIJ printing in the fabrication of not only metal

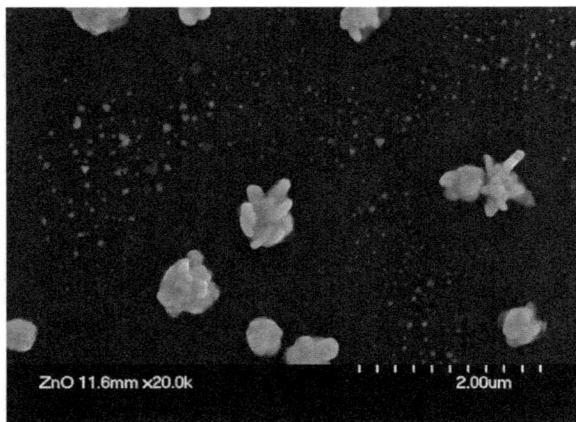

Figure 6.25   SEM of ZnO nanorod structure on top of Au seed layer for fully RIJ printing of both Au seed and ZnO starting materials. The scale bar is 2 μm.

based nanoparticles, but also oxide compound nanostructures, such as ZnO nanorods and nanowalls. The triangular shaped region in the background of Figure 6.25 represents nucleated Au sheets within the area of the heat-treated printed droplets.

6.11   RIJ Printing of Quantum Dots

Given the demonstrated RIJ fabrication of NPs of various materials, sizes, and shapes, we have no doubt that it is possible to extend such an approach to the synthesis of QDs. Here we present the first steps towards that goal. In this regard, we present the RIJ growth of QDs of lead sulfide (PbS), on a ZnO nanowall structure fabricated by pulsed laser deposition on silicon (100) substrates.[67]

Two inks were prepared, an anionic ink composed of $PbCl_2$ (at 0.6 mmol) dissolved in 10 ml of DMSO (ink A), and a cationic ink was formulated from a solution of Na_2S (0.6 mmol) in 10 ml of methanol (ink B). The two inks were loaded in separate inkjet printing cartridges. Ink A was then printed in the form rectangular areas of 1 cm^2, on the top of the ZnO honeycomb nanowall structure. Immediately, we printed ink B directly on top of ink A. At the end of the printing experiment, the substrates were rinsed gently with DMSO for 30 s, then dried in nitrogen.

Figure 6.26 is an SEM scan of the ZnO nanowall structure before, and after, the growth of NPs. The image (right) reveals clearly the growth of PdS QDs, with an average size of around 9–14 nm, on the ZnO nanowall structure. The Bohr radius of PdS is around 18 nm[68] indicating that the above size range falls within the PbS QD set, although the confinement effects might not be as pronounced as QDs having smaller size than 9 nm. Further assessment, using TEM, of the size of the RIJ printed PdS confirms the SEM results.

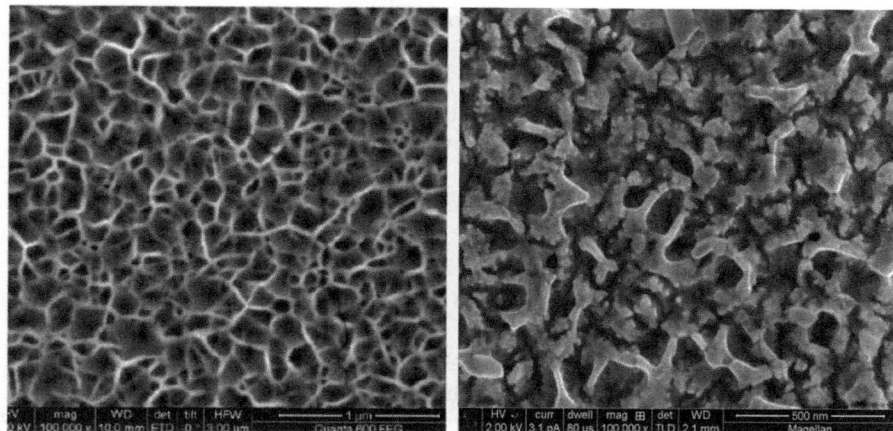

Figure 6.26 SEM image of ZnO honeycomb nanowall (left) predeposited on silicon substrate, and SEM image of ZnO honeycomb nanostructure after RIJ growth of PdS QDs. Scale bar is 1 μm (left image), and 500 nm (right image).

XRD was used to confirm the growth of PdS, the results of which will be published elswhere.

6.12   Conclusion

Our aim in this introductory chapter was to demonstrate that one could use inkjet printing not only as passive materials dispensing technique, but also as an active tool in materials fabrication. In this regard, we surveyed our developments in this exciting area. Examples demonstrating the potential of RIJ printing in nanomaterials fabrication were given. With such approach, we demonstrated the creation of grey-scale sheet resistivity in conducting polymer of PEDOT:PSS, and the self-assembled synthesis of uniform size Au NPs. We presented initial results demonstrating the extension of RIJ approach to other material systems including alloys of Au : Ag NPs, ZnO NPs and nanorods, and PdS QDs. Given what has been achieved using RIJ, there is no doubt that it will see increasing utility in various fields of science and engineering.

References

1. S. E. Shaheen, R. Radspinner, N. Peyghambarian and G. E. Jabbour, Fabrication of bulk heterojunction plastic solar cells by screen printing, *Appl. Phys. Lett.*, 2001, **79**, 2996–2998.
2. P. Kopola, T. Aernouts, R. Sliz, S. Guillerez, M. Ylikunnari, D. Cheyns, M. Välimäki, M. Tuomikoski, J. Hast, G. Jabbour, R. Myllylä and A. Maaninen, Gravure printed flexible organic photovoltaic modules, *Sol. Energy Mater. Sol. Cells*, 2011, **95**, 1344–1347.

3. R. Søndergaard, M. Hösel, D. Angmo, T. T. Larsen-Olsen and F. C. Krebs, Roll-to-roll fabrication of polymer solar cells, *Mater. Today*, 2012, **15**, 36–49.

4. C. N. Hoth, S. A. Choulis, P. Schilinsky and C. J. Brabec, High photovoltaic performance of inkjet printed polymer: fullerene blends, *Adv. Mater.*, 2007, **19**, 3973–3978.

5. K. A. Baert, J. Roggen, J. F. Nijs and R. P. Mertens, Amorphous-Silicon Solar Cells with Screen-Printed Metallization, *IEEE Trans. Electron Devices*, 1990, **37**, 702–707.

6. M. M. Hilali, A. Rohatgi and S. Asher, Development of screen-printed silicon solar cells with high fill factors on 100 Ohm/sq emitters, *IEEE Trans. Electron Devices*, 2004, **51**, 948–955.

7. G. E. Jabbour, R. Radspinner and N. Peyghambarian, Screen printing for the fabrication of organic light-emitting devices, *IEEE J. Sel. Top. Quantum Electron.*, 2001, 7, 769–773.

8. S. Liu, W. Liu, W. Ji, J. Yu, W. Zhang, L. Zhang and W. Xie, Top-emitting quantum dots light-emitting devices employing microcontact printing with electricfield-independent emission, *Sci. Rep.*, 2016, **6**, 1–9, article number: 22530.

9. C. C. Ho, K. Murata, D. A. Steingart, J. W. Evans and P. K. Wright, A super ink jet printed zinc–silver 3D microbattery, *J. Micromech. Microeng.*, 2009, **19**, 094013–094117.

10. A. G. Kelly, D. Finn, A. Harvey, T. Hallam and J. N. Coleman, All-printed capacitors from graphene-BN-graphene nanosheet heterostructures, *Appl. Phys. Lett.*, 2016, **109**, 023107–023109.

11. Y. Gao, H. Li and J. Liu, Directly Writing Resistor, Inductor and Capacitor to Composite Functional Circuits: A Super-Simple Way for Alternative Electronics, *PLoS One*, 2013, **8**(8), 69761.

12. A. Sawhney, A. Agrawal, P. Patra and P. Calvert, Piezoresistive sensors on textiles by inkjet printing and electroless plating, *MRS Proc.*, 2006, **920**, 0920–S05-04.

13. B. Weng, A. Morrin, R. Shepherd, K. Crowley, A. J. Killard, P. C. Innis and G. G. Wallace, Wholly printed polypyrrole nanoparticle-based biosensors on flexible substrate, *J. Mater. Chem. B*, 2014, **2**(7), 793–799.

14. R. S. Deol, H. W. Choi, M. Singh and G. E. Jabbour, Printable Displays and Light Sources for Sensor Applications: A Review, *IEEE Sens. J.*, 2015, **15**, 3186–3195.

15. J. Li, M. Chen, X. Fan and H. Zhou, Recent advances in bioprinting techniques: approaches, applications and future prospects, *J. Transl. Med.*, 2016, **14**, 271, (Open Access).

16. R. Schirhagl, Bioapplications for Molecularly Imprinted Polymers, *Anal. Chem.*, 2014, **86**, 250–261.

17. T. Xu, J. Jin, C. Gregory, J. J. Hickman and T. Boland, Inkjet printing of viable mammalian cells, *Biomaterials*, 2005, **26**, 93–99.

18. H. W. Choi, T. Zhou, M. Singh and G. E. Jabbour, Recent developments and directions in printed nanomaterials, *Nanoscale*, 2015, 7, 3338–3355.

19. P. J. Smith and A. Morrin, Reactive inkjet printing, *J. Mater. Chem.*, 2012, **22**, 10965–10970.
20. Y. Yoshioka, P. Calvert and G. E. Jabbour, Simple modification of sheet resistivity of conducting polymeric anodes *via* combinatorial ink-jet techniques, *Macromol. Rapid Commun.*, 2005, **26**, 238–246.
21. Y. Yoshioka and G. E. Jabbour, Inkjet printing of oxidants for patterning of nanothick conducting polymer electrodes, *Adv. Mater.*, 2006, **18**, 1307–1312.
22. S. Lin and R. M. Carlson, Susceptibility of Environmentally Important Heterocycles to Chemical Disinfection: Reactions with Aqueous Chlorine, Chlorine Dioxide, and Chloramine, *Environ. Sci. Technol.*, 1984, **18**, 743–748.
23. B. Jiang and T. D. Tilley, General, efficient route to thiophene-1-oxides and welldefined, mixed thiophene- thiophene-1-oxide oligomers, *J. Am. Chem. Soc.*, 1999, **121**, 9744–9745.
24. R. A. Weiss, A. Sen, C. L. Willis and L. A. Pottick, Block copolymer ionomers: 1. synthesis and physical properties of sulphonated poly(styrene-ethylene/butylenesstyrene), *Polymer*, 1999, **32**, 1867–1874.
25. G. Socrates, *Infrared Characteristics Group Frequencies*, John Wiley & Son, New York, 1980.
26. R. Servaty, J. Schiller, H. Binder, B. Kohlstrunk and K. Arnold, IR and NMR studies on the action of hypochlorous acid on chondroitin sulfate and taurine, *Bioorg. Chem.*, 1998, **1998**(26), 33–43.
27. V. Hulea, F. Fajula and J. Bousquet, Mild Oxidation with H2O2 over Ti-Containing Molecular Sieves, A very Efficient Method for Removing Aromatic Sulfur Compounds from Fuels, *J. Catal.*, 2001, **198**, 179–186.
28. V. Maurino, P. Calza, C. Minero, E. Pelizzetti and M. Vincenti, Light-assisted 1,4-dioxane degradation, *Chemosphere*, 1997, **35**, 2675–2688.
29. S. Kirchmeyer and K. Reuter, Scientific importance, properties and growing applications of poly(3,4-ethylenedioxythiophene), *J. Mater. Chem.*, 2005, **15**, 2077–2088.
30. M. Zhou, B. Wang, Z. Rozynek, Z. Xie, J. O. Fossum, X. Yu and S. Raaen, Minute synthesis of extremely stable gold nanoparticles, *Nanotechnology*, 2009, **20**, 505606.
31. D. A. Giljohann, D. S. Seferos, W. L. Daniel, M. D. Massich, P. C. Patel and C. A. Mirkin, Gold Nanoparticles for Biology and Medicine, *Angew. Chem., Int. Ed.*, 2010, **49**, 3280–3294.
32. X. Huang, P. K. Jain, I. H. El-Sayed and M. A. El-Sayed, Gold nanoparticles: interesting optical properties and recent applications in cancer diagnostics and therapy, *Nanomedicine*, 2007, **2**, 681–693.
33. Y. Chen, Gold nanoparticles for applications in energy and environment: synthesis and characterization, *Rare Met.*, 2011, **30**(suppl. 1), 116.
34. M. Notarianni, K. Vernon, A. Chou, M. Aljada, J. Liu and N. Motta, Plasmonic effect of gold nanoparticles in organic solar cells, *Sol. Energy*, 2014, **106**, 23–37.
35. H. A. Atwater and A. Polman, Plasmonics for improved photovoltaic Devices, *Nat. Mater.*, 2010, **9**, 205–213.

36. Y. Chen, Gold nanoparticles for applications in energy and environment: Synthesis and characterization, *Rare Met.*, 2011, **30**, 116–120.

37. M. Notarianni, K. Vernon, A. Chou, M. Aljada, J. Liu and N. Motta, Plasmonic effect of gold nanoparticles in organic solar cells, *Sol. Energy*, 2014, **106**, 23–37.

38. M. C. Oliveira, A. L. S. Fraga, A. Thesing, R. L. Andrade, J. F. L. Santos and J. L. Santos, Interface dependent plasmon induced enhancement in dye-sensitized solar cells using gold nanoparticles, *J. Nanomater.*, 2015, **2015**, 1–9.

39. M. Yao, X. Jia, Y. Liu, W. Guo, L. Shen and S. Ruan, Surface plasmon resonance enhanced polymer solar cells by thermally evaporating Au into buffer layer, *ACS Appl. Mater. Interfaces*, 2015, **7**, 18866–18871.

40. I. Hussain, S. Z. Hussain, H. U. Rehman, A. Ihsan, A. Rehman, Z. M. Khalid, M. Brust and A. I. Cooper, *In situ* growth of gold nanoparticles on latent fingerprints—from forensic applications to inkjet printed nanoparticle patterns, *Nanoscale*, 2010, **2**, 2575–2578.

41. G. C. Jensen, C. E. Krause, G. A. Sotzing and J. F. Rusling, Inkjet-printed gold nanoparticle electrochemical arrays on plastic. Application to immunodetection of a cancer biomarker protein, *Phys. Chem. Chem. Phys.*, 2011, **13**, 4888–4894.

42. S. A. Maier, P. G. Kik and H. A. Atwater, Optical pulse propagation in metal nanoparticle chain waveguides, *Phys. Rev. B*, 2003, **67**, 205402.

43. A. Grubisic, S. Mukherjee, N. Halas and D. J. Nesbitt, Anomalously strong electric near-field enhancements at defect sites on Au nanoshells observed by ultrafast scanning photoemission imaging microscopy, *J. Phys. Chem. C*, 2013, **117**, 22545–22559.

44. M. C. Daniel and D. Astruc, Gold Nanoparticles: Assembly, Supramolecular Chemistry, Quantum-Size-Related Properties, and Applications toward Biology, Catalysis, and Nanotechnology, *Chem. Rev.*, 2004, **104**, 293–346.

45. A. Kamyshny and S. Magdassi, Conductive nanomaterials for printed electronics, *Small*, 2014, **10**, 3515–3535.

46. B. J. Y. Tan, C. H. Sow, T. S. Koh, K. C. Chin, A. T. S. Wee and C. K. Ong, Fabrication of size-tunable gold nanoparticles array with nanosphere lithography, reactive ion etching, and thermal annealing, *J. Phys. Chem. B*, 2005, **109**, 11100–11109.

47. M. K. Corbierre, J. Beerens and R. B. Lennox, Gold nanoparticles generated by electron beam lithography of gold(I)–thiolate thin films, *Chem. Mater.*, 2005, **17**, 5774–5779.

48. A. M. Hung, C. M. Micheel, L. D. Bozano, L. W. Osterbur, G. M. Wallraff and J. N. Cha, Large-area spatially ordered arrays of gold nanoparticles directed by lithographically confined DNA origami, *Nat. Nanotechnol.*, 2010, **5**, 121–126.

49. J. M. Tour, L. Cheng, D. P. Nackashi, Y. Yao, A. K. Flatt, S. K. S. Angelo, T. E. Mallouk and P. D. Franzon, Nanocell electronic memories, *J. Am. Chem. Soc.*, 2003, **125**, 13279–13283.

50. F. Meriaudeau, T. R. Downey, A. Passian, A. Wig and T. L. Ferrell, Environment effects on surface-plasmon spectra in gold-island films potential for sensing applications, *Appl. Opt.*, 1998, **37**, 8030–8037.
51. J. Turkevich, P. C. Stevenson and J. Hillier, A study of the nucleation and growth processes in the synthesis of colloidal gold, *Discuss. Faraday Soc.*, 1951, **11**, 55–75.
52. G. Frens, Controlled nucleation for the regulation of the particle size in monodisperse gold suspensions, *Nat. Phys. Sci.*, 1973, **241**, 20–22.
53. J. Kimling, M. Maier, B. Okenve, V. Kotaidis, H. Ballot and A. Plech, Turkevich method for gold nanoparticle synthesis revisited, *J. Phys. Chem. B*, 2006, **110**, 15700–15707.
54. J. Benson, C. M. Fung, J. S. Lloyd, D. Deganello, N. A. Smith and K. S. Teng, Direct patterning of gold nanoparticles using flexographic printing for biosensing applications, *Nanoscale Res. Lett.*, 2015, **10**, 127.
55. S. R. Samarasinghe, I. Pastoriza-Santos, M. J. Edirisinghe, M. J. Reece and L. M. Liz-Marzan, Printing gold nanoparticles with an electrohydrodynamic direct-write device, *Gold Bull.*, 2006, **39**(2), 48–53.
56. G. C. Jensen, C. E. Krause, G. A. Sotzingab and J. F. Rusling, Inkjet-printed gold nanoparticle electrochemical arrays on plastic. Application to immunodetection of a cancer biomarker protein, *Phys. Chem. Chem. Phys.*, 2011, **13**, 4888–4894.
57. M. Abulikemu, E. Daas, H. Haverinen and G. E. Jabbour, *In situ* synthesis of Au nanoparticles using reactive inkjet printing, *Angew. Chem., Int. Ed.*, 2014, **53**, 420–423.
58. M. Abulikemu, and G. E. Jabbour, *In situ* synthesis of nanoparticles on substrates by inkjet printing, US Patent 8916457, 2014.
59. J. He, P. Kanjanaboos, N. L. Frazer, A. Weis, X. M. Lin and H. M. Jaeger, Fabrication and Mechanical Properties of Large-Scale Freestanding Nanoparticle Membranes, *Small*, 2010, **6**, 1449–1456.
60. Q. F. Zhou, J. C. Bao and Z. Xu, Shape-controlled synthesis of nanostructured gold by a protection–reduction technique, *J. Mater. Chem.*, 2002, **12**, 384–387.
61. S. M. Taheri, S. Fischer and S. Förster, Routes to nanoparticle-polymer superlattices, *Polymers*, 2011, **3**, 662–673.
62. J. K. Young, N. A. Lewinski, R. J. Langsner, L. C. Kennedy, A. Satyanarayan, V. Nammalcar, A. Y. Lin and R. A. Drezek, Size-controlled synthesis of monodispersed gold nanoparticles via carbon monoxide gas reduction, *Nanoscale Res. Lett.*, 2011, **6**, 428.
63. M. Zhou, B. Wang, Z. Rozynek, Z. Xie, J. O. Fossum, X. Yu and S. Raaen, Minute synthesis of extremely stable gold nanoparticles, *Nanotechnology*, 2009, **20**, 505606.
64. S. Link, Z. L. Wang and M. A. El-Sayed, Alloy Formation of Gold-Silver Nanoparticles and the Dependence of the Plasmon Absorption on Their Composition, *J. Phys. Chem. B*, 1999, **103**, 3529–3533.
65. S. Baruah and J. Dutta, Hydrothermal growth of ZnO nanostructures, *Sci. Technol. Adv. Mater.*, 2009, **10**(1), 013001.

66. A. B. Djurišić, X. Chen, Y. H. Leung and A. M. C. Ng, ZnO nanostructures: growth, properties and applications, *J. Mater. Chem.*, 2012, **22**, 6526–6535.

67. B. El Zein, S. Boulfrad, G. E. Jabbour and E. Dogheche, Parametric study of self-forming ZnO Nanowall network with honeycomb structure by Pulsed Laser Deposition, *Applied Surface Science*, 2014, **292**, 598–607.

68. M. A. Hines and G. D. Scholes, Colloidal PbS Nanocrystals with Size-Tunable Near-Infrared Emission: Observation of Post-Synthesis Self-Narrowing of the Particle Size Distribution, *Adv. Mater.*, 2003, **15**, 1844–1849.

CHAPTER 7

Reactive Inkjet Printing of Silk Barrier Membranes for Dental Applications

P. M. RIDER*[a], I. M. BROOK[a], P. J. SMITH[b] AND C. A. MILLER[a]

[a]School of Clinical Dentistry, The University of Sheffield, S10 2TA, Sheffield, UK; [b]Department of Mechanical Engineering, The University of Sheffield, S1 3JD, Sheffield, UK
*E-mail: prider1@sheffield.ac.uk

7.1   Introduction

Inkjet printing is an important and versatile manufacturing technique which has been used for producing tissue engineering constructs.[1] With its layer by layer approach to producing structures, it has the ability to produce precise, repeatable, surface topographies and graduated structures. Reliably controlling structural characteristics and the repeatability of samples could be useful for producing barrier membranes.

7.1.1   Guided Bone Regeneration

Guided bone regeneration (GBR) is a procedure in which a bone filler is commonly used in conjunction with a barrier membrane to augment alveolar bone for dental restoration in the treatment of pathogenic lesions. The role of the barrier membrane is to support bone growth within the defect site

Smart Materials No. 32
Reactive Inkjet Printing: A Chemical Synthesis Tool
Edited by Patrick J. Smith and Aoife Morrin
© The Royal Society of Chemistry 2018
Published by the Royal Society of Chemistry, www.rsc.org

and prevent the infiltration of fibrous tissues from the surrounding environment. Periodontitis is a common chronic pathogenic infection which leads to the inflammation of the gingiva (gum) and supporting structures of the teeth. Periodontal diseases with different degrees of severity affect 90% of people aged 70 years and over.[2] The disease is caused by prolonged exposure to bacteria resulting in the breakdown of surrounding tissues such as the alveolar bone, which provides support for the teeth. If untreated it will lead to eventual tooth loss.

Surgical intervention is used to treat severe cases, with 40% of surgical procedures requiring the use of a dental barrier membrane to help restore defect sites.[3] The membranes are positioned between the alveolar bone and gingiva (Figure 7.1). A bone filler or scaffold is placed within the bony defect space and a barrier membrane is used to seclude the defect site from surrounding tissues. Other properties required of the barrier membrane include; tissue integration, cell occlusivity, clinical manageability, space provision and biocompatibility.[3]

7.1.2 Dental Barrier Membrane Materials

Common materials used for producing the barrier membranes are either non-resorbable such as titanium or resorbable such as polyesters and collagen.[2,4] Non-resorbable materials are able to provide mechanical support for the duration of the regeneration and can remain in position for as long as required. However, to extract the membrane requires increased surgical costs as well as site morbidity.

Resorbable membranes are able to circumvent a second surgery, and an idealised version would have a controllable degradation rate to suit patient requirements. Collagen is one of the most commonly used membrane materials due to its excellent biocompatibility.[5] Yet collagen has a very poor

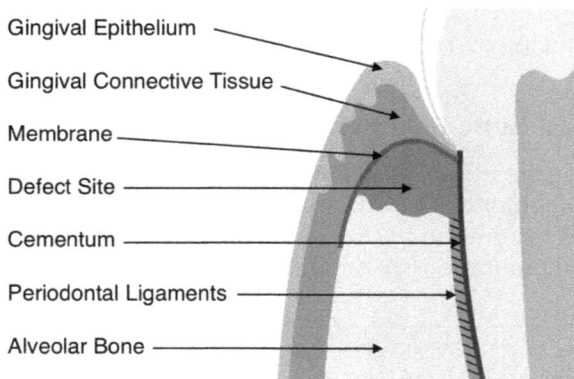

Figure 7.1 Barrier membrane positioned over a defect site, supporting regeneration of the periodontium.

tensile strength when wet and can disintegrate when exposed to the open oral environment.[6] Moreover, collagen is expensive to use due to the difficulty involved with sterilisation, and the source of collagen can also be an issue for certain religious groups.

Synthetic materials such as polyesters have a better structural integrity than that of collagen. Degradation rates are more easily controlled. However, during degradation, polyesters degrade into acidic by-products. If these by-products build up in the surrounding tissues they can elicit a chronic inflammatory reaction which can prevent complete successful regeneration.[7]

7.1.3 Silk

Silk is a material which has the potential to address all of the above issues and still provide the necessary requirements to be a successful dental barrier membrane. Silk does not provoke an immunogenic response upon implantation, has controllable degradation rates and robust mechanical properties, can be sterilised with a variety of different techniques and is easily processed. Silk has had a long history as a medical material, being used for over 2000 years as a suture material due to its superior mechanical and biocompatibility properties. Silk is still used as a suture material for dental operations. More recently regenerated silk fibroin (RSF) has been used as a building block to create scaffolds for tissue engineering.[8]

Silk has several polymorphs of its secondary protein structure. Silk I and II are the important polymorphs used in this study. Silk I is amorphic, consisting of α-helical and random coil structures which are water soluble, whilst silk II is crystalline and consists mainly of a β-sheet structure which is non-water soluble. Silk I can be transformed into silk II through a variety of methods such as stretching,[9] exposure to methanol[10] and changes in pH. The ability to change silk I to silk II make it ideal for processing and manufacture.

RSF films have already been explored for use in ocular,[11] hepatic tissue,[12] bladder,[13] skin[14] and eardrum[15] regeneration. The most commonly used method for producing an RSF film is by casting and more recently by electrospinning.[16] The mechanical properties of the RSF films can be adapted using various processing techniques to increase β-sheet content. Controlled drying or water annealing, are used during the setting of films. Stretching and alcohol immersion are used after the film has set. Different processes give variations in the overall mechanical properties and degradation rates.[17,18]

However, as yet, there hasn't been much research into the inkjet printing of silk films. The first instance of RSF processing with an inkjet printer was demonstrated by Suntivich *et al.* in 2014 who created silk nests for the entrapment of cells.[19] Their approach required the fibroin side chains to be modified with poly-(L-lysine) and poly-(L-glutamic acid) to produce oppositely charged polyelectrolytes, which when printed and paired together in bilayers, produced a silk II structure. Another study was performed by Tao *et al.* in 2015 which demonstrated a range of applications for inkjet printed RSF, as well as printing RSF constructs with active components such as anti-biotics.[20] The study postulated that a composite RSF ink would be required to control mechanical properties, however they had not considered the use of reactive inkjet printing. Reactive inkjet printing could be used in the production of RSF constructs by printing RSF solution alongside methanol to induce β-sheet formation.

Inkjet printing, specifically reactive, has the potential to reduce processing steps by combining film production with an alcohol treatment to tailor mechanical properties. Inkjet printing offers the ability to produce complex samples with control over shape, surface topography as well as the capability to produce graduated compositional structures. Therefore, inkjet printing may offer an alternative manufacturing method for the production of barrier membranes.

7.2 Materials and Methods

7.2.1 RSF Synthesis

In order for constructs to be made using an inkjet printer, silk first needs to be in its silk I water soluble state. Reconstituted silk fibroin (RSF) was produced using Bombyx Mori silk worm cocoons (Wild Fibres, Birmingham, UK), following the protocol outlined by D. N. Rockwood *et al.*[21] Silk cocoons were boiled for 30 minutes in a 0.02 molar solution of sodium carbonate (≥99.5% purity, ACS reagent, Sigma Aldrich, UK) (Figure 7.2(a–b)). The alkaline solution removes the sericin, which acts as a glue holding the cocoon together, from the silk. Once the sericin was removed the fibres were released. The fibres were thoroughly rinsed and dried before the next procedural step (Figure 7.2(c)).

To produce a solution of water soluble silk, the fibres were heated in a chaotropic solvent. The chaotropic solvent breaks down the hydrogen bonds holding the crystalline regions together giving the silk an amorphous water soluble structure. The fibres were put in a 9.3 M solution of lithium bromide (≥99% purity, ReagentPlus®, Sigma Aldrich, UK) and heated in an oven at 70 °C for 3 hours and 30 minutes (Figure 7.2(d)). This step differs from that outlined by D. N. Rockwood *et al.* as they recommended using a temperature of 60 °C for 4 hours. However, after 4 hours at this temperature there was incomplete dissolution of the fibres. Instead a temperature of 70 °C was chosen, as recommended by M. K. Sah *et al.*[22]

Once the fibres had dissolved, the resulting solution was added to dialysis cassettes with a 3500 molecular weight pore size and dialysed against distilled water (Figure 7.2(e–f)). The solution was dialysed for a period of 72 hours after which the lithium bromide had been removed from the RSF solution. The solution was then centrifuged to remove any large contaminants such as remnants of the silk worm (Figure 7.2(g)). At this point the RSF solution has a concentration of about 70–75 mg mL^{-1}. To get a more concentrated

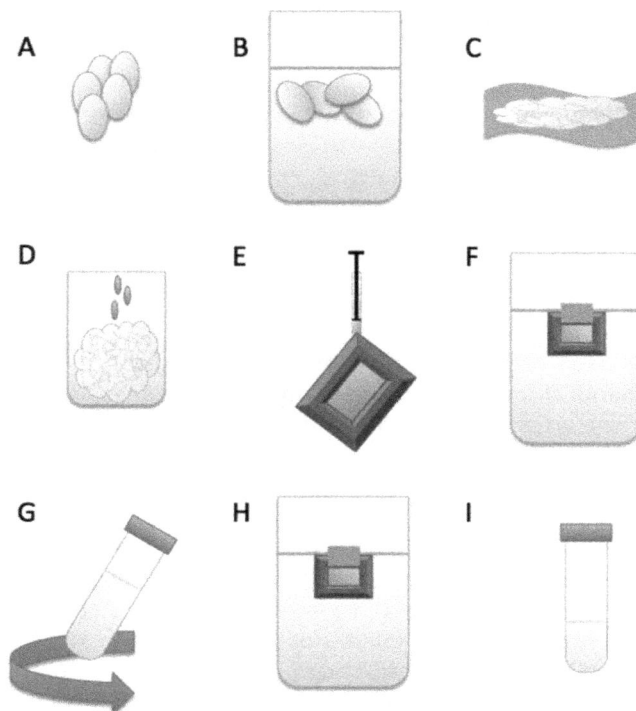

Figure 7.2 Processing of Bombyx Mori silkworm cocoons into a silk fibroin solution. (A) Silk cocoons, (B) cocoons boiled to remove sericin and release fibres, (C) silk fibres rinsed and dried, (D) silk fibres dissolved in 9.3 M LiBr at 70 °C for 3 hours, (E) RSF/LiBr solution added to dialysis cassette, (F) dialysed against D.I. water for 48 hours, (G) RSF solution centrifuged, (H) RSF solution dialysed against a 10 wt% PEG solution, (I) concentrated RSF solution stored at 4 °C.

solution, the RSF solution was dialysed a 10 wt% polyethylene glycol (Av. mol. wt 10 000, Sigma Aldrich, UK) solution for a period of 20 hours (Figure 7.2(h–i)). The final concentration of the RSF solution was then adjusted by the addition of distilled water.

7.2.2 nHA Synthesis

Nano-hydroxyapatite (nHA) was synthesised using a wet precipitation method. nHA is directly precipitated from an aqueous solution at room temperature using a method based upon the Fluidinova process.[23] To produce the nHA, 250 mL of 0.12 M potassium phosphate monobasic (≥99.0%, powder, Sigma Aldrich, UK) with a pH of 13 was added to a 500 mL solution of 0.13 M calcium chloride (USP testing specification, Sigma Aldrich, UK) with a pH of 12.8 and stirred continuously for 1 hour. The resultant solution was left to settle over a period of 7 hours after which the nHA slurry sank to the bottom of the beaker leaving a clear solution floating on top. To wash the nHA the clear solution was syphoned off and the beaker topped up with water. The solution was stirred continuously for 20 minutes and then left to settle for 24 hours. The washing process was repeated a further 2 times.

After the washing process the final nHA solution was used directly for printing. The solution was determined to have a concentration of 23.49 mg mL^{-1} through thermogravimetric analysis.

7.2.3 Film Production

Printing was performed using a MicroFab drop on demand piezoelectric ink-jet printer using JetLab4 software. A piezo printhead with a nozzle aperture of 80 μm was used to print the RSF inks; and a 60 μm printhead was used to print methanol (MeOH). Films were produced by printing a layer of RSF ink followed by a layer of methanol. Films were printed to a height of 20 RSF layers.

To calculate droplet sizes, photographs were taken of fully detached spherical droplets using the built in camera. Droplet diameters were measured using ImageJ2[24] software and volumes calculated using the equation for a sphere. Calculated volumes, as well as step sizes, were used to calculate the volume of ink printed per unit area. Therefore, films of different ratios of RSF to MeOH were able to be produced. Ratios used were; 1:0, 3:1, 2:1, 1:1, 1:2 and 1:3 millilitres of RSF solution (1 mg mL^{-1}) to MeOH.

nHA/RSF films were also produced. nHA solution, resulting from the synthesis process, was mixed with RSF solution which had a concentration of 100 mg mL^{-1}. nHA/RSF ink concentrations were calculated based upon the dry weights of each component solution. Inks were made which had dried weights equivalent to 100%, 75%, 50% and 25% nHA content. From this point on, each of these inks will be referred to based upon their nHA content. Films were produced as described above, with a 1:1 ratio between the nHA/RSF ink and methanol.

Films were also made using a casting method. 0.5 mL of RSF was pipetted onto glass coverslips. The RSF solution was spread over the coverslip to give a film. Films were then left for 48 hours to dry. Some of the films were then subjected to a methanol treatment to induce β-sheet formation. Films were submerged in methanol and left in an enclosed petri dish for 4 days to induce maximum β-sheet formation.[25]

7.2.4 Degradation

Films were degraded in either an enzymatic solution consisting of Protease XIV (3.5 units/mg, from *Streptomyces griseus*, powder, Sigma Aldrich, UK) or in phosphate buffered saline (PBS) (Dulbecco's Phosphate Buffered Saline, without calcium chloride and magnesium chloride, Sigma Aldrich, UK). The enzymatic solution had a concentration of 0.35 U mL^{-1} protease XIV to provide a slow degradation rate. Due to the small mass of the RSF films produced with the inkjet printer, a low concentration of protease solution was used to provide a gradual degradation which would give a comparison between the films. RSF films were placed in 24-well plates and submerged in either a 1 mL solution of PBS or protease. Well plates were then incubated for up to 8 days, with degradation solutions replenished daily. At designated time points of 1, 2, 3, 5 and 8 days, a subset ($n = 3$) of films were removed and analysed. When removing films, each film was rinsed three times by immersion in 1 mL PBS for 2 minutes before being dried in a drying oven at 60 °C for 1 hour. Films were weighed before and after the degradation test.

7.2.5   Infrared Spectroscopy

Infrared spectroscopy was performed on B. Mori silkworm cocoons before and after the degumming procedure, as well as on the RSF films. RSF films measured after degradation test were first dried as described above. Samples were then used for attenuated total reflection Fourier transform infrared spectroscopy (FTIR-ATR) (Frontier FTIR, PerkinElmer). For each measurement, 16 scans were co-added with a resolution of 4 cm^{-1}, wavenumbers within the range 800–4000 cm^{-1}.

Fourier self-deconvolution was performed on the amide I region (1705–1595 cm^{-1}) by using the second derivative to identify hidden peaks within the spectra. Peaks were then identified as belonging to different secondary protein structures; beta sheets, alpha helix, random coil, turns and side chains. Using Origin Pro software peaks were fitted to an accumulative curve which fitted the original spectra. Deconvolution was performed using a nine-point Savitsky–Golay smoothing filter and Gaussian line shape profiles. To avoid possible artefacts, the peaks were fitted with different values for each band half width. These bands were 15, 20, 25 and 30 cm^{-1}; the variation of these results are represented as error bars.

7.2.6   Cell Viability

Human osteosarcoma MG-63 cells were maintained at a subconfluent density in minimum essential medium (α-MEM, Sigma Aldrich, UK), supplemented with 5% (v/v) foetal bovine serum (biosera), 50 µg mL^{-1} penicillin-streptomycin and 2 mMol L-glutamine. Before seeding, RSF and nHA/RSF films were sterilized using UV radiation as this has been shown to preserve the secondary protein structure.[26] Each film was placed within a 24-well plate and seeded with 1×10^4 cells in 500 µL of media. These were then incubated and the media replaced every 2 days.

Cell metabolic activity was monitored over a 2 week period. PrestoBlue® assays were performed on the samples after 1, 3, 7 and 14 days in culture. PrestoBlue® was mixed with cell media to a ratio of 1 : 10. 700 µL of the solution was added to each well and incubated at 37 °C for 1 hour. After incubation, 200 µL aliquots were taken from each well in triplicate and placed into a 96-well plate. Fluorescence intensity was measured using a fluorescence reader (Tecan Spectrophotometer) at an excitation wavelength of 570 nm and an emission wavelength of 600 nm.

7.2.7 Statistical Analysis

Statistical comparisons were performed using a two-way ANOVA test to compare features of groups, followed by a Tukey's test *post hoc* for multiple comparisons to determine significant differences. Significance of *p* values are as stated.

7.3 Removal of Sericin

To determine if sericin had been successfully removed during the degumming procedure, FTIR-ATR was performed on the B. Mori cocoons before and after the degumming procedure. Sericin has bands positioned at 1650, 1530, 1400 and 1070 cm^{-1}, however it is its band positioned at 1400 cm^{-1} which is its signature peak.[27] This band is positioned away from any silk fibroin bands which makes it easy to distinguish. Figure 7.3 is of the amide I (1600–1690 cm^{-1}), amide II (1480–1575 cm^{-1}) and amide III (1229–1301 cm^{-1}) spectral regions[28] of silk fibres before and after the degumming procedure. The unprocessed silk fibres have narrower bands than that of the degummed fibres. The broadening of the peaks is representative of a loss in crystallinity induced by the degumming procedure. The broadest band for the unprocessed fibres is positioned at 1400 cm^{-1}, symbolic of sericin. Sericin is a non-crystalline protein which also helps to distinguish it from the crystalline silk fibroin present in the B. Mori silkworm cocoon. After the

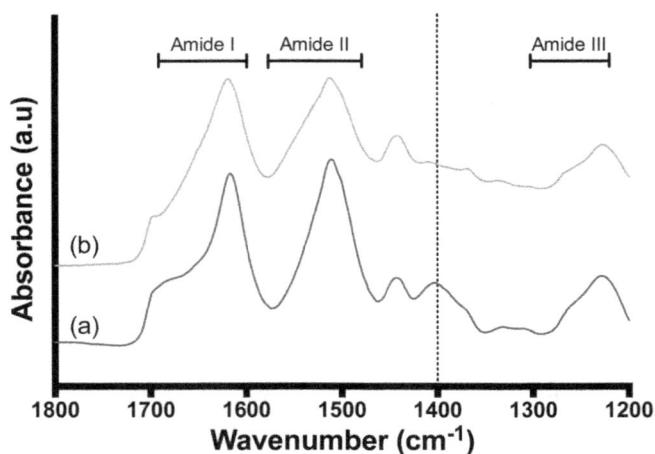

Figure 7.3   FTIR-ATR of amide I, II and III regions for (a) Bombyx Mori silk worm cocoon and (b) silk fibres after the degumming process. The vertical dotted line at 1400 cm^{-1} is positioned at the signature sericin peak.

degumming procedure the band at 1400 cm^{-1} disappears indicative that sericin has been removed.

7.4 Results and Discussion

7.4.1 Controlling β-sheet Formation

A key concept for using reactive inkjet printing to produce RSF films over other more common methods, is the condensing multiple procedural steps. This condensing is achieved by printing RSF and methanol together, thereby producing the film and improving the mechanical properties all at once. Control of film crystallinity is important as it is related to film degradation rates and mechanical stability. Figure 7.4 is the FTIR-ATR spectra of the RSF films, with different ratios of RSF to methanol, within the amide I and II regions. Figure 7.3(a) represents the spectrum of unprocessed B. Mori silkworm cocoons, Figure 7.4(b) is the spectrum of RSF solution which has been cast by being dispensed onto a glass coverslip and left to dry. Spectra (d–h) (Figure 7.4) are of films with RSF:MeOH ratios 3:1, 2:1, 1:1, 1:2 and 1:3 respectively. Figure 7.4(i) is the spectrum of an RSF cast film which has been submerged in methanol for 4 days to induce maximum β-sheet content. Table 7.1 compares the ratios of band intensities representative of unordered silk I (1640 cm^{-1} and 1535 cm^{-1}) and silk II (1619 cm^{-1} and 1515 cm^{-1}) within the amide I and amide II regions.

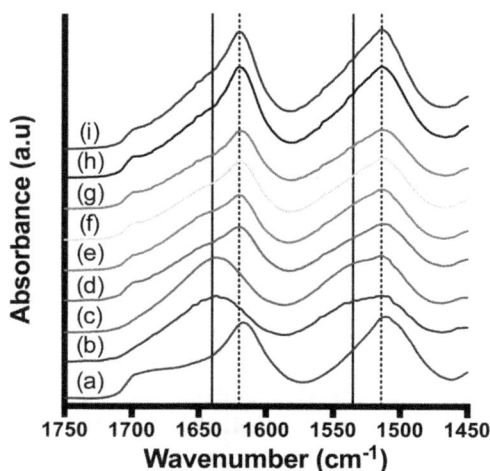

Figure 7.4   FTIR-ATR spectra of the amide I and amide II regions for silk films printed with different ratios of RSF (100 mg mL^{-1}):MeOH (1 mL); (c) 1:0, (d) 3:1, (e) 2:1, (f) 1:1, (g) 1:2 and (h) 1:3. Also shown are the spectra for; (a) B. Mori silkworm cocoon (b) cast RSF, (i) cast RSF with 4 day MeOH treatment. Solid vertical lines represent band positions for a silk I structure and dotted vertical lines represent bands for a silk II structure.

Table 7.1 Ratio of silk II to silk I band intensities within the amide I and amide II region.

Ratio	A_{1619}/A_{1640} Silk II/Silk I	A_{1515}/A_{1535} Silk II/Silk I
NP	0.83	1.08
1:0	0.79	1.11
3:1	1.24	1.22
2:1	1.25	1.22
1:1	1.36	1.27
1:2	1.41	1.26
1:3	1.53	1.31
NP + M	1.58	1.34
SC	1.85	1.76

Comparing spectra (b) and (c) (Figure 7.4) determines if β-sheet formation has been induced during printing. Each spectrum appears to have a similar shape with a prominent peak positioned for silk I at 1640 cm^{-1} and no peaks associated with silk II (1619 or 1515 cm^{-1}). This indicates that they remain in a non-crystalline state. A comparison of silk II to silk I band intensities made in Table 7.1 show that a similar ratio of silk I and silk II is present for each film. It can therefore be deduced that printing of the RSF solution through a 60 μm nozzle does not induce crystallisation of the silk fibroin.

For RSF films which have had methanol printed between sequential layers, bands appear positioned at 1619 cm^{-1} and a shift of peak occurs within the amide II region from 1535 cm^{-1} to 1515 cm^{-1} indicating the formation of β-sheet. As the volume of methanol is increased, the intensity of these bands also increases due to the formation of β-sheet structures with denser inter-sheet packing. As the amount of silk II increases, bands associated with silk I, positioned at 1640 cm^{-1} and 1535 cm^{-1}, begin to decrease. There is a substantial increase of intensity for silk II peaks observed between ratios of 1:2 and 1:3, with the peaks similar in size to that of the cast RSF film with induced maximum β-sheet formation. This suggests that films with a printed ratio of 1:3 have reached a maximum crystallinity.

Table 7.1 confirms the observations made of the spectra. Ratios of silk II to silk I within the amide I and amide II regions support a gradual increase of silk II with increasing volumes of methanol. For both amides I and II regions, a significant increase of silk II is measured between the ratios of 1:2 and 1:3.

β-sheet crystallinity of the films was calculated by performing Fourier self-deconvolution (FSD) on the amide I region of each spectrum. By performing FSD, each contributing structure: β-sheet, α-helix, random coils, β-turns and side chains, of silk fibroin can be calculated for each RSF film. Due to the position of their bands, percentage contributions from α-helix and random coils were combined as it is difficult to accurately distinguish

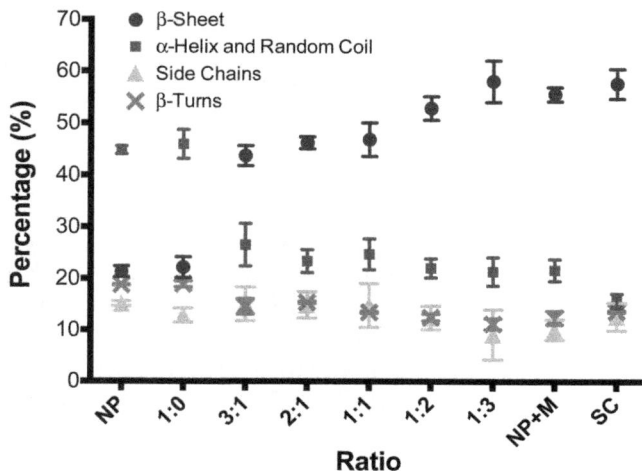

Figure 7.5 FSD data showing percentage contribution of protein structures for RSF films.

between the two. Figure 7.5 shows the FSD data of the calculated percentage contributions from each structure.

Figure 7.5 confirm that crystallisation is not induced during printing and demonstrates that with the addition of a small volume of methanol there is an immediate increase of β-sheet structure. β-sheet content rises from ~20% to ~45%, when comparing RSF films 1:0 and 3:1. β-sheet content remains relatively consistent for RSF films 3:1, 2:1 and 1:1 with no significant difference between them. However, when the volume of methanol increases beyond the 1:1 ratio, significant differences are observed between sequential films. β-sheet crystallinity increases by 6% between RSF films 1:1 and 1:2 ($p < 0.001$), and 5% between RSF films 1:2 and 1:3 ($p < 0.01$). There was no significant difference of β-sheet content between RSF films 1:3, NP + M and that of the unprocessed silk cocoon (SC) with average crystallinities of ~58%, ~56% and ~58% respectively. NP + M has the same crystallinity as that observed previously by Xiao Hu *et al.* confirming maximum β-sheet crystallinity.[25]

α-Helix and random coil contributions significantly reduces between RSF films 1:0 and 3:1, dropping from ~45% to ~25% of the structure. Their contribution to the overall structure remains consistently around 25% for RSF films 3:1, 2:1 and 1:1, and drops to ~20% for RSF films 1:2, 1:3 and NP + M. α-Helix and random coil contributions were significantly lower for the unprocessed silk cocoons at just below 16% ($p < 0.001$).

The FSD data in Figure 7.5 show that small volumes of methanol are able to affect the RSF structure significantly by inducing a crystalline structure. The crystallinity of the RSF structure can then be gradually increased to a maximum potential β-sheet content, similar to that of unprocessed silk cocoons. These observations are obtained with an average volume of 12 μL of RSF printed per square centimetre, or 1.2 mg of silk fibroin per square centimetre, for each layer of RSF. Any changes in the volumes printed could change the influence of methanol on the RSF structure for each ratio.

7.4.2 Degradation

Degradation of the RSF films were studied and compared by immersing them in an enzymatic solution of protease XIV and in a phosphate buffered saline (PBS) over an 8 day period. Studying the degradation of the films helps to predict how the films will perform when *in vivo*. Figure 7.6(a) shows mass loss of the RSF films submerged in an enzymatic solution. From the degradation profiles it is apparent that small differences in crystallinity, produced by reactive inkjet printing influences the degradation rates of the RSF films.

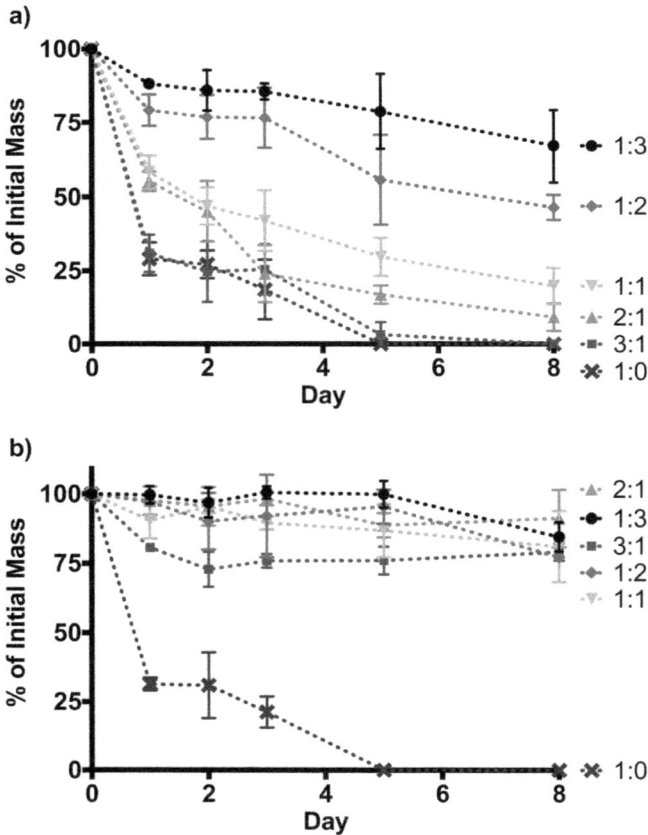

Figure 7.6   Degradation mass loss for RSF films printed with different ratios of RSF to MeOH Films degraded in (a) Protease XIV or (b) PBS over an 8 day period.

All films experienced the largest mass loss over the first 24 hours with the initial mass loss directly related to the crystallinity of the films. RSF films which were exposed to the smallest volumes of methanol (therefore less crystalline) experienced the largest mass loss. The RSF film without any methanol exposure (1:0) had a similar mass loss to that of the RSF film printed with the smallest volume of methanol (3:1), losing around 70% of their mass over the first 24 hours. RSF films 2:1 and 1:1 also experienced a similar initial mass loss to each other, losing around 45% mass. Films 1:2 and 1:3 lost about 20% and 10% mass respectively.

Over the following 7 days, for films submerged in the enzymatic solution, degradation rates remained related to initial film crystallinity. By day 8, films 1:0 and 3:1 had completely degraded. RSF film 1:0 had completely degraded by day 5 followed by 3:1 by day 8. On the final day, the amount that each film had degraded by was related to its crystallinity. Films with the smallest to highest remaining masses were as follows 2:1 < 1:1 < 1:2 < 1:3, each losing around 90%, 80%, 55% and 35% respectively of their initial mass.

RSF films 1:0 and 3:1 remained similar to each other for every time point of the experiment. RSF films 2:1 and 1:1 remained significantly similar apart from on day 3 ($p < 0.05$) when 2:1 had an average mass 22% of its initial value, whereas 1:1 had a mass 44% of its initial value. The degradation profiles of RSF films 1:2 and 1:3 were only significantly different on the final day ($p < 0.01$).

Figure 7.6(b) shows the profiles of the RSF films degraded in PBS. The PBS solution was used as a control to monitor dissolution of the films due to the presence of soluble unordered silk. As expected the mass lost by the films in the enzymatic solution was not matched by the films submerged in PBS. A larger mass loss for films within the protease XIV solution would be expected as the protease facilitates in the degradation and dissolution of silk fibroin. However, there seemed to be no pattern between crystallinity and mass loss for films degraded in a PBS solution. The RSF film without methanol treatment (1:0) was the only film to have a similar degradation profile in both solutions. There was no pattern to the degradation profiles of the remaining RSF films in PBS, which is similar to previous degradation studies performed using PBS where mass loss has been minimal.[29]

The large mass loss experienced over the first 24 hours for films in the enzymatic solution could be linked to the dissolution of unordered silk present within the films. Comparing the initial mass loss of films submerged in PBS to that of the enzymatic solution, the only film to experience a large drop in mass was the RSF film which had not been exposed to methanol. A potential reason for only this film experiencing a large mass loss, could be due to the manner in which the films are produced. By using inkjet printing to produce the RSF films, multiple layers had to be printed in order to build up the depth and mass of the films. This layer-by-layer approach meant that layers of methanol were printed between sequential layers of RSF. This could have produced a structure where layers of unordered soluble silk are encapsulated under layers of silk II. Films which have had larger volumes of methanol printed between layers of RSF have a greater crystallinity than those with lower volumes of methanol. Larger volumes of methanol would take longer to evaporate on the substrate and therefore have longer to penetrate into the film. Thereby films which have had a longer exposure to methanol will have thicker layers of silk II with denser crystal packing encapsulating the unordered silk beneath. Films which have been submerged in an enzymatic solution have had the silk II layers degraded, enabling the silk I to be dissolved. This could also explain how after the initial drop in mass, rate of degradation is similar for most of the films due to the enzyme digesting the silk II structure. Subsequently, films submerged in PBS have an outer layer of silk II which protects the silk I and prevents it from dissolving in solution. Hence RSF film 1:0 is only the film without any methanol treatment and therefore experienced substantial mass loss over the initial 24 hour period within the PBS solution.

IR spectra of the films were measured using FTIR-ATR for each time point. FSD was then performed on the spectra to examine how the RSF structure changed during degradation. Figures 7.7 and 7.8 are of the FSD data for the RSF films degraded in an enzymatic or PBS solution.

Figure 7.7(a) shows that RSF film 1:0 had a significant increase in crystallinity over the first 24 hour period when degraded in the enzymatic solution with an increase of β-sheet content by 30%. This was accompanied with a loss of α-helix and random coil of 19%. Over the following 24 hours there is a slight loss of crystallinity as the β-sheet content drops by 5%, whilst the proportion of α-helix and random coil remained similar with no significant difference between days. In comparison (Figure 7.8(a)) there is also a significant increase in β-sheet content for the 1:0 RSF film over the first 24 hours submerged in PBS, however, the relative increase is only about half as much at 15%, increasing to ~38% content and does not significantly change the following day. There is a steady decline in α-helix and random coil structure dropping by 14% over the first 24 hours, and then by another 5% over the following 24 hours.

Over the first 2 days there were no significant differences in β-sheet content for RSF film 3:1 when degraded in an enzymatic solution which increases with an average of 2% (Figure 7.7(b)). There is an initial drop of 8% for α-helix and random coil contributions, however by day 2 they account for ~24% of the structure which is significantly similar to the starting contributing value. In PBS (Figure 7.8(b)), RSF film 3:1 averaged around 42% total content for the experiment with the lowest β-sheet content on day 2 at ~39%. An increase of 6% of α-helix and random coil structure was measured over the first 24 hours and then remained around 32% total contribution.

RSF film 2:1 (Figure 7.7(c)) had a slight increase in β-sheet content over the first 48 hours in the enzymatic solution, increasing by 6%, however by the final day the proportion of β-sheet content was similar to that of its initial value on day 0. The percentage of α-helix and random coil structures within the films fluctuated around an average of 22%. However, in the PBS there is an initial dip in β-sheet content by 7% giving a structural contribution of ~40% before increasing to ~47% by day 8 (Figure 7.8(c)). The α-helix and

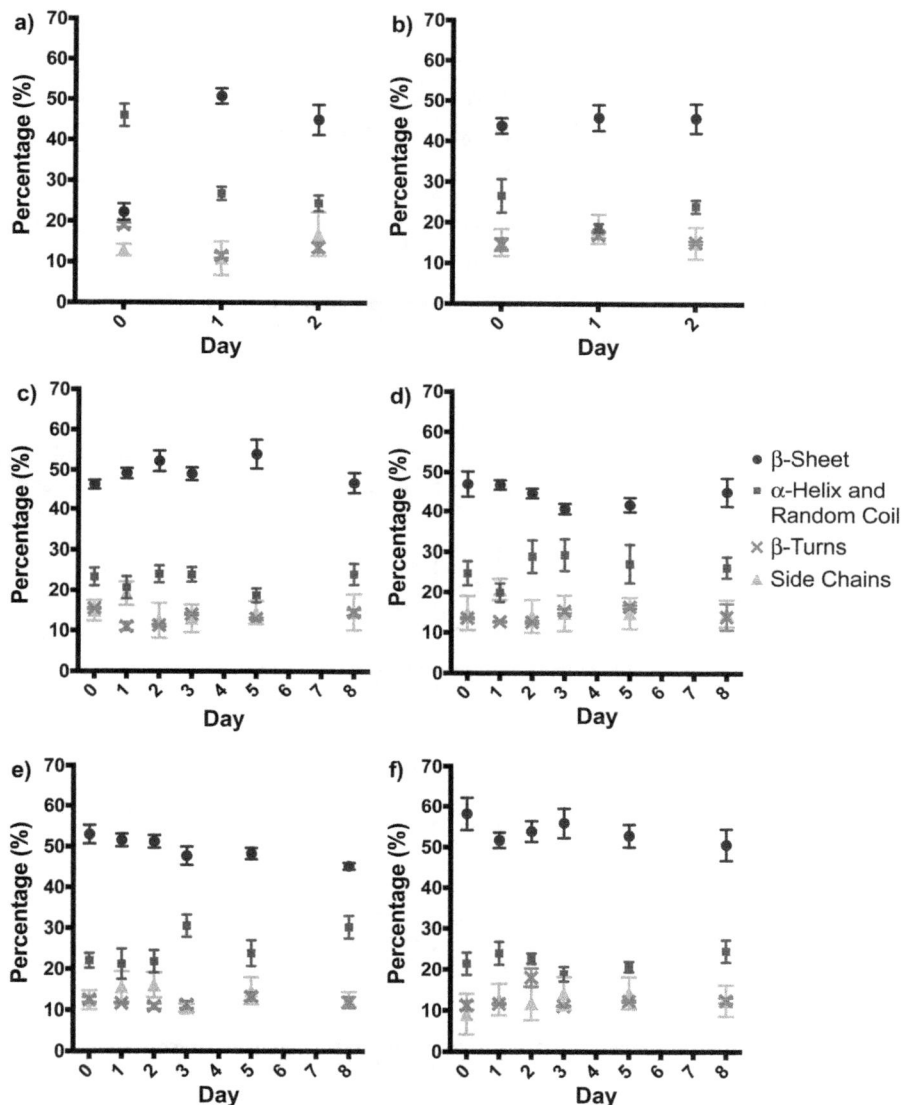

Figure 7.7 FSD data showing percentage of component secondary structures for each RSF film degraded in Protease XIV at different degradation time points. RSF film ratios of RSF (100 mg mL^{-1}) : MeOH (1 mL) shown in; (a) 1:0, (b) 3:1, (c) 2:1, (d) 1:1, (e) 1:2 and (f) 1:3.

random coil content increases over the first 48 hours in PBS by 10% to ~33% total content, and reduces to ~28% total content by day 8.

RSF film 1:1 (Figure 7.7(d)) had an initial loss of β-sheet content over the first 3 days in the Protease XIV solution, accumulating to a total loss of 5%, however, by day 8, none of its contributing structures were significantly different to that of the film on day 0. When in PBS (Figure 7.8(d)), β-sheet content drops by

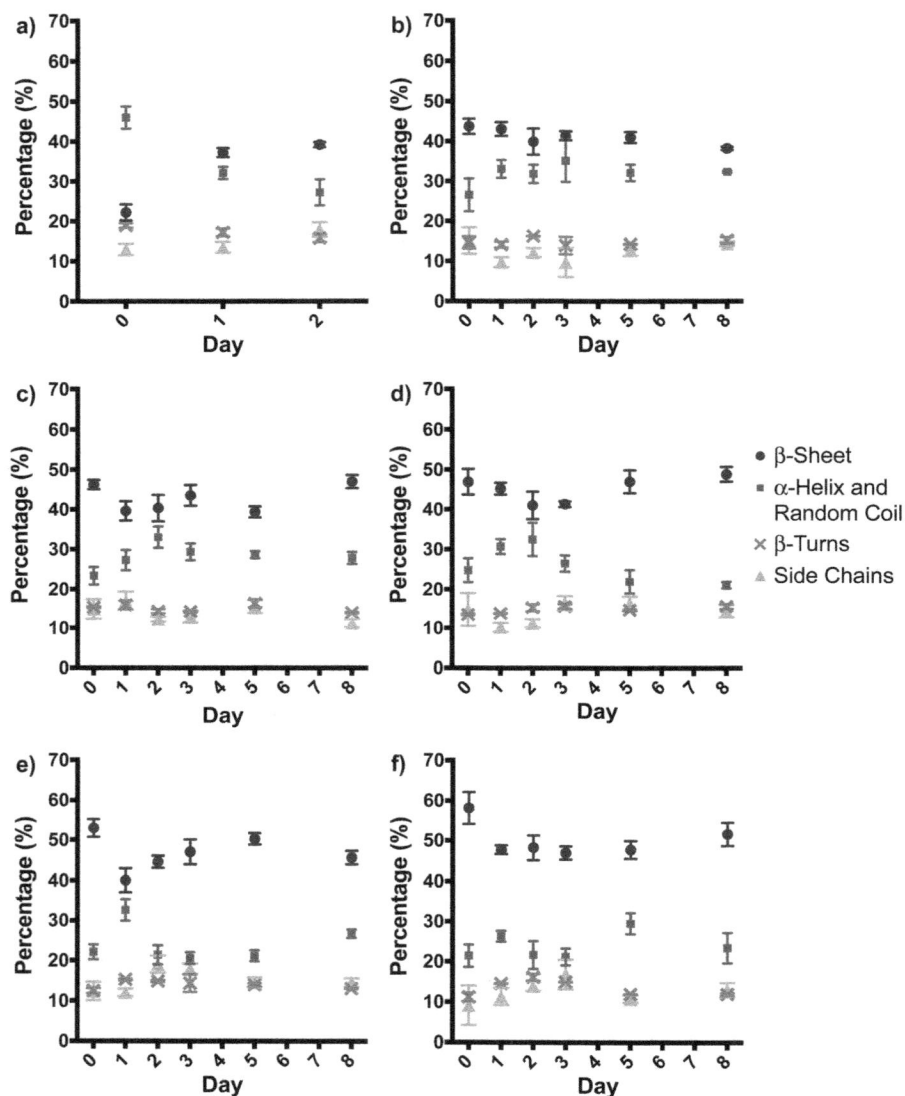

Figure 7.8 FSD data showing percentage of component secondary structures for each RSF film degraded in PBS at different degradation time points. RSF film ratios of RSF (100 mg mL^{-1}):MeOH (1 mL) shown in; (a) 1:0, (b) 3:1, (c) 2:1, (d) 1:1, (e) 1:2 and (f) 1:3.

6% after 48 hours in solution, which was accompanied by an 8% rise in α-helix and random coil structure. However, as in the protease solution, by day 8 there was no significant difference between contributing structures.

RSF film 1:2 (Figure 7.7(e)) had no significant change in β-sheet content between sequential days in the protease solution, however over the 8 day period, a general loss in β-sheet content is observed so that on the final day there has been around an 8% drop in crystallinity with a corresponding 8% increase in α-helix and random coil contributions. In PBS (Figure 7.8(e)), there is an initial sharp drop in β-sheet content dropping by 13% to provide ~40% of its structure whereas a rise in α-helix and random coil content of 11% was observed so that it contributes to ~33% of the RSF structure. However, by day 5, β-sheet content had increased by 10% before reducing slightly on day 8 to ~46%. α-helix and random coil content decreases to ~21% by day 5 and increases to ~27% by day 8.

RSF film 1:3 (Figure 7.7(f)) has an initial drop in β-sheet content by 6% and averages out around 52% of the overall structure until day 8. α-Helix and random coils contributed to between ~19% and ~25% of the film structure over the 8 day period, reducing to its lowest contribution on day 3 and its maximum on day 8. In PBS (Figure 7.8(f)) there was an initial drop in β-sheet content of 10% after the first 24 hours, after which, β-sheet content continued to average around 48% of the structure. α-Helix and random coil structures increased by 5% over the first 24 hours in solution, and then over the following days contributed to between 20–30% of the structure.

Minimal changes in overall β-sheet crystallinity for films degraded in the protease could support the notion that the scaffolds have a layered structure of silk II and unordered silk. As the enzymes only work on the surface of the films and do not penetrate the structure, the layers of silk II would have to be degraded to expose the unordered silk below which can then be dissolved by the solution. In a previous study the β-sheet peak at 1693 cm^{-1} was almost lost after 24 days in an protease XIV solution as the enzyme hydrolysed the silk β-sheet crsytals,[29] however β-sheet content remains prominent in our study. Only films with the highest crystallinity showed a loss of β-sheet after 8 days in protease XIV, both RSF films 1:2 and 1:3 lost about 8% β-sheet content. An increase of 23% was measured for RSF film 1:0, however its structure began with a high proportion of soluble unordered silk which could have simply dissolved in solution. RSF films 3:1, 2:1 and 1:1 experienced no significant changes in β-sheet content. RSF films 1:0, 3:1, 2:1 and 1:1 all experienced a drop in average α-helix and random coil content over the first 24 hours, which could potentially be linked to the large mass loss experienced by these samples.

Films degraded in PBS show greater fluctuations in crystallinity. This could be the result of amorphic regions of the silk II structure hydrolysing and secreting β-sheet crystals into the surrounding solution. Overall the greatest loss in crystallinity within the PBS solution occurred for films; 3:1 which lost 10% of its β-sheet content, 1:2 which lost 7% and 1:3 which lost 7%. RSF film 1:0 was again observed to increase in crystallinity, this time by 17%, and again possibly because of its starting high content of soluble unordered silk within its structure. There was no significant change in crystallinity of RSF films 2:1 and 1:1 by the end of the 8 day period. All of the methanol treated films experienced an increase of α-helix and random coil content over the first 24 hours.

7.4.3 Cell Viability

Human osteosarcoma MG-63 cells were used to evaluate bone cell viability and proliferation on the RSF and nHA/RSF films using PrestoBlue® assays. Controls of tissue culture plastic (TCP), glass coverslips and Poly(L-lactide) (PLLA) were used to compare proliferation on other common cell substrates.

Figure 7.9 shows the fluoresce for MG-63 cells seeded upon either RSF (Figure 7.9(a)) or nHA/RSF (Figure 7.9(b)) films. Both Figures 7.9(a) and (b) show that the cells survived and proliferated on all of the samples, with increased metabolic activity over the 14 day test period. Growth between each time point was significant for all samples ($p < 0.0001$) other than between days 1 and 3. After 24 hours, most of the inkjet printed samples experienced no significant increase in fluorescence except RSF film 3 : 1 ($p < 0.05$), however all of the controls experienced a significant increase over the same period.

After 1 day in culture the fluorescence of all RSF and nHA/RSF films were significantly similar to each other and the controls. After 3 days in culture the RSF films had increased in fluorescence by similar amounts, and therefore had no significant difference between each film (Figure 7.9(a)). However, in comparison to the controls, all RSF films were similar to PLLA, except RSF film 1 : 3 ($p < 0.05$), which had a lower fluorescence. TCP was significantly higher than all of the RSF films ($p < 0.01$). Fluorescence of the glass coverslips was only significantly greater than RSF films 1 : 1 ($p < 0.05$) and 1 : 3 ($p < 0.01$).

nHA/RSF films had no significant difference between each other by day 3 (Figure 7.9(b)). PLLA was significantly greater than nHA/RSF films 100% ($p < 0.05$) and 50% ($p < 0.01$). TCP was significantly higher than all of the nHA/RSF films by a significance of $p < 0.0001$. The glass coverslips were significantly higher than three of the nHA/RSF films; 100% ($p < 0.01$), 50% ($p < 0.001$) and 25% ($p < 0.05$).

By day 7, a difference in fluorescence was measured between the RSF film samples (Figure 7.9(a)). All of the RSF films were significantly similar to each other apart from RSF film 1 : 3 which had a lower fluorescence ($p < 0.001$). RSF film 1 : 3 had the lowest fluorescence of all the films and was significantly less than that of the controls ($p < 0.0001$). PLLA had a significantly higher fluorescence than RSF films 2 : 1 ($p < 0.05$) and 1 : 1 ($p < 0.001$). Both TCP and the glass coverslips had significantly higher fluorescence than RSF film 1 : 1 ($p < 0.05$).

The nHA/RSF films had begun to differentiate in fluorescence after 7 days in culture (Figure 7.9(b)). Only nHA/RSF films 100% and 25% remained similar. 75% had the highest recorded fluorescence and was significantly higher than all nHA/RSF films ($p < 0.0001$). nHA/RSF film 50% had the lowest fluorescence which was significantly lower than nHA/RSF films 100% and 25% ($p < 0.0001$). Fluorescence of nHA/RSF film 50% was also significantly lower than all of the controls ($p < 0.0001$). PLLA was the only control to be significantly different to any of the other samples; 100% ($p < 0.001$) and 25% ($p < 0.001$).

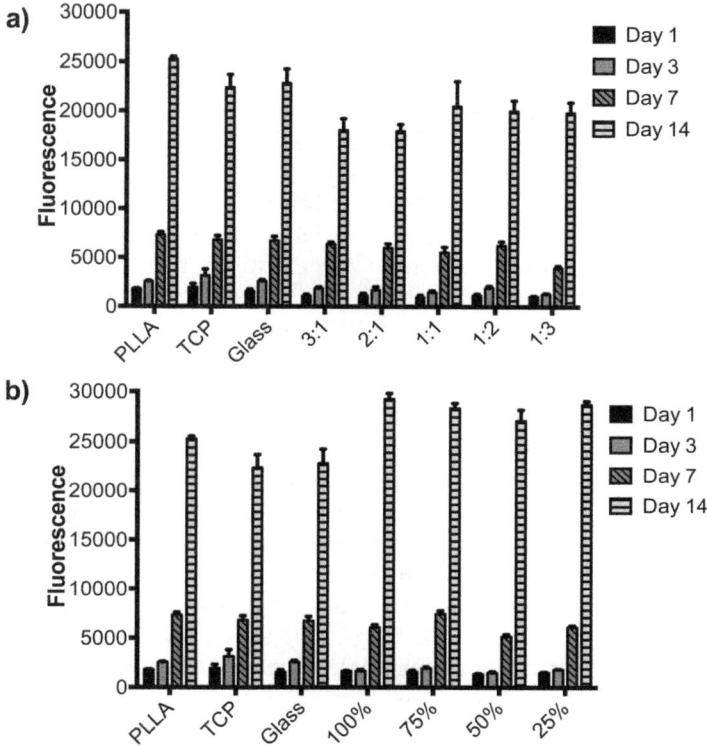

Figure 7.9 Fluorescence of (a) RSF and (b) nHA/RSF films over a 2 week period.

By the final day of culture, the RSF films with the highest crystallinity $1:1$, $1:2$ and $1:3$ had significantly higher crystallinity than that of the less crystalline samples $3:1$ and $2:1$ ($p < 0.0001$) (Figure 7.9(a)). RSF films $1:1$, $1:2$ and $1:3$ all had similar fluorescence, as did RSF samples $3:1$ and $2:1$. All of the RSF samples had significantly less fluorescence than that of the controls ($p < 0.0001$).

In contrast, all of the nHA/RSF films had significantly higher fluorescence than that of the controls after 14 days in culture ($p < 0.0001$) (Figure 7.9(b)). nHA/RSF film 100% had the highest fluorescence value, however was not significantly higher than that of nHA/RSF film 25%. nHA/RSF film 25% was also similar to film 75%. The lowest fluorescence value of the nHA/RSF films was for film 50%, which was also significantly lower than the other nHA samples ($p < 0.0001$).

Over the first 3 days in culture there was not much of a difference in fluorescence values to distinguish between the RSF samples. It was only by day 7 that differences began to arise between the RSF samples. On day 7 it was the most crystalline sample which had the lowest fluorescence value indicating lower levels of cell metabolism. However, by day 14, it was the more crystalline samples which had a significantly higher fluorescence than that of the lower crystalline RSF films 3 : 1 and 2 : 1. Therefore, the crystallinity of the RSF samples does not seem to influence cell attachment and proliferation of MG-63 cells over the first week of culture, although by the end of a 2 week period, higher crystallinities have a more positive affect on the cells. Using reactive inkjet printing to control the crystallinity of the RSF films could be useful for improving cellular responses.

The inclusion of nHA within the RSF films was shown to be beneficial. By the end of the first week, nHA/RSF film 75% had a similar fluorescence to that of the controls, and by the end of 2 weeks all of the nHA/RSF films had a significantly higher fluorescence than that of the controls. Therefore, printing a composite scaffold of nHA and RSF could be beneficial for promoting bony regeneration of defect sites.

7.5 Conclusion

Reactive inkjet printing has successfully been utilised to produce RSF films with controllable crystallinity and degradation rates. Inkjet printing enables precise volumes of RSF and methanol to be dispensed onto a substrate to produce a film. Due to the reliability of constant repeatable droplet volumes, films can be produced with varying ratios of RSF to methanol and thereby control overall crystallinity of the films. Due to the layer by layer method of producing the films, it is hypothesized that the structure of the films is a combination of alternate layers of unordered silk I and silk II. Larger volumes of methanol increase the depth of the silk II layers and density of β-sheet crystals. The silk II layers protect the unordered silk, which is why even the least crystalline RSF film printed with methanol has minimal mass loss when submerged in PBS over 8 days. When the RSF films were exposed to an enzymatic solution containing protease XIV, crystallinity was seen to affect degradation rates. However, due to layers of silk II encapsulating unordered silk, β-sheet, α-helix and random coil structural contributions remained relatively consistent during degradation. Higher crystallinities of the RSF films were shown to improve cellular responses over a 2 week period. It was also shown that the inclusion of nHA was beneficial in promoting cellular proliferation of MG-63 osteosarcoma cells.

The ability to control the mechanical properties and degradation rates is important for all tissue engineering constructs. Optimal tissue regeneration can be achieved by matching the mechanical properties of the surrounding tissues and degrading once the damaged tissue has been repaired. As reactively inkjet printed RSF has been shown to have control over β-sheet content, which is linked to degradation and mechanical properties, reactively inkjet printed RSF could be an ideal material for tissue engineering. Cellular interactions have been shown to be influenced by substrate elasticity[30] and crystallinity.[31] Regulating RSF elasticity and crystallinity to match cellular requirements may make reactively inkjet printed RSF appealing for both hard and soft tissue regenerations. Inkjet printing can also be used to produce graduated structures.[32] Thus a structure could be designed to interact with multiple tissue types by printing gradients of bioactive components such as nHA.

Control over RSF film β-sheet content could also be useful for other applications such as for controlled drug release or in optics. Wang *et al.* showed that β-sheet content of RSF coatings affect drug release,[33] however to build up layers of RSF in the reported study required a dip-drying process which could be more accurately controlled through reactive inkjet printing. Manipulating the transport of light by RSF crystallinity may be of use for optical wave guides, where silk has already been considered as a suitable material for their production.[34] Overall the ability of reactive inkjet printing to control many of the characteristics of RSF, and the capability and accuracy of inkjet printing to build complex patterns, could have potential uses for a range of different applications.

Acknowledgements

The authors would like to thank Nobel Biocare UK for their support in funding this research.

References

1. Y. Zhang, C. Tse, D. Rouholamin and P. J. Smith, *Cent. Eur. J. Eng.*, 2012, **2**, 325–335.
2. M. C. Bottino, V. Thomas, G. Schmidt, Y. K. Vohra, T.-M. G. Chu, M. J. Kowolik and G. M. Janowski, *Dent. Mater.*, 2012, **28**, 703–721.
3. T. Scantlebury and J. Ambruster, *J. Evid. Based Dent. Pract.*, 2012, **12**, 101–117.
4. Y. D. Rakhmatia, Y. Ayukawa, A. Furuhashi and K. Koyano, *J. Prosthodont. Res.*, 2013, **57**, 3–14.
5. P. Bunyaratavej and H. L. Wang, *J. Periodontol.*, 2001, **72**, 215–229.
6. H. Tal, A. Kozlovsky, Z. Artzi, C. E. Nemcovsky and O. Moses, *Clin. Oral Implants Res.*, 2008, **19**, 760–766.
7. G. Polimeni, K.-T. Koo, G. A. Pringle, A. Agelan, F. F. Safadi and U. M. E. Wikesjo, *Clin. Implant Dent. Relat. Res.*, 2008, **10**, 99–105.
8. P. Rider, Y. Zhang, C. C. W. Tse, Y. Zhang, D. Jayawardane, J. Stringer, J. Callaghan, I. M. Brook, C. A. Miller, X. Zhao and P. J. Smith, *J. Mater. Sci.*, 2016, **51**, 8625–8630.
9. I. Greving, M. Cai, F. Vollrath and H. C. Schniepp, *Biomacromolecules*, 2012, **13**, 676–682.
10. D. Huemmerich, U. Slotta and T. Scheibel, *Appl. Phys. A*, 2005, **82**, 219–222.
11. B. D. Lawrence, M. Cronin-Golomb, I. Georgakoudi, D. L. Kaplan and F. G. Omenetto, *Biomacromolecules*, 2008, **9**, 1214–1220.
12. B. Cirillo, M. Morra and G. Catapano, *Int. J. Artif. Organs*, 2004, **27**, 60–68.
13. Y. G. Chung, D. Tu, D. Franck, E. S. Gil, K. Algarrahi, R. M. Adam, D. L. Kaplan, C. R. Estrada Jr and J. R. Mauney, *PLoS One*, 2014, **9**, e91592.

14. T.-L. Liu, J.-C. Miao, W.-H. Sheng, Y.-F. Xie, Q. Huang, Y.-B. Shan and J.-C. Yang, *J. Zhejiang Univ., Sci., B*, 2010, **11**, 10–16.
15. R. Ghassemifar, S. Redmond, Zainuddin and T. V. Chirila, *J Biomater Appl*, 2010, **24**, 591–606.
16. P. X. Ma and J. W. Choi, *Tissue Eng.*, 2001, **7**, 23–33.
17. H. J. Jin, J. Park, V. Karageorgiou, U. J. Kim, R. Valluzzi, P. Cebe and D. L. Kaplan, *Adv. Funct. Mater.*, 2005, **15**, 1241–1247.
18. Q. Lu, B. Zhang, M. Li, B. Zuo, D. L. Kaplan, Y. Huang and H. Zhu, *Biomacromolecules*, 2011, **12**, 1080–1086.
19. R. Suntivich, I. Drachuk, R. Calabrese, D. L. Kaplan and V. V. Tsukruk, *Biomacromolecules*, 2014, **15**, 1428–1435.
20. H. Tao, B. Marelli, M. Yang, B. An, M. S. Onses, J. A. Rogers, D. L. Kaplan and F. G. Omenetto, *Adv. Mater.*, 2015, **27**, 4273–4279.
21. D. N. Rockwood, R. C. Preda, T. Yucel, X. Wang, M. L. Lovett and D. L. Kaplan, *Nat. Protoc.*, 2011, **6**, 1612–1631.
22. M. K. Sah and K. Pramanik, *Int. J. Environ. Sci. Dev.*, 2010, **1**, 404–408.
23. L. J. C. Brito, D. Q. D. M. Gomes, M. Da Silva Vivia Tenedorio, D. O. E. S. P. Quadros, M. F. J. Mendes, C. G. P. J. Da and M. A. Y. Pataquiva, U.S. Patent, WO 2008/007992 A2, 2008.
24. J. Schindelin, C. T. Rueden, M. C. Hiner and K. W. Eliceiri, *Mol. Reprod. Dev.*, 2015, **82**, 518–529.
25. X. Hu, K. Shmelev, L. Sun, E. S. Gil, S.-H. Park, P. Cebe and D. L. Kaplan, *Biomacromolecules*, 2011, **12**, 1686–1696.
26. M. A. de Moraes, R. F. Weska and M. M. Beppu, *J. Biomed. Mater. Res., Part B*, 2014, **102**, 869–876.
27. X. Zhang and P. Wyeth, *Sci. China: Chem.*, 2010, **53**, 626–631.
28. J. Kong and S. Yu, *Acta Biochim. Biophys. Sin. (Shanghai)*, 2007, **39**, 549–559.
29. J. Zhou, C. Cao, X. Ma, L. Hu, L. Chen and C. Wang, *Polym. Degrad. Stab.*, 2010, **95**, 1679–1685.
30. F. Han, C. Zhu, Q. Guo, H. Yang and B. Li, *J. Mater. Chem. B*, 2015, **4**, 9–26.
31. H. Cui and P. J. Sinko, *Front. Mater. Sci.*, 2011, **6**, 47–59.
32. M. S. Khan, D. Fon, X. Li, J. Tian, J. Forsythe, G. Garnier and W. Shen, *Colloids Surf., B*, 2010, **75**, 441–447.
33. X. Wang, X. Hu, A. Daley, O. Rabotyagova, P. Cebe and D. L. Kaplan, *J. Controlled Release*, 2007, **121**, 190–199.
34. S. T. Parker, P. Domachuk, J. Amsden, J. Bressner, J. A. Lewis, D. L. Kaplan and F. G. Omenetto, *Adv. Mater.*, 2009, **21**, 2411–2415.

CHAPTER 8

Reactive Inkjet Printing of Regenerated Silk Fibroin as a 3D Scaffold for Autonomous Swimming Devices (Micro-rockets)

DAVID A. GREGORY*, YU ZHANG, STEPHEN J. EBBENS AND XIUBO ZHAO

The University of Sheffield, Department of Chemical and Biological Engineering, South Yorkshire, UK
*E-mail: d.a.gregory@sheffield.ac.uk

8.1 Introduction

In recent years the printing of biomaterials has become of increasing interest, in particular the ability to generate biological scaffold materials has attracted substantial media attention. Here we use a new kind of inkjet printing called *Reactive Inkjet Printing* (RIJ)[1] as a method to print three dimensional silk scaffolds for use in biomedical applications. RIJ refers to the use of two or more different kinds of inks that are printed in the sample location and react with each other to form a new substance or generate conformational changes in one of the inks. In the case of a silk ink this means that a water soluble silk

Smart Materials No. 32
Reactive Inkjet Printing: A Chemical Synthesis Tool
Edited by Patrick J. Smith and Aoife Morrin
© The Royal Society of Chemistry 2018
Published by the Royal Society of Chemistry, www.rsc.org

is converted by a curing ink to a water insoluble silk scaffold structure. Using regenerated silk fibroin as ink provides an opportunity to encapsulate different moieties into the scaffold structure during printing without the need for complex chemical reactions taking place, as is the case for the frequently used covalent immobilisation processes.[2] For RIJ of silk it has been shown, for example, that enzyme molecules can simply be mixed into *regenerated silk fibroin* (RSF), which is water soluble, but that the conformational structure change during RIJ will encapsulate these enzyme molecules, immobilising them into the scaffold lattice structure.

In this chapter particular focus is given to the use of highly biocompatible silk inks and the production of 3D silk scaffolds to produce self-motile micromotors ('rockets') *via* a *layer-by-layer* (LBL) printing approach. Initially this chapter will look at what self-motile micromotors are and their potential benefits and application areas and how current fabrication methods of current devices is affecting the ability to produce complex structures. Following on from this, silk as a biocompatible ink will be discussed along with what major advantages it bestows. The fabrication of silk-based microrockets is then discussed in detail, considering some of the challenges that have been reported in recent publications.

8.1.1 What Are Micromotors/Autonomous 'Swimming' Devices

Over the last ten years the production of small-scale devices that are able to generate autonomous motion *via* catalytic reactions within fluidic environments has become an increasingly active field of research.[3-7] There are numerous potential applications for these devices, including environmental monitoring, water remediation,[8-12] Lab-on-a-Chip diagnostics,[13] as well as *in vivo* drug delivery and repair,[14] which have been some of the key aspects as to why these devices have received increasing attention and substantial media coverage. In order to understand the challenges that are faced when fabricating these devices, it is important to look at how different length scales affect particles in fluids. When viewing small particles (micro- or nanometre range) suspended in a solution (*e.g.* in water) under a microscope the particles are seen to move around randomly. This motion is called Brownian motion and is due to the thermal energy of the surrounding fluid.[15] In view of this ubiquitous motion, it is clear that useful propulsion generating mechanisms must result in velocity magnitudes that exceed Brownian motion velocity. Based on our macroscopic experiences the most obvious mechanisms for propulsion generation use reciprocal deformations, *e.g.* when humans swim they do so by moving their arms and legs backwards and forwards. However, deformations do not have the same effect on small scale devices due to the dominance of viscous forces over inertia. In order to determine the dominating forces at small length scales the Reynolds number (Re) can be calculated. This is given by dividing the inertial forces by the viscous forces as shown in eqn (8.1), where 'a' is the dimension of the object, 'v' is its velocity, 'ρ' is the density of the liquid and 'μ' the liquid's viscosity.

$$\text{Re} = \frac{av\rho}{\mu} \tag{8.1}$$

Purcell[16] looked into how the size of an object affects the Reynolds number. For an average-sized human swimming in water the Reynolds number may be of the order of 10^4, whereas for a guppy (small fish) Purcell suggests it could be 10^2. However, for microorganisms such as bacteria and algae, which have a size of about one micron, the Reynolds number is of the order of 10^{-4} or 10^{-5}. For such organisms inertia is of no relevance. To illustrate, when looking at one of these organisms in water with a kinematic viscosity of 10^{-2} cm s^{-1} and an average speed of 30 μm s^{-1}, if it is necessary to push the organism in order for it to move, upon removal of the force the organism will travel around 0.1 Å before it stops. This makes it obvious that inertia plays a vanishing role at this scale and only forces that are exerted precisely at the time of movement determine any movement, while conventional propulsion systems using reciprocal motion fail to cause propulsion.[3,4] Despite this, there are various natural microorganisms that have evolved active propulsion systems and are able to move at the micro scale. These systems build on the principle of non-reversible motion. For example, in order to achieve propulsion *Spirillum volutans* (a bacterium) moves its body in a waving fashion and then a spiral wave flows down the tail, similar to a corkscrew twisting into a cork.[17] In another example *Escherichia coli*, a very well-known microbe, uses a flagellum to propel itself forward. The flagellum in simple terms works like a flexible oar.[18] In all these cases however, very complex processes take place which have taken millions of years of evolution to be created. There are some examples where researchers have attached synthetic flagella to cells and used external fields to generate motion,[19] however these cannot be defined as self-motile particles, as without the external fields motion will cease. Therefore in order to synthetically produce particles that can move on their own in a liquid at the micro- and nano-scale, a simpler approach has emerged. Either an interfacial gradient is generated over a particle *via* surface chemical reactions,[20] or particles are propelled *via* the production of gas bubbles, which are expelled from the surface of the object, again due to chemical reactions.[21] Self-motile particles such as these are often referred to as 'swimmers', micro-jets or micro-motors. In order to be called self-motile particles, they must have no need for any external forces or fields to power propulsion, but simply rely on chemical reactions/effects on their surface to act as the motor.[22] Presently there are two main mechanistic sources of motion for these micro-motors used in the current research field: firstly, phoretic phenomena (*e.g.* self-diffusiophoresis,[23,24] self-electrophoresis[25–27]) and secondly, bubble propulsion.[10,21,28]

8.1.1.1   Phoretic Mechanisms

Motion generated *via* a phoretic mechanism such as self-diffusiophoresis is based on the principle that a concentration gradient, for example of H_2O_2, is generated across a PS/Pt colloid. In the case of spherical Janus-particles, for example, the chemical reaction (eqn (8.2),[29] where the reaction rate limiting constraints are shown as k_1 and k_2), takes place on the platinum catalyst surface and produces more product molecules than reactants. The resulting asymmetric distribution of the reaction products across the entire colloid propels the particle,[30] in this case usually away from the platinum-coated side, see schematic in Figure 8.1. This effect is analogous to diffusiophoresis: the mechanism by which colloids show enhanced motion in response to an external solute concentration gradient.[31,32]

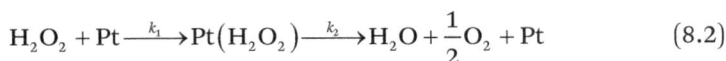

$$H_2O_2 + Pt \xrightarrow{k_1} Pt(H_2O_2) \xrightarrow{k_2} H_2O + \frac{1}{2}O_2 + Pt \qquad (8.2)$$

8.1.1.2   Bubble-propelled Motion

Another motion producing mechanism for larger catalytic swimming devices is bubble propulsion.[21] Active particles undergoing motion *via* bubble release can be primarily divided into two geometrical groups, tube-like particles and spherical particles. Bubble propelled devices can therefore be divided up into two general types; tube-like micro-swimmers and spherical micro-swimmers.

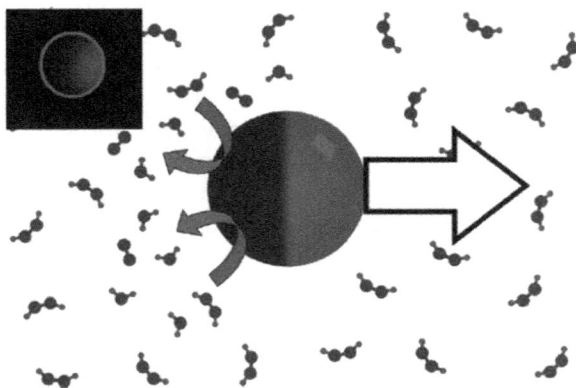

Figure 8.1   Schematic of a Janus-particle, the dark side represents the platinum coating and the lighter (right side) the uncoated part, which is also fluorescent. The molecules represent hydrogen peroxide, water and oxygen; hydrogen peroxide is asymmetrically decomposed by the Janus-particle. The concentration gradient of the fuel is thought to cause the propulsion as indicated by the arrow. Reprinted with permission from ref. 24. Copyright (2011) American Chemical Society.

8.1.1.2.1 Tube-like Micro-swimmers. Rolled-up nanotube swimmers based on the decomposition of hydrogen peroxide by platinum localised on the inside of the tube have been described in several papers.[8,28,33–40] Rolled-up nanotubes, or micro-tubes are also referred to as micro-jets or micro-motors. However, not all micro-tubes are based on platinum and hydrogen peroxide as their reaction mechanism. Thus, micro-tubes based on zinc and generating hydrogen bubbles in acidic media are described by Gao *et al.*,[41] but unlike Pt-based micro-tubes, these zinc-based ones use up zinc as part of the reaction to produce hydrogen gas. The mechanism by which tube-like micro-motors propel themselves, *i.e.* bubble propulsion, is described in detail by Li *et al.*[33] To date, it appears that all tube-like micro-motors that utilize reaction materials on the inside of the tubes need the addition of surfactants such as Triton X 100 or sodium dodecyl sulfate (SDS) to lower surface tension in order to allow bubble propulsion to take place.[42] This in turn leads to important challenges in the pursuit of biocompatibility.

8.1.1.2.2 Spherical Micro-motors. Spherical micro-motors can also produce propulsion based on bubble release. One such system is described by Gibbs *et al.*,[21] where particular detail was given to how the forces of bubble propulsion work on a spherical particle and how the surface tension of the bulk solution changes the propulsion. Surprisingly, there are very few examples of confirmed spherical, bubble-propelled micro-motors based on platinum as a catalyst; the vast majority of spherical micro-motors are based on chemical reactions that tend to use up the particle body, meaning these particles have a finite lifetime. The following are some examples of bubble swimmers in this category: Gao *et al.*[43] demonstrated a novel, multi-fuel driven swimmer, made from palladium and an aluminium core producing bubble-propelled motion in hydrogen peroxide and in strong acidic or basic solutions such as HCl and NaOH.[43] A different water-driven, spherical, bubble-propelled micro-motor is also described by Gao *et al.*[44] In this case the particle core consists of an aluminium-gallium (Al–Ga) alloy made *via* micro-contact mixing and coated on one side with titanium *via* electron-beam evaporation, as illustrated in Figure 8.2.

Other water-driven micro-motors described by Gao *et al.*[9] are bubble-propelled *via* magnesium. In this case magnesium reacts in water to form magnesium hydroxide and hydrogen gas. The possible use of these Janus swimmers in oil drop capture and transport, for applications such as cleaning up oil spills at sea is demonstrated in Figure 8.3. The gold layer can be modified with self-assembled monolayers (SAMs) of long chain alkanethiols to create strong surface hydrophobicity in order to collect oil droplets.

Li *et al.*[11] also produced water-driven photoreactive spherical bubble swimmers that were made of $TiO_2/Au/Mg$. These micro-motors create a highly reactive oxygen species that efficiently destroys the cell membranes of the anthrax simulant Bacillus globigii spores demonstrating another potential future application area of micro-motors.

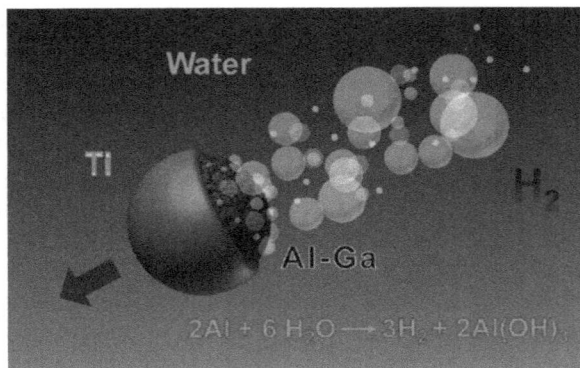

Figure 8.2   Schematic of water-driven hydrogen propelled Al–Ga/Ti micro-motor. The dark hemisphere (right side) is the Al–Ga alloy and the left hemisphere the Ti coating. Reprinted with permission from ref. 44. Copyright (2012) American Chemical Society.

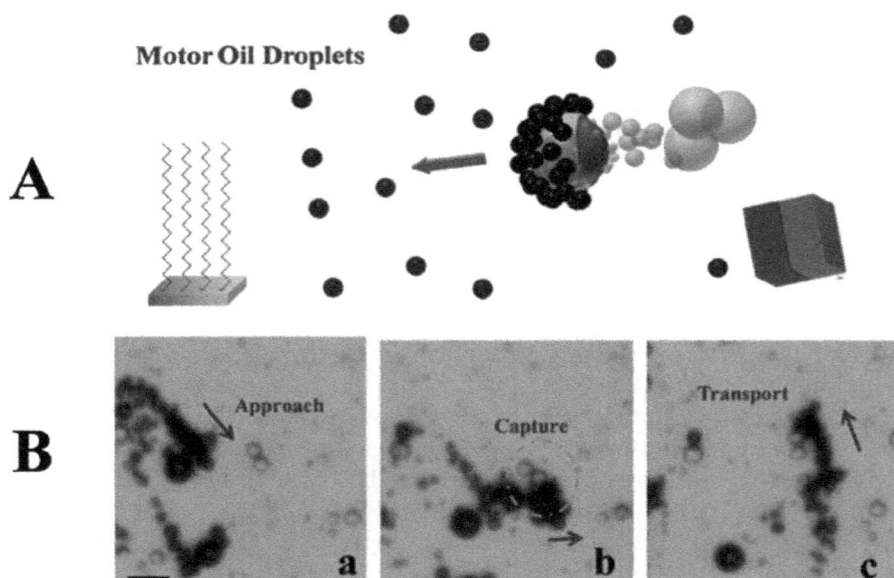

Figure 8.3   (A) Schematic of the seawater-driven Janus micro-motors capturing motor oil droplets. (B) Time lapse images taken from a movie (a) approach of the micro-motor to oil droplets (b) capture of droplets on motor (c) transport of captured oil droplets in seawater. Scale bar 50 μm. Reproduced from ref. 9 with permission from The Royal Society of Chemistry.

Finally, the Ebbens group has recently produced PS/Pt/Cr based spherical micro-motors that undergo bubble-propulsion and show how it is possible to alter trajectory behaviour according to different fabrication methods.[45] As shown in Figure 8.4 this entailed using different masking techniques to confine the catalytically active area of the swimmers *via* E-beam and sputter coating.

8.1.2 Current Issues Concerning the Uses of Micro-motors

Many micro-motors, to date, are based on materials such as polystyrene and silicon dioxide and use metals such as gold,[46] silver,[47] titanium[11,44] and platinum.[28] None of these materials are biocompatible; they are also costly. For example, platinum used as a catalytic material to power the motion has been shown to have strong biofouling issues, where proteins adsorb onto the metal surface inhibiting the catalytic reaction and therefore stopping the motion of these particles.[48]

Currently, the vast majority of autonomous, bubble-propelled micro-motors need to have surfactants added to the working medium in order to

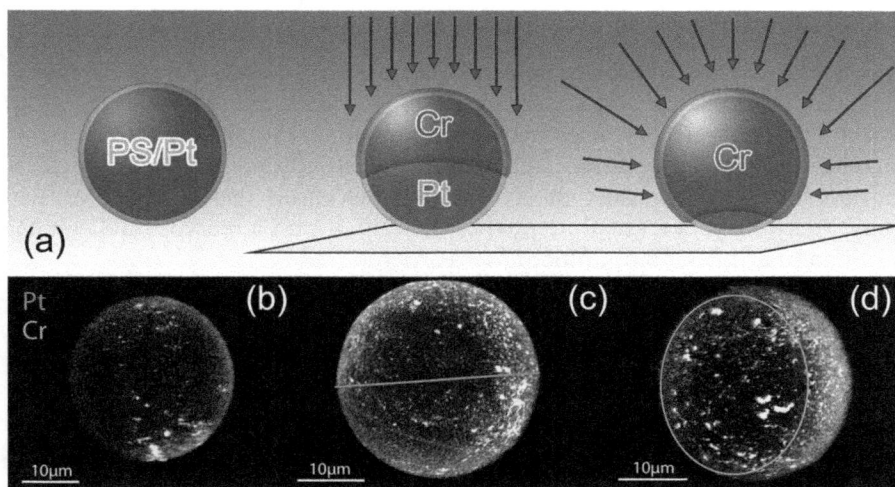

Figure 8.4 (a) Schematic for the three different bubble swimmer geometries targeted (left to right: symmetrical, Janus, and pore activity); arrows indicate directionality of metal deposition. E-beam shows strong directionality (middle), while sputter coating is less directional (right). Surface (below middle and right) indicates glass slide substrate. (b–d) Overlaid EDS images for Pt peak intensity (energy range of 1.93–2.17 eV) and Cr peak intensity (energy range of 5.24–5.58 eV) distribution for typical platinum-coated polystyrene colloids: (b) unmasked; (c) masked by Cr evaporation; (d) masked by Cr sputter coating. Lines indicate approximate boundary between the catalytically active and Cr masked regions. Reprinted with permission from ref. 45. Copyright (2015) American Chemical Society.

achieve any appreciable propulsion.[41,42,49-51] However, surfactants, *e.g.* SDS, also cause unwanted side effects, such as enzyme denaturation, which will render the systems useless for many applications.[52] Simple bubble-propulsive spherical devices can produce motion without surfactants in water,[45] however our experiments showed that motion in biological fluids does require surfactant. The role of surfactant in this case was found to be to reduce biofouling of the platinum catalyst and thereby maintain surface reactivity, rather than to reduce surface tension as had been previously suggested. Fouling of catalytic platinum in human serum illustrates the problem that this expensive catalyst is very far from optimal for most biological applications and therefore there is a clear impetus to move to more biologically based devices that use enzymes as chemically active species and away from metals such as platinum.

Finally, the ability to fabricate micro-motors which can be tuned for specific trajectory behaviour is critical for future challenges, but as shown in the last part of Section 8.2.1.2.2 the current fabrication methods such as evaporation make the generation of complex structures very hard to achieve and therefore the development of new devices is limited.

8.1.3 Immobilization of Enzymes

A variety of devices have been reported that use enzymes as motors to power the catalytic reactions and cause motion of particles. As the catalase (CAT) enzyme complex decomposes hydrogen peroxide[53] this facilitates comparison with the previously described Pt based micro-motors. However, it is clear that for future applications alternative enzymes could also be used to drive these micro-motors instead, removing the reliance on a reactive fuel, which it is incompatible with *e.g.* blood.

Immobilization methods can be categorized into three groups:[2] *adsorption*,[54] *covalent binding* (cross-linking)[55,56] and *incorporation*.[57] Adsorption is when the enzymes (proteins) adsorb onto the surface and hold an affinity to stay immobilised *via* forces such as van der Waals and hydrophobic forces.[54,58,59] Covalent binding means the chemical attachment of a part of the enzyme to the surface, or a molecule attached to the surface. Incorporation refers to enzyme entrapment or encapsulation, which is similar to the method used here during RIJ. Some examples of enzyme-powered propulsive devices have been reported in the literature. Orozco *et al.*[46] created enzyme-powered micro-jets using covalent attachment of the highly reactive enzyme catalase to the inner surface of a tube coated in gold. The covalent attachment method used for these micro-jets was a chemistry using self-assembled monolayers (SAM) of a mixture of 11-mercaptoundecanoic acid (MUA)/6-mercaptohexanol (MCH) onto the gold surface to which CAT was attached *via* carbodiimide chemistry (EDC stabilised with *N*-hydroxysuccinimide (NHS)).

An example of a multi-enzyme self-propelling ensemble is shown in Pantarotto *et al.*,[60] where the enzymes glucose oxidase (GOx) and CAT are both immobilised onto a multi-walled carbon nanotube (MWCNT), as shown in Figure 8.5. Here the two enzyme reactions are linked *via* hydrogen peroxide. Under the catalysis of GOx oxygen reacts with glucose to form gluconolactone and hydrogen peroxide while CAT catalyses the breakdown of hydrogen peroxide to form water and molecular oxygen. However, the amount of oxygen produced by CAT is not sufficient to keep the reaction with GOx going and to do this it is necessary to have a constant flow of oxygen over the sample.

CAT enzyme as a catalyst for hydrogen peroxide has also been mimicked by Vicario *et al.*,[61] where a manganese synthetic catalyst was immobilized covalently onto an object *via* a tether. The catalyst decomposed hydrogen peroxide in a similar way as CAT, into water and oxygen, causing movement of the object.

8.1.4 Why Use Silk as a Bio-ink?

As can be seen from the previously described examples many of these processes involving enzyme immobilisation are often costly, involve complex multistage chemical processes and are very inefficient. In addition, the surfaces needed for covalent immobilization are not always considered to be highly biocompatible and must also be particularly clean upon initial chemical activation. It is therefore desirable to immobilise enzymes in a simple, cheap yet biocompatible way, and where possible also with beneficial side effects such as enzyme stabilisation. One such material which shows great potential for this sort of application is silk.[62-64] It has already been shown by Zhang[65] that natural silk fibroin can be used as a support for enzyme immobilization and increased stability,[66-68] and there have been various studies into using silk for drug delivery challenges.[69-74]

As biocompatibility is essential for many prospective applications of self-motile micro-motors, the use of silk together with an enzyme as a propulsion motor is likely to be of great benefit and enzymes are by nature biocompatible. In addition to this, silk is already an FDA approved biomaterial

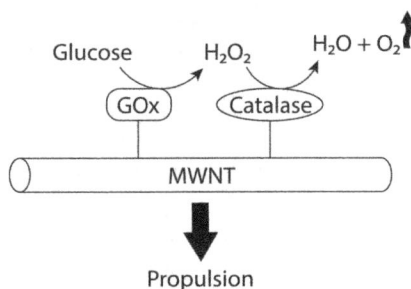

Figure 8.5 Biohybrid propulsion system. GOx and CAT immobilized onto a multi-walled carbon nanotube. Adapted from ref. 60 with permission from The Royal Society of Chemistry.

and has been used for many biomedical applications.[62] It is a versatile material due to its strong mechanical properties,[75] excellent biocompatibility,[76] adaptable biodegradability,[77] and easy processing.[78] It is for these reasons that silk from the silkworm (*B. mori*) has been tested as the base for generating particles *via* silk scaffolds.

Silk Fibroin (SF) has three different conformations, which are also known as polymorphs. Silk I is water soluble with a random coil conformation, Silk II is the state which consists of β-sheet secondary structure (spun silk state), and Silk III is an air/water assembled interfacial silk consisting of a helical structure.[62,79,80] Secreted silk (Silk II) is commonly processed into water soluble regenerated silk fibroin (RSF) (Silk I) to generate the silk ink. In order to generate 3D scaffold silk structures, as needed here, a further conversion stage is therefore required converting the printed material back into the water insoluble rigid scaffold. Exposing Silk I to chemicals such as methanol or potassium chloride, or heat, or shear stress converts it to a β-sheet secondary structure (Silk II). This phenomenon has been widely used to make silk scaffolds for different biomedical applications.[62]

In a recent publication Suntivich *et al.*[81] reported the printing of RSF for silk nest arrays for cell hosting. In this case the authors claim the shear force during printing was enough to cause conformation changes of the silk. This was done for only very few layers and no high 3D structures were required.

Our data shows that in order to produce 3D objects from silk the shear force due to the jetting of inkjet printing is not enough to produce an adequate β-sheet secondary structure.[82] Therefore RIJ was employed for the first time, which included chemical treatment with methanol, in order to ensure a rigid detachable silk scaffold was formed.

8.1.5 Advanced Fabrication of Micro-motors

Spherical micro-motors undergoing phoretic mechanisms have in general been fabricated by means of spin coating colloids onto a surface and then evaporating metals such as platinum onto one hemisphere, thus producing Janus particles (Figure 8.1).[30,83,84] Bimetallic nanorods on the other hand require porous templates and multistage electrochemical processes for their manufacture.[85] The synthesis/fabrication processes are in general not useful for mass production techniques.

There have been some recent examples utilising screen printing and micro-scale continuous optical printing (μCOP) in order to generate swimming devices. Two examples currently exist. In the first case Kumar *et al.*[86] used a layer-by-layer screen printing approach in order to generate large scale self-motile 'fish' that can 'swim' by bubble propulsion in hydrogen peroxide fuel. By using multiple stencils together with different inks they could generate 'fish' in the region of 1 cm in length with specific materials in different regions of the fish, *i.e.* a chitosan/Pt tail to drive the fish *via* bubble propulsion, an acrylic body and a mid-body made of carbon/nickel which allows for magnetic guidance of the fish *via* external magnetic fields (see Figure 8.6). This method however still requires the generation of micro-motors *via* a multiple stage process, so is not ideal for large scale manufacturing.

The second optical printing method by Zhu *et al.*,[87] achieves the fabrication of much smaller, multi-factionalized fish-shaped devices *via* μCOP using a UV light source which is focused in the correct locations *via* a Computer-Aided Design (CAD) software and a digital micro-mirror device (DMD), see Figure 8.7. For this method it is important to note that all inks need to contain UV curable precursors, which in this case are poly(ethylene glycol) diacrylate (PEGDA) based hydrogels. This allows for the incorporation of other functional particles such as platinum and magnetic nanoparticles during the curing process and thus by changing ink solutions above the substrate it is possible to generate multi-functional PEGDA based particles. This method is said to have a resolution of ~1 μm. For this method there is once again the need for a multiple stage process and inks are restricted to being UV curable.

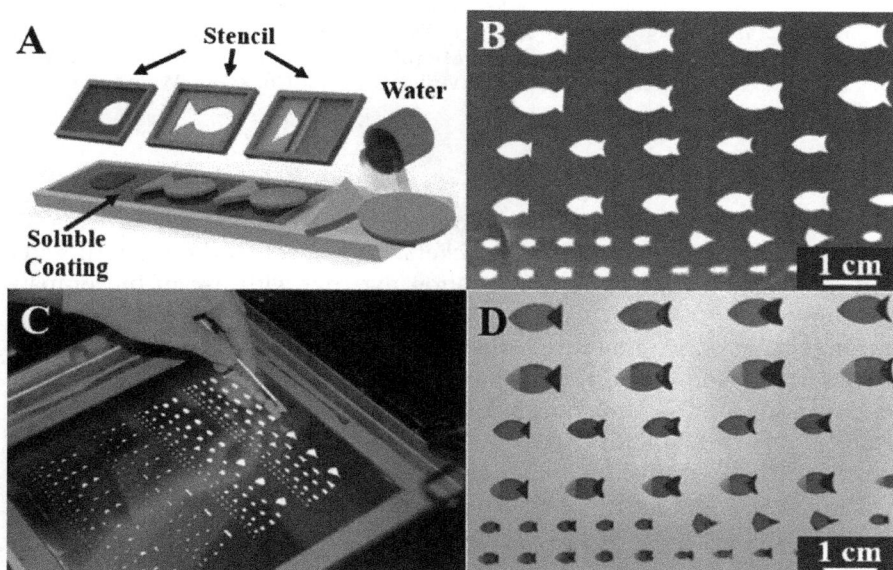

Figure 8.6 (A) Schematic illustration of the layer-by-layer screen-printing micro-fabrication of the synthetic catalytic fish: sequential printing of specific layers based on different modified inks for localizing different functionalities at specific sections of the printed fish. (B) Image of the stainless steel stencil containing the pre-cut design of the entire fish shape using different dimensions and shapes. (C) Image of mass fabrication procedure; coloured acrylic ink is physically applied onto the stencil above the pattern before a squeegee or doctor blade pushes the ink across the design. (D) Image of an array of mass-printed fish on a water-dissolvable coated substrate using the stencil shown in (B). Reproduced from ref. 86 with permission from The Royal Society of Chemistry.

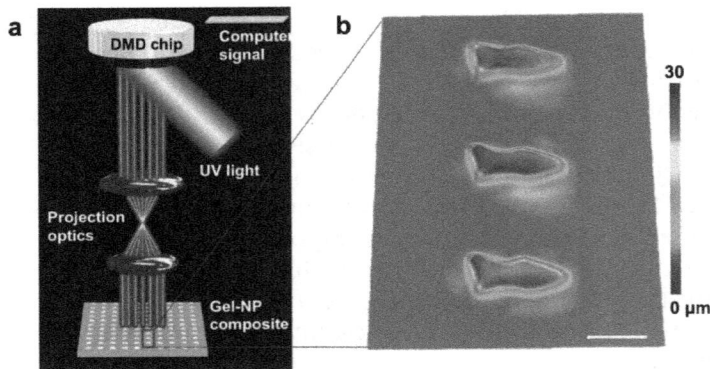

Figure 8.7   (a) Schematic illustration of the μCOP method to fabricate micro-fish. UV light illuminates the DMD mirrors, generating an optical pattern specified by the control computer. The pattern is projected through optics onto the photosensitive monomer solution to fabricate the fish layer-by-layer. (b) 3D microscopy image of an array of printed micro-fish. Scale bar, 100 μm. Reproduced with permission from W. Zhu, J. Li, Y. J. Leong, I. Rozen, X. Qu, R. Dong, Z. Wu, W. Gao, P. H. Chung and J. Wang, 3D-Printed Artificial Microfish, *Advanced Materials*,[87] John Wiley & Sons, © 2015 WILEY-VCH Verlag GmbH & Co. KGaA, Weinheim.

8.1.5.1   *Reactive Inkjet Printing*

Based on these examples we have developed a novel approach utilising RIJ of silk material to generate micro-rockets. This novel method incorporates both the capability of swimming motion together with use of biocompatible components. Inkjet printing overcomes the need for production in multiple stages and allows for the simple printing of various inks digitally-predefined by software. RIJ also shares the advantages of conventional inkjet printing to allow the straightforward manufacture of 3D objects with well controlled shape and size, for example by utilising a layer-by-layer approach.[88–90]

RIJ printing allows the generation of self-motile swimmers with digitally-defined compositions, structures and shapes without the need for multiple step methods to generate complex structures, which is the problem with many other methods currently used. This new method therefore allows for the rapid development and understanding of how different structures will affect the directionality of these particles. RIJ therefore enables highly efficient structures to be rapidly fabricated and investigated. Further to this, in the future additional functionalities can be simply added into silk-ink solutions and immobilized, as with CAT in the silk scaffolds *via* RIJ.[1] These additions could be other proteins or even components such as magnetic nanoparticles. This method of immobilizing enzymes *via* silk is much cheaper, easier and quicker than, for example some of the previously mentioned covalent immobilisation techniques on metal surfaces. This is because coupling agents use complicated chemistry and often cause high loss of enzyme activity as they, in general, attack several areas of an enzyme which can lead to enzyme denaturation during the immobilization process.[2,91,92] In contrast, in the RIJ process enzymes are immobilized *via* chemisorption and encapsulation into the silk scaffolds during the formation of the β-sheet structure of silk[93] from Silk I to Silk II.

At present, features smaller than 10 μm in size have been produced *via* inkjet printing,[94] which gives the chance of generating highly defined structures on the micron scale. One recently published paper by Tao *et al.*[93] has demonstrated some of the benefits of printing silk scaffolds doped with enzymes onto its surfaces to generate *e.g.* silk biosensors, which can show contamination levels on surfaces *via e.g.* colour change reactions. In Gregory *et al.*[45] it was demonstrated how the size of the exposed catalytically-active area on a masked spherical particle can affect the trajectories of bubble-propelled particles. This is a general feature for many swimming devices, *i.e.* that controlling the location of the catalyst determines performance, and so a key advantage for a print based manufacturing method is the potential ease with which catalyst location can be defined and modified.

The following part of this chapter describes how it is possible to use LBL inkjet printing of RSF ink, containing CAT enzyme molecules, to generate different silk micro-rockets that undergo different motion patterns dependent on their structure, as predefined by the printing process.

8.2 Production of Silk-based Enzyme-powered Micro-rockets

In the following section the process by which RIJ is used to create self-motile enzyme-powered silk micro-rockets is explained.[82]

8.2.1 Preparation of Silk Ink Solution

In order to prepare the silk ink for printing, *i.e.* producing RSF, there are four main steps that need to be followed:

- Degumming of silk
- Dissolution of silk fibroin fibres
- Dialysis
- Preparation of ink solution

8.2.1.1 Silk Degumming

In order to generate RSF, silk from *Bombyx mori is* degummed, in order to remove sericin, by briefly boiling the raw silk in 0.02 M sodium carbonate (Na_2CO_3). After degumming, the resulting silk materials are rinsed with deionized water until the solution turns clear and are then dried at 30 °C overnight in a drying oven, as shown representatively in Figure 8.8.

Degumming   Dissolving silk fibroin

| Raw silk | 0.02M Na₂CO₃ Boiling water | Degummed silk | 15%-20% of silk dissolved in Ajisava's reagent | Regenerated silk fibroin solution |

Figure 8.8   Preparation of RSF solution contains three main steps which are degumming, dissolving and dialysis. Reprinted from Zhang *et al.*[97] with permission of IS&T: The Society for Imaging Science and Technology, sole copyright owners of the *Printing for Fabrication 2016 32nd International Conference on Digital Printing Technologies (NIP) Technical Program, Abstracts, and USB Proceedings*.

8.2.1.2   Dissolution of Silk Fibroin Fibre

For the dissolution of silk fibroin fibre (SF) Ajisawa's reagent is used: $CaCl_2$/ethanol/water in a $1:2:8$ molar ratio: as suggested by Ajisawa *et al.*[95,96] Ajisawa's reagent is added to the degummed fibroin and stirred at 75 °C for 3 hours after which it is left to cool down to room temperature and dialysed against deionized water to remove the salts, see Figure 8.8. Finally any particles in the RSF are then removed by centrifugation.

8.2.1.3   Preparation of Ink Solution

A stock solution of highly concentrated enzyme, in our case catalase, is made in deionised water, which is then filtered with a 0.7 μm glass filter to remove any particles that might block the nozzle during printing. The desired concentration of RSF and catalase are then mixed together by gentle inversion of the sample. It is important at this stage that the solution does not undergo any high temperatures or shear forces otherwise the RSF solution will form a gel. If strong salts are used in the enzyme solution this may also lead to the final ink turning to gel in a short time and can therefore make printing a challenge.

8.2.2   Inkjet Printing Process of Silk Micro-rockets

As previously described in Section 8.1.4, silk can undergo different polymorphic states, RSF, which is Silk I is water soluble and can therefore be used as ink, which is printable. Different ink solutions containing RSF as the scaffold material are used to print 3D enzyme-driven self-motile particles. Different types of silk micro-rocket designs were made by Gregory *et al.*,[82] where the main designs were aimed at rockets being either fully active all through the

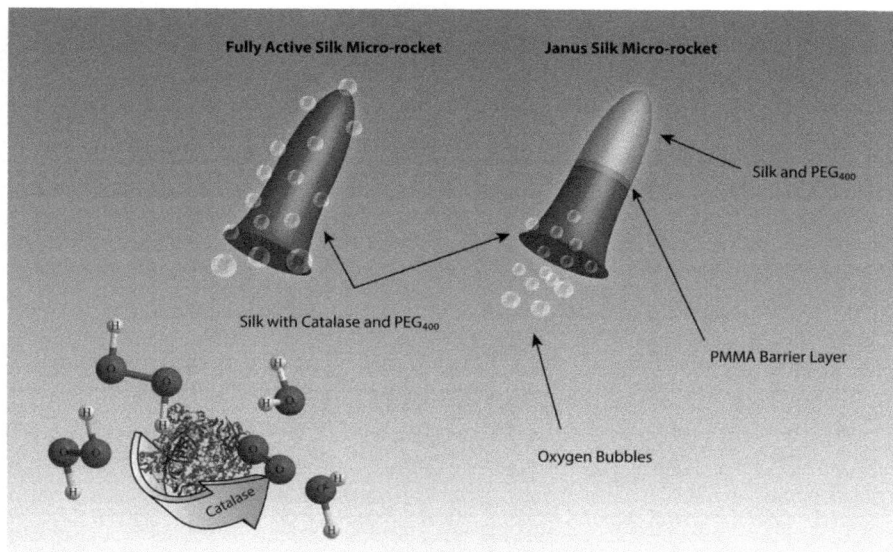

Figure 8.9   Schematic showing the two designed silk-micro-rockets, (left) fully active rockets, (right) Janus rocket.

particle or Janus, where the bottom half contained encapsulated catalase enzymes, as shown in the schematic in Figure 8.9.

The silk scaffold turns into a solid β-sheet structure (Silk II) by printing a drop of methanol *via* a second print head and switching between these two print heads for every layer, as shown in Figure 8.10. It is well known that organic solvents such as methanol are able to denature enzymes causing them to stop functioning.[98] The drops of methanol printed on top of the silk layer containing the enzyme are very small (droplet diameter ~80 μm resulting in ~3000 pL), and the evaporation and beta sheet formation of the RSF appear to happen before the methanol can denature any significant amount of enzyme. The example schematic shown in Figure 8.10 is a representation of Janus silk micro-rockets containing a barrier layer of PMMA in the middle. This barrier layer is explained later. In order to make fully active rockets, as depicted by the left rocket in Figure 8.9, the stages 2 to 4 are left out and simply more layers of stage 1 are printed.

In order to print the silk micro-rockets depicted in Figure 8.9 it is therefore necessary to use four different kinds of ink, for which in the case of the micro-rockets described here nozzles with a pore diameter of 60 μm were used. The inks used were made up of: RSF solution blended with polyethylene glycol (PEG$_{400}$), SRF blended with catalase and PEG$_{400}$ and pure methanol. In the case of Janus particles a barrier layer of 10 layers of poly-methyl methacrylate (PMMA) in Dimethylformamide (DMF) is required. This is necessary in order to decrease the amount of bubbles leaking from the active half to the inactive half.

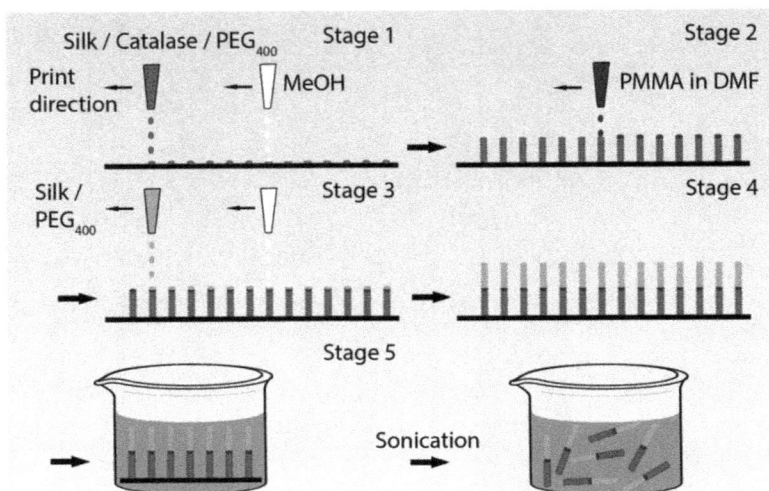

Figure 8.10   Schematic representing the layer-by-layer printing procedure of silk scaffolds with (dots indicate enzyme) and without catalase enzyme molecules, the darker colour represents the PMMA barrier layer: generation of Janus silk micro-rockets. Reprinted under CC BY 4.0 (https://creativecommons.org/licenses/by/4.0/) from D. A. Gregory, Y. Zhang, P. J. Smith, X. Zhao and S. J. Ebbens, Reactive Inkjet Printing of Biocompatible Enzyme Powered Silk Micro-rockets, *Small*, 2016, **12**, 4048–4055, DOI: 10.1002/smll.20160092.[82] © 2016 The Authors. Published by WILEY-VCH Verlag GmbH & Co. KGaA, Weinheim.

Optimum results can be obtained when printing LBL by choosing jetting parameters that generate droplets that do not contain satellites; example droplets are shown for different concentrations of RSF in Figure 8.11. Droplet formation is improved with higher RSF concentration.

8.2.3   Optimisation of Silk Micro-rockets and Printing Process

It may be advantageous during LBL printing to add brilliant blue FCF $(C_{37}H_{34}N_2Na_2O_9S_3)$ into the methanol to accurately establish the locations where methanol is printed. During the RIJ process the silk scaffold will turn blue, as can be seen in Figure 8.12 where a silk rocket has been dyed blue. This method enables the user to make sure methanol reaches the silk material and reacts with it to form the scaffold.

8.2.3.1   Influence of PEG$_{400}$

Printing of RSF ink containing enzyme molecules requires frequent cleaning of the print head as enzyme accumulation occurs. On the other hand, for inks containing an additional component, PEG$_{400}$, this is less of an issue allowing for longer printing times before cleaning of the nozzle is necessary.

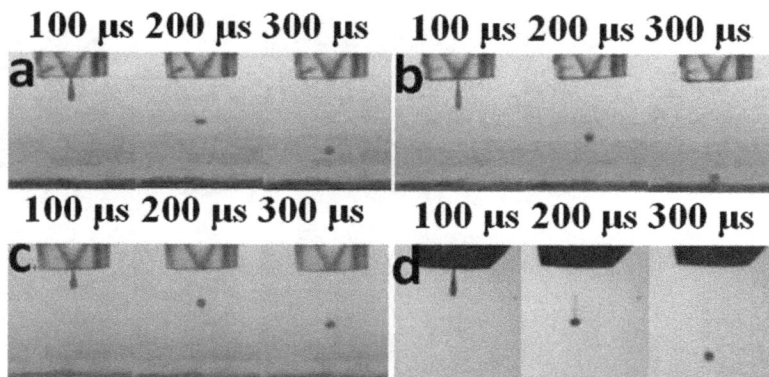

Figure 8.11 CCD camera images, depicting silk droplets ejected from 60 μm Micro-fab nozzles, showing the formation of a single droplet from different concentrations of RSF solutions, (a), 10 mg ml^{-1}, (b), 20 mg ml^{-1}, (c), 30 mg ml^{-1} and (d), 40 mg ml^{-1} respectively. Reprinted from Zhang *et al.*[97] with permission of IS&T: The Society for Imaging Science and Technology, sole copyright owners of the *Printing for Fabrication 2016 32nd International Conference on Digital Printing Technologies (NIP) Technical Program, Abstracts, and USB Proceedings*.

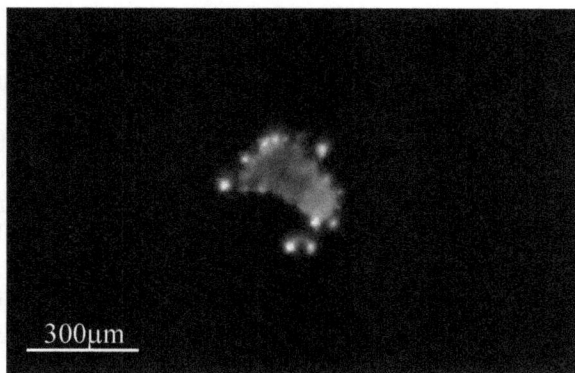

Figure 8.12 Fully active RIJ printed silk-micro-rocket dyed with brilliant blue.

The addition of PEG$_{400}$ within the silk scaffold also has a large impact on the bubble release from the silk micro-rockets. For micro-rockets that do not contain PEG$_{400}$ as shown in Figure 8.13 left, a large bubble can be seen forming on the surface of the rocket. This only pops when very large, meaning that rockets are stopped from moving freely, but rather rotate around the forming bubble. If, however, PEG is blended into the ink then, as shown in Figure 8.13 (right), bubbles are rapidly released, allowing for the rapid bubble-propelled motion of the rocket. Altering the concentration of PEG within the rocket influences the bubble release directly.

In a further development of the effect PEG has on the bubble release on silk rockets experimental data shows that if rockets are made up of two different

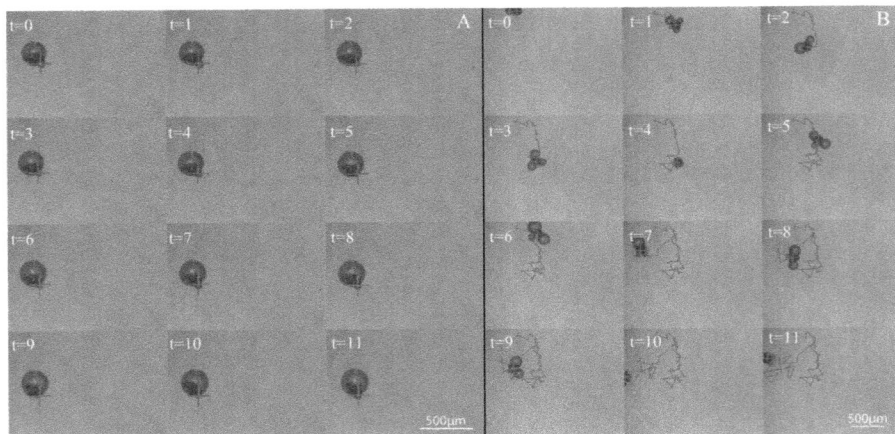

Figure 8.13   Silk swimmer containing catalase enzymes but no blended in PEG_{400} (A) showing the bubble detachment/popping issue—image is taken from above so looking down onto the particles, (B) silk swimmer containing PEG_{400}. Reprinted from Gregory *et al.*[99] with permission of IS&T: The Society for Imaging Science and Technology, sole copyright owners of *the Printing for Fabrication 2016 32nd International Conference on Digital Printing Technologies (NIP) Technical Program, Abstracts, and USB Proceedings.*

halves then it appears that a higher concentration of PEG in the inactive half is of benefit to discourage bubbles becoming attached to the inactive half. This might be due to silk being hydrophobic and PEG altering the hydrophobicity of the silk rocket, as was described by Gregory *et al.*[99] Contact angle measurements reported by Gregory *et al.*[99] show that with increasing PEG concentration the silk surface becomes more hydrophilic which would agree with the previous observations. This means that PEG allows tuning of the bubble release of micro-rockets and has the additional beneficial effect of lowering the enzyme accumulation on the nozzle.

8.2.4   The Silk Rocket Barrier Layer

As the silk scaffold is a porous structure it is possible for the fuel solution to permeate through the entire rocket meaning the total surface area where the enzyme can catalyse the reaction is very large. This however also means that bubbles are produced throughout the rocket and need to escape from within the rocket. For Janus rockets, which are made up of one half of silk scaffold that does not contain any enzyme, it is therefore possible for bubbles to migrate through to the inactive half of the rockets, which leads to bubbles detaching from the inactive half. This in turn alters the trajectory behaviour of the rockets. If it is desired to simply have bubbles detaching from the rockets on one half, a barrier layer of a mere 5 to 10 layers of PMMA is sufficient to stop oxygen gas leaking into the inactive half of the

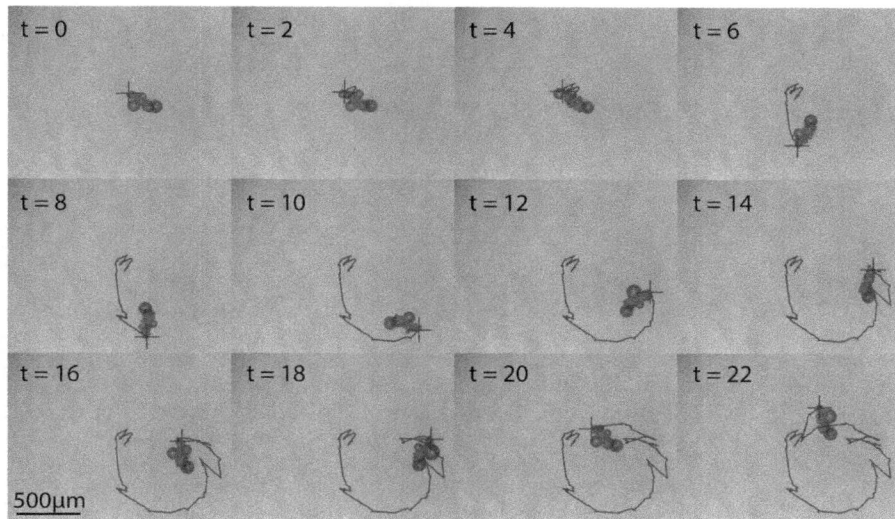

Figure 8.14 Example silk Janus micro-rocket without PMMA barrier layer, bubbles appear all over the rocket and trajectory is affected accordingly.

rocket and thus bubbles are simply released on the active side. The leaking of bubbles to the inactive half of the silk particles can be seen in Figure 8.14. There may be other inks that are better suited for this type of barrier layer, such as possibly wax, and this will need further research in the near future.

8.3   Characterisation of Micro-motors

When producing printed silk-micro-rockets it can be assumed that for future mass production of these particles *via* RIJ it is necessary to look at how the different inks can affect different parameters of the rockets, therefore it is important to investigate the printed structures before these particles are suspended in fuel solutions.

As every micro-rocket starts off with one droplet printed on a surface such as a silicon wafer one might want to see how the RSF concentration might affect the droplet diameter and Z number. The following optical profiler microscope images (Contour GT-K, USA) shown in Figure 8.15 display the morphology of (A) 10, (B) 20, (C) 30 and (D) 40 mg ml^{-1} RSF solution printed on Silicon wafer substrates. The Z values of these RSF solutions are 40.9, 32.1, 25.3 and 20.3 respectively. It is worth noting that the Z values of all the inks are above 14, which means that they require extra pressure in order to form stable single droplets.[100] As the RSF concentration goes up the diameter of droplets decreases down to a value of ~125 μm for 40 mg ml^{-1} RSF, see Figure 8.15.[82] This means that higher concentration is desirable if rockets are supposed to be narrower.

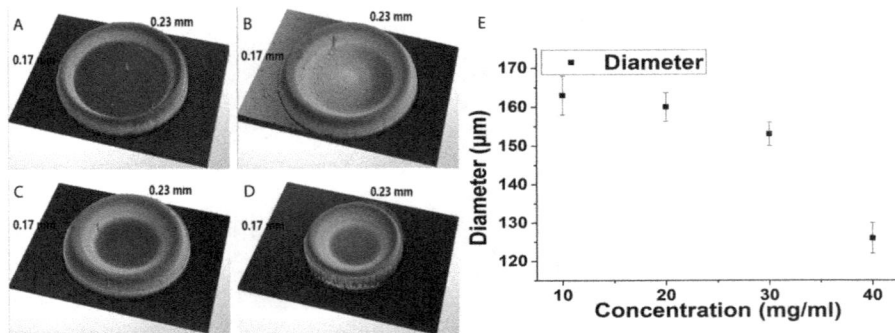

Figure 8.15 Optical profiler microscope images show different concentrations of RSF solution printed dots. The concentrations are (A), 10 mg ml^{-1}, (B), 20 mg ml^{-1}, (C), 30 mg ml^{-1} and (D), 40 mg ml^{-1} respectively. (E) Printed droplet diameter on Si-wafer substrates *versus* silk ink concentrations. Reprinted from Zhang *et al.*[97] with permission of IS&T: The Society for Imaging Science and Technology, sole copyright owners of the *Printing for Fabrication 2016 32nd International Conference on Digital Printing Technologies (NIP) Technical Program, Abstracts, and USB Proceedings*.

8.3.1 How Does Layer Thickness Affect Average Column Height

When designing silk-rockets consisting of different inks, such as the Janus ones presented here, it is important to understand how the additional enzyme molecules might affect the average layer height. This is important in order to compensate for any significant changes in layer thickness such as to print more or less layers. In order to investigate this issue, samples of varying height can be printed onto silicon wafers and then measured in the optical profiler microscope. In the case of RSF, this has been done with and without catalase enzyme and average heights measured for a varying number of layers. The analysis suggests that the average height of layers printed containing catalase enzyme is slightly lower than those without enzyme. As predicted there is a clear relationship between height and number of layers printed as shown in Figure 8.16.

8.3.2 Final Structures

Lastly, one may ask oneself how the final printed rockets may look under the microscope. For experimental results it is obviously important to be able to test the printing for reproducibility and how comparable rockets may be when comparing fully active silk micro-rockets, which contain enzyme throughout their structure, to Janus ones. The SEM images below show the initial arrays of printed columns for both fully active (Figure 8.17 (left)) and Janus (Figure 8.18 (left)) silk particles on silicon wafer surfaces. As can be seen the particles all appear very similar in height and diameter. The enlarged views of both the side of the columns (Figure 8.17 (right)) and the top Figure 8.18 (right) show

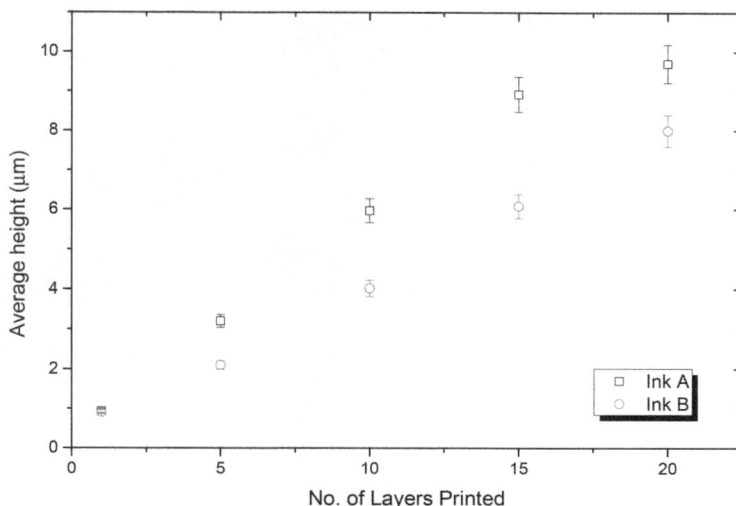

Figure 8.16 Comparison of average height measurements of silk printed columns as used for the active particles ink A (30 mg ml^{-1} silk, 12 mg ml^{-1} PEG$_{400}$), ink B (30 mg ml^{-1} silk, 4 mg ml^{-1} catalase, 10 mg ml^{-1} PEG$_{400}$). Reprinted under CC BY 4.0 (https://creativecommons.org/licenses/by/4.0/) from D. A. Gregory, Y. Zhang, P. J. Smith, X. Zhao and S. J. Ebbens, Reactive Inkjet Printing of Biocompatible Enzyme Powered Silk Micro-rockets, *Small*, 2016, **12**, 4048–4055, DOI: 10.1002/smll.20160092.[82] © 2016 The Authors. Published by WILEY-VCH Verlag GmbH & Co. KGaA, Weinheim.

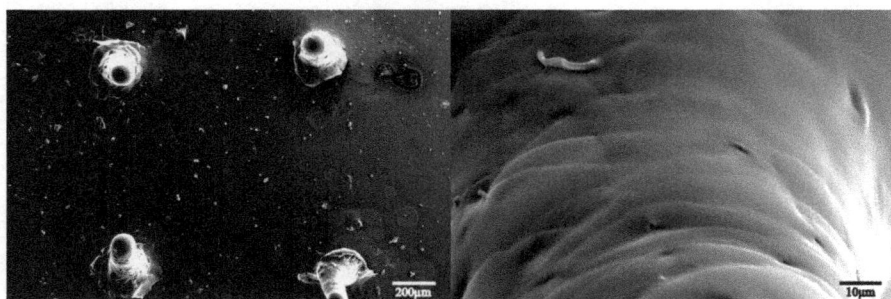

Figure 8.17 Two SEM (secondary electron) images at 15 KeV; (left) array of fully active silk particles made up of 500 layers containing catalase enzyme (4 mg ml^{-1}), RSF (30 mg ml^{-1}) and PEG$_{400}$; (right) enlarged view of the side of a column.

that at this resolution there were no noticeable pores, but rather the surfaces of the particles were smooth and the diameter of the columns near the top measured ~100 μm.

Closer comparison of fully active particles and Janus particles shows that despite the difference in the distribution of enzyme, the overall size and shape remain comparable, as shown representatively by two fully active particles in

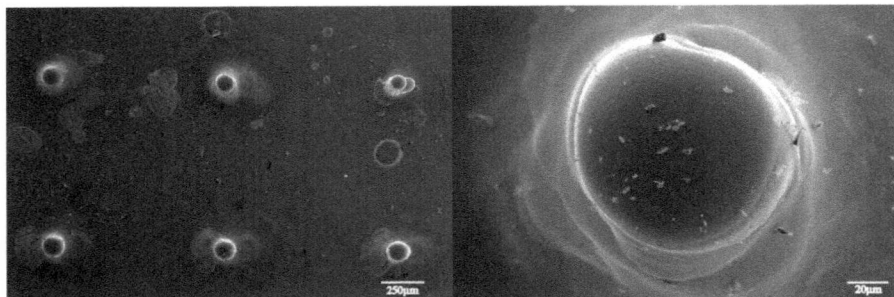

Figure 8.18   Two SEM (secondary electron) images at 10 KeV; (left) array of Janus printed particles viewed from above consisting of 10 layers of PMMA as a barrier layer; (right) close up image of the top surface of a column of a printed silk swimmer, the diameter of the column is around 100 μm. Adapted with permission under CC BY 4.0 (https://creativecommons.org/licenses/by/4.0/) from D. A. Gregory, Y. Zhang, P. J. Smith, X. Zhao and S. J. Ebbens, Reactive Inkjet Printing of Biocompatible Enzyme Powered Silk Micro-rockets, *Small*, 2016, **12**, 4048–4055, DOI: 10.1002/smll.20160092.[82] © 2016 The Authors. Published by WILEY-VCH Verlag GmbH & Co. KGaA, Weinheim.

Figure 8.19   Two secondary electron images (at 12 KeV) of fully active silk swimmer particles containing CAT (4 mg ml^{-1}), Silk (30 mg ml^{-1}) and PEG$_{400}$ (10 mg ml^{-1}). Adapted with permission under CC BY 4.0 (https://creativecommons.org/licenses/by/4.0/) from D. A. Gregory, Y. Zhang, P. J. Smith, X. Zhao and S. J. Ebbens, Reactive Inkjet Printing of Biocompatible Enzyme Powered Silk Micro-rockets, *Small*, 2016, **12**, 4048–4055, DOI: 10.1002/smll.20160092.[82] © 2016 The Authors. Published by WILEY-VCH Verlag GmbH & Co. KGaA, Weinheim.

Figure 8.19 and two Janus particles in Figure 8.20 (which also contain 5 layers of PMMA as a barrier layer between the active and inactive half). The images show that during the printing process a little material sometimes flows over the sides down the particles; *i.e.* during LBL printing of the silk-inks some ink can run along the outer side of the forming rockets. This suggests that printing first the ink containing the enzyme catalyst could result in better defined Janus structures than first printing the inactive rocket half, as some of the enzyme containing silk-ink might flow over the inactive side.

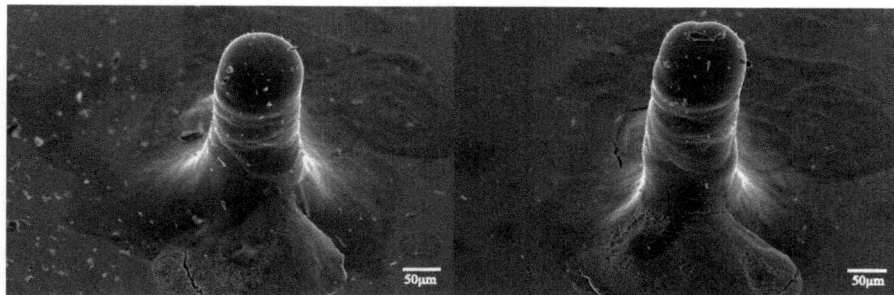

Figure 8.20   Two SEM (secondary electron) images at 14 KeV of representative Janus silk swimmer particles containing CAT (4 mg ml^{-1}), Silk (30 mg ml^{-1}) and PEG$_{400}$ (10 mg ml^{-1}) with 10 layers of PMMA barrier and an inactive part containing silk (30 mg ml^{-1}) and PEG$_{400}$ (12 mg ml^{-1}). Adapted with permission under CC BY 4.0 (https://creativecommons.org/licenses/by/4.0/) from D. A. Gregory, Y. Zhang, P. J. Smith, X. Zhao and S. J. Ebbens, Reactive Inkjet Printing of Biocompatible Enzyme Powered Silk Micro-rockets, *Small*, 2016, **12**, 4048–4055, DOI: 10.1002/smll.20160092.[82] © 2016 The Authors. Published by WILEY-VCH Verlag GmbH & Co. KGaA, Weinheim.

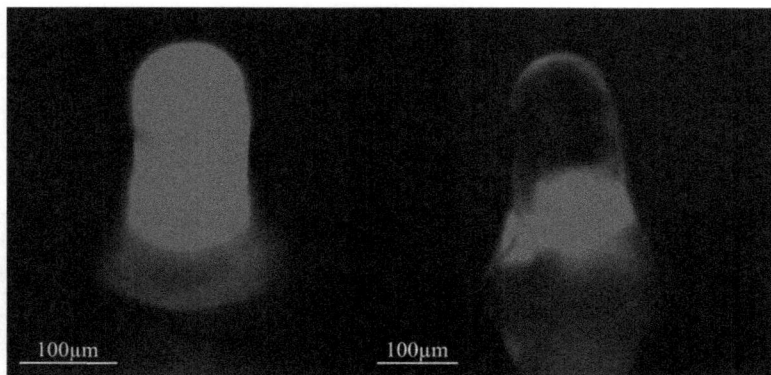

Figure 8.21   Fluorescent microscopy images of FITC labelled catalase (brighter region) in silk rockets, (left) fully active, (right) Janus. Reprinted under CC BY 4.0 (https://creativecommons.org/licenses/by/4.0/) from D. A. Gregory, Y. Zhang, P. J. Smith, X. Zhao and S. J. Ebbens, Reactive Inkjet Printing of Biocompatible Enzyme Powered Silk Micro-rockets, *Small*, 2016, **12**, 4048–4055, DOI: 10.1002/smll.20160092.[82] © 2016 The Authors. Published by WILEY-VCH Verlag GmbH & Co. KGaA, Weinheim.

In order to visualise the localisation of the catalase enzyme in fully active and Janus rockets the enzyme can be labelled with fluorescein isothiocyanate (FITC) as shown in Figure 8.21, where the location of the enzyme within the rockets fluoresces. The fully active rocket clearly contains catalase all over and the Janus rocket only contains catalase in the bottom half of the particle, see Figure 8.21. Finally, both SEM and fluorescence microscope images show that the printed columns are slightly wider towards the bottom than at the top, giving rise to a rocket-like shape. The difference in size between top and bottom of the particle is relatively small, of the order of ~20 μm (maximum), but this ultimately gives the particles more of a rocket structure than a simple rod/column shape.

8.4 Analysing the Trajectory Behaviour of Symmetrical and Janus Silk Micro-rockets

In order to investigate what effects result from the two previously described catalytic distributions within the silk micro-rockets, the particles need to be detached from their silicon wafer substrates. This is usually done by immersing the sample in water and then giving the sample a quick sonication burst. This ensures the safe detachment of the rockets from the substrate without damaging the structure.

After detachment the rockets are then placed into a petri dish containing an aqueous sample solution which also contains the fuel, in this case hydrogen peroxide at concentrations of 1% to 10%. Representative images taken over time of fully active and Janus micro-rockets in 5% hydrogen peroxide are shown in Figures 8.22 and 8.23 respectively. Clearly there is a strong difference between the trajectory behaviours of these two different devices. In the following section this behaviour will be more closely analysed.

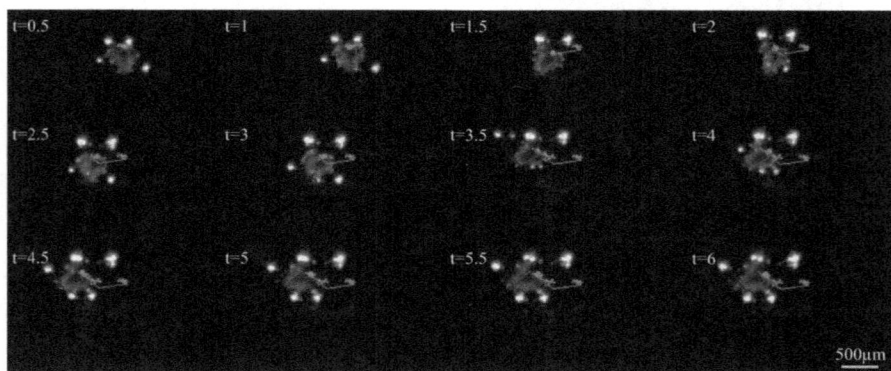

Figure 8.22   Example image captures of a fully active silk rocket swimming in 5% H$_2$O$_2$ solution – dark line indicates top and light line bottom of the rockets. Reprinted under CC BY 4.0 (https://creativecommons.org/licenses/by/4.0/) from D. A. Gregory, Y. Zhang, P. J. Smith, X. Zhao and S. J. Ebbens, Reactive Inkjet Printing of Biocompatible Enzyme Powered Silk Micro-rockets, *Small*, 2016, **12**, 4048–4055, DOI: 10.1002/smll.20160092.[82] © 2016 The Authors. Published by WILEY-VCH Verlag GmbH & Co. KGaA, Weinheim.

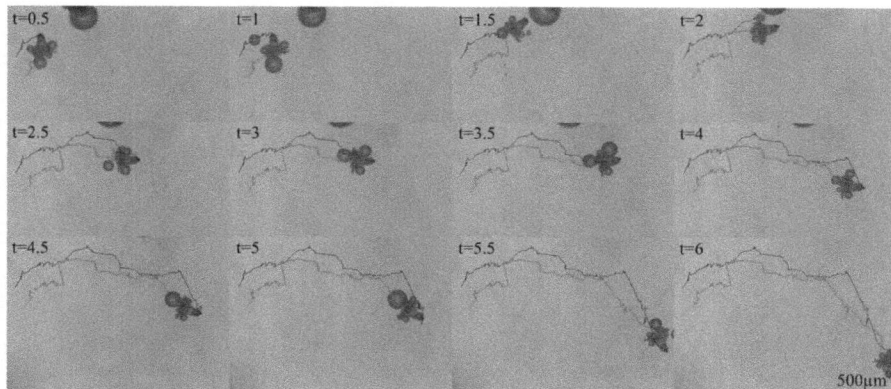

Figure 8.23 Example image captures of a Janus silk rocket swimming in 5% H_2O_2 solution – dark line indicates top and lighter line bottom of the rockets. Reprinted under CC BY 4.0 (https://creativecommons.org/licenses/by/4.0/) from D. A. Gregory, Y. Zhang, P. J. Smith, X. Zhao and S. J. Ebbens, Reactive Inkjet Printing of Biocompatible Enzyme Powered Silk Micro-rockets, *Small*, 2016, **12**, 4048–4055, DOI: 10.1002/smll.20160092.[82] © 2016 The Authors. Published by WILEY-VCH Verlag GmbH & Co. KGaA, Weinheim.

8.4.1 Directionality Analysis – Alignment of Particle to its Direction of Motion

When investigating the trajectories of micro-motors it is important to first look qualitatively at the raw trajectories, which are plotted in Figure 8.24. Fully active micro-rockets show trajectories that had more turns and twists in them, whereas Janus micro-rocket trajectories on the whole are more directional resembling straighter lines.

Gregory *et al.*[82] describes in detail how to correlate the orientation of these micro-rockets *versus* their trajectory direction where it is reported that the correlation between these two angles is $r^2 = 0.003 \pm 0.017$ for symmetrically active rockets but $r^2 = 0.656 \pm 0.023$ for Janus rockets. This correlation analysis emphasises that Janus rockets tend to travel on more linear paths.

Two other data analysis techniques employed are instantaneous velocity measurements and persistence length measurements. Persistence length is a form of measurement where the linearity of the trajectory is investigated and given a real dimension, it is a technique often used for calculating the stiffness or bending properties of a polymer chain.

Based on the calculated centre of mass trajectory from the dual tracked micro-rockets (see Figures 8.22 and 8.23), trajectories were analysed for their mean instantaneous velocities and their persistence lengths. Instantaneous velocities were calculated for every frame and averaged over the whole trajectory. The results are shown in Table 8.1, and it can be seen that despite the lower amount of catalase present in the Janus micro-rockets, on average these rockets moved ~1.4 times faster than fully active ones. A decrease in velocity of ~1/3rd

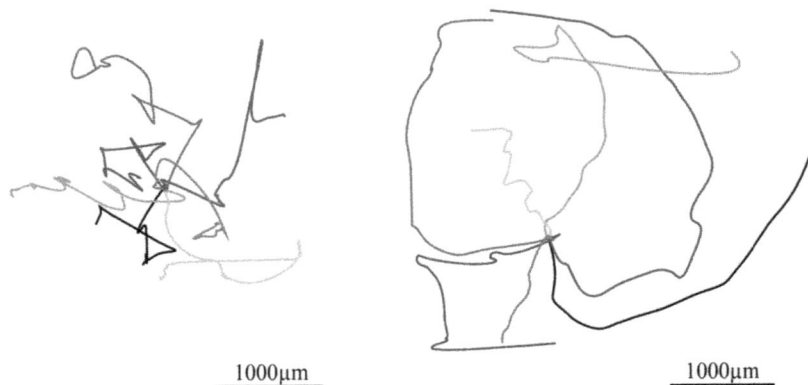

1000µm   1000µm

Figure 8.24   Comparison of raw trajectories, (left) fully active particles over a period of ~10 seconds, (right) Janus particles over a period of ~5–10 seconds. Reprinted under CC BY 4.0 (https://creativecommons.org/licenses/by/4.0/) from D. A. Gregory, Y. Zhang, P. J. Smith, X. Zhao and S. J. Ebbens, Reactive Inkjet Printing of Biocompatible Enzyme Powered Silk Micro-rockets, *Small*, 2016, **12**, 4048–4055, DOI: 10.1002/smll.20160092.[82] © 2016 The Authors. Published by WILEY-VCH Verlag GmbH & Co. KGaA, Weinheim.

Table 8.1   Velocity and persistence length data for silk swimmers. Reprinted under CC BY 4.0 (https://creativecommons.org/licenses/by/4.0/) from D. A. Gregory, Y. Zhang, P. J. Smith, X. Zhao and S. J. Ebbens, Reactive Inkjet Printing of Biocompatible Enzyme Powered Silk Micro-rockets, *Small*, 2016, **12**, 4048–4055, DOI: 10.1002/smll.20160092.[82] © 2016 The Authors. Published by WILEY-VCH Verlag GmbH & Co. KGaA, Weinheim.

Type of Micro-motor	Average velocity [μm s^{-1}]	Persistence length [μm]	L_P/L_C [arb. units]
Fully active rocket	370 ± 31	26 ± 6	0.014 ± 0.002
Janus rocket	511 ± 93	423 ± 184	0.191 ± 0.063
Fully active rocket In 2% serum	282 ± 13	40 ± 4	NA
Janus rocket In 2% serum	338 ± 17	135 ± 14	NA

over the course of 1 hour was observed during the experiment for both fully active and Janus rockets. This velocity decrease is comparable to the decrease in the amount of bubbles observed to be released from the micro-rockets over this time course (comparing the initial and final bubble releases). The velocities of silk-micro-rockets seen here in the region of 370–500 µm s^{-1} are comparable to the velocities reported for other bubble- propelled micro-jets.[101,102]

Persistence length analysis (shown in Table 8.1), of the centre of mass trajectories fitted in with the orientation correlation analysis and confirms that printed Janus silk micro-rockets had on average a ~ 16 times higher persistence length than fully active ones (persistence length of Janus micro-rockets divided by fully active micro-rockets). Because the trajectories of Janus and fully active particles were of different lengths, the persistence length data was also corrected for the trajectory length (longer length on average for fully active particles) by dividing the persistence length by the total track length. The relative persistence length calculated this way also agreed with the previous results showing $a \sim 13$ times higher relative persistence length of the Janus micro-rockets compared to the fully active micro-rockets.

8.5 Biocompatibility of Enzyme-powered Micro-motors—Ability to Swim in Biological Solutions

As was discussed earlier it is highly important to be able to produce micro-swimmers that function in biological solutions such as human serum, PBS and other complex media. The silk micro-motors described here were tested in 2% human serum samples containing 3% w/v hydrogen peroxide as a fuel source. Both symmetrically active (Figure 8.25) and Janus micro-rockets showed continuous motion. More directionality can be objectively seen for Janus rockets in 2% human serum, Figure 8.26 corresponding to the previously reported increase in persistence length (Table 8.1). In view of the amount

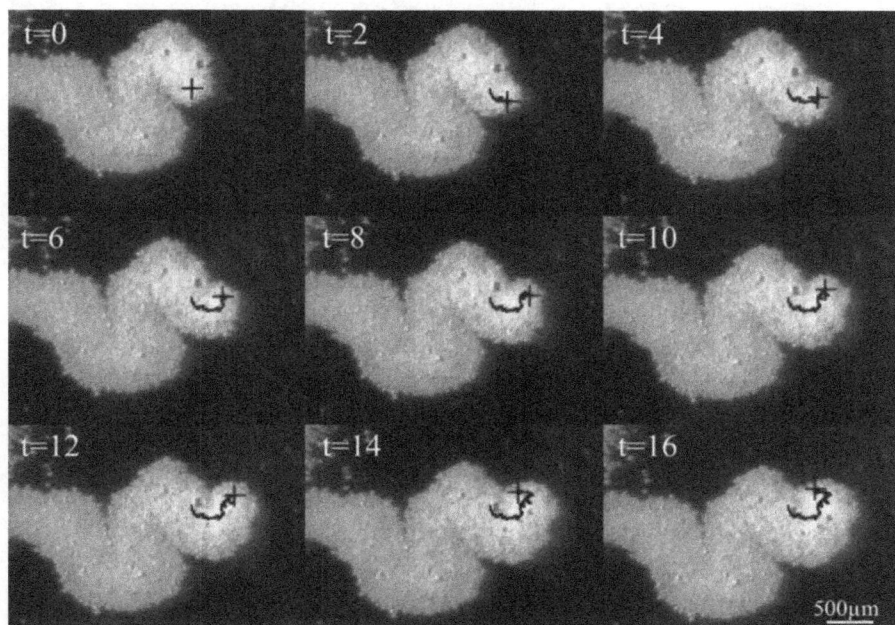

Figure 8.25 Example of a fully active silk swimmer swimming in 2% serum containing 3% w/v hydrogen peroxide as a fuel source. Reprinted under CC BY 4.0 (https://creativecommons.org/licenses/by/4.0/) from D. A. Gregory, Y. Zhang, P. J. Smith, X. Zhao and S. J. Ebbens, Reactive Inkjet Printing of Biocompatible Enzyme Powered Silk Micro-rockets, *Small*, 2016, **12**, 4048–4055, DOI: 10.1002/smll.20160092.[82] © 2016 The Authors. Published by WILEY-VCH Verlag GmbH & Co. KGaA, Weinheim.

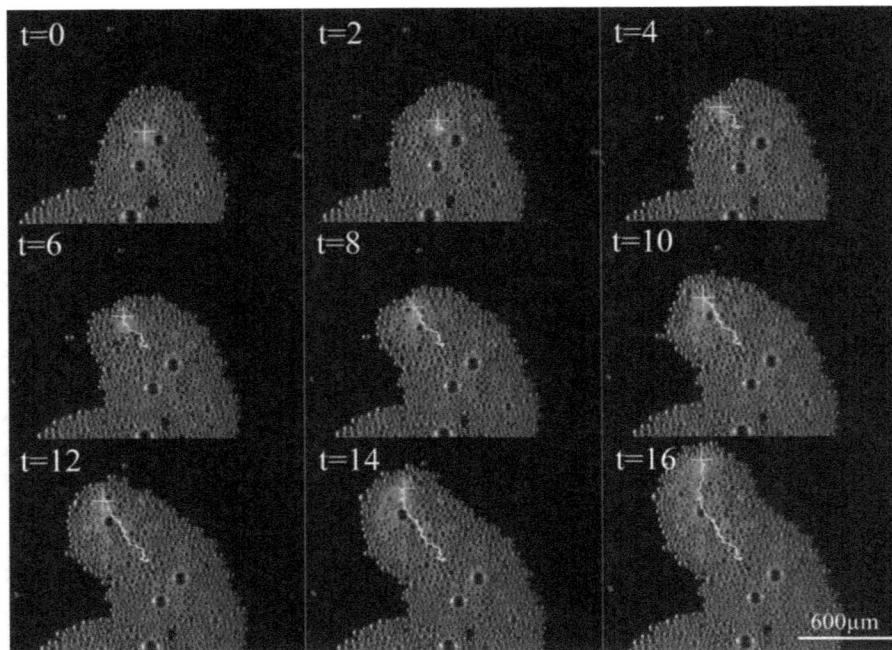

Figure 8.26   Janus silk rocket swimming in 2% Human Serum with 3% H$_2$O$_2$.
Reprinted under CC BY 4.0 (https://creativecommons.org/licenses/
by/4.0/) from D. A. Gregory, Y. Zhang, P. J. Smith, X. Zhao and S. J.
Ebbens, Reactive Inkjet Printing of Biocompatible Enzyme Powered Silk Micro-rockets, *Small*, 2016, **12**, 4048–4055, DOI: 10.1002/
smll.20160092.[82] © 2016 The Authors. Published by WILEY-VCH Verlag GmbH & Co. KGaA, Weinheim.

of bubbles and velocity of the rockets over the course of 30 minutes there
was no indication that the reaction rate decreased by any substantial amount.
This therefore strongly suggests that the activity barely decreased throughout
the experiment, but quantitative measurements were not possible. No biofouling of the catalyst was detected, furthermore it is very important to note
that propulsion was present despite no addition of surfactants to the bulk
solution. This is indeed a vast improvement over many of the other self-motile
bubble-propelled swimmers previously reported in publications.[11,44,86,87]

8.6   Lifetime of Enzyme Incorporated in Silk Structure *Versus* Free Enzyme

Finally, it is important to look at the stability of catalase enzyme within the silk
scaffold structure. It is well known that enzymes will denature if exposed to
high or low pH extremes, therefore the silk-micro-rockets were tested in high
and low pH media. The pH of 5% hydrogen peroxide can be measured to be
pH ~ 4.2. This pH is not very suitable for long-term enzyme activity. Activity

measurements of free catalase enzyme in 5% H_2O_2 show 90% enzyme activity is lost within just 10 minutes. In contrast the motion of silk micro-rockets only drops by $1/3^{rd}$ of the initial velocity after one hour of exposure in 5% H_2O_2. A possible explanation for this is that the loss in activity observed is due to enzymes on the outermost side of the rockets being denatured over time while enzyme molecules within the inner part of the silk scaffold stay stabilized despite the acidic pH. Similar results are also reported at high pH values. These results indicate that the use of silk may allow prolongation of enzyme lifetime and extension of the possible environments in which they can be used.

8.7   Conclusions

In conclusion, it has become clear that enzyme immobilization within silk scaffolds to generate self-motile particles *via* RIJ is a convenient manufacturing process without the need for complicated covalent coupling chemistry or multistage printing processes. The immobilization of catalase enzymes in RIJ layer-by-layer printed silk scaffolds is very easy to achieve without any major loss of enzyme activity. The generated structures are porous enough to allow fuel solutions to ingress into the micro-rockets and therefore induce bubble-propelled motion of the printed micro-rockets, as has been demonstrated using catalase as the motor molecule for hydrogen peroxide fuel solutions. With the addition of PEG_{400} into silk inks it is possible to help tune the release of bubbles from the silk micro-rockets by altering the hydrophobicity of the silk surface. As the silk rocket scaffold structure is porous Janus silk micro-rockets benefit from having a barrier layer of PMMA printed between the active and inactive halves in order to discourage oxygen bubbles produced within the rocket from diffusing into the inactive half and altering desired trajectory paths.

Trajectory analysis using persistence length and comparison of particle orientation and direction show that trajectories are altered dependent on the printed structure and position of the enzyme molecules. This means that it is possible to exert a considerable degree of control over the trajectories *via* the digitally defined printed structure of these micro-rockets. Thus for fully active micro-rockets the trajectories are of random diffusive behaviour, whereas for Janus printed micro-rockets, trajectories mainly follow straight lines, which greatly increases the directionality and overall velocity of the micro-rockets. This is the first step towards producing higher efficiency of micro-rockets as the Janus rockets have a higher velocity than their symmetrical counterparts. Therefore it is possible to conclude that RIJ appears to offer a chance to rapidly design, manufacture and test a wide range of swimming devices with different catalyst distributions, shapes and sizes.

Additional advantages arising from the use of RSF with enzymes entrapped in the scaffold structure are that the enzymes are stabilized and keep their activity longer in otherwise unsuitable pH environments (*e.g.* pH ~ 4), where free enzyme molecules survive only briefly. Thus, silk swimmers are shown to swim for time periods of over 60 minutes with a loss of only around $1/3^{rd}$ of their initial velocity whereas in comparison, free catalase does not survive longer than 15 minutes in the same conditions (90% activity loss).

Finally it has been shown that silk printed micro-rockets show excellent biocompatibility: they do not need any addition of surfactant to the bulk solution they swim in, and further show good swimming capabilities in biological solutions, such as 1–10% human serum without any issues of biofouling such as seen with platinum-based micro-motors.[48]

References

1. P. J. Smith and A. Morrin, *J. Mater. Chem.*, 2012, **22**, 10965–10970.
2. R. A. Sheldon, *Adv. Synth. Catal.*, 2007, **349**, 1289–1307.
3. M. Leoni, J. Kotar, B. Bassetti, P. Cicuta and M. C. Lagomarsino, *Soft Matter*, 2009, **5**, 472–476.
4. A. Najafi and R. Golestanian, *J. Phys.: Condens. Matter*, 2005, **17**, S1203–S1208.
5. S. Duhr and D. Braun, *Proc. Natl. Acad. Sci. U. S. A.*, 2006, **103**, 19678–19682.
6. F. Julicher and J. Prost, *Eur. Phys. J. E*, 2009, **29**, 27–36.
7. S. Ebbens, D. A. Gregory, G. Dunderdale, J. R. Howse, Y. Ibrahim, T. B. Liverpool and R. Golestanian, *EPL (Europhys. Lett.)*, 2014, **106**, 58003.
8. L. Soler and S. Sanchez, *Nanoscale*, 2014, **6**, 7175–7182.
9. W. Gao, X. Feng, A. Pei, Y. Gu, J. Li and J. Wang, *Nanoscale*, 2013, **5**, 4696–4700.
10. L. Soler, V. Magdanz, V. M. Fomin, S. Sanchez and O. G. Schmidt, *ACS Nano*, 2013, **7**, 9611–9620.
11. J. Li, V. V. Singh, S. Sattayasamitsathit, J. Orozco, K. Kaufmann, R. Dong, W. Gao, B. Jurado-Sanchez, Y. Fedorak and J. Wang, *ACS Nano*, 2014, **8**, 11118–11125.
12. J. Orozco, G. Cheng, D. Vilela, S. Sattayasamitsathit, R. Vazquez-Duhalt, G. Valdes-Ramirez, O. S. Pak, A. Escarpa, C. Kan and J. Wang, *Angew. Chem., Int. Ed.*, 2013, **52**, 13276–13279.
13. L. Baraban, D. Makarov, R. Streubel, I. Mönch, D. Grimm, S. Sanchez and O. G. Schmidt, *ACS Nano*, 2012, **6**, 3383–3389.
14. Z. Ghalanbor, S. A. Marashi and B. Ranjbar, *Med. Hypotheses*, 2005, **65**, 198–199.
15. A. Einstein, *Ann. Phys.*, 1906, **324**, 371–381.
16. E. M. Purcell, *Am. J. Phys.*, 1977, **45**, 3–11.
17. K. F. Jarrell and M. J. McBride, *Nat. Rev. Microbiol.*, 2008, **6**, 466–476.
18. E. M. Purcell, *Proc. Natl. Acad. Sci. U. S. A.*, 1997, **94**, 11307–11311.
19. R. Dreyfus, J. Baudry, M. L. Roper, M. Fermigier, H. A. Stone and J. Bibette, *Nature*, 2005, **437**, 862–865.
20. S. J. Wang and N. Wu, *Langmuir*, 2014, **30**, 3477–3486.
21. J. G. Gibbs and Y. P. Zhao, *Appl. Phys. Lett.*, 2009, **94**, 163104.
22. R. Kapral, *J. Chem. Phys.*, 2013, **138**, 020901.
23. J. de Graaf, G. Rempfer and C. Holm, *IEEE Trans. Nanobioscience*, 2015, **14**, 272–288.
24. S. J. Ebbens and J. R. Howse, *Langmuir*, 2011, **27**, 12293–12296.

25. Y. Wang, R. M. Hernandez, D. J. Bartlett Jr, J. M. Bingham, T. R. Kline, A. Sen and T. E. Mallouk, *Langmuir*, 2006, **22**, 10451–10456.

26. P. M. Wheat, N. A. Marine, J. L. Moran and J. D. Posner, *Langmuir*, 2010, **26**, 13052–13055.

27. J. L. Moran, P. M. Wheat and J. D. Posner, *Phys. Rev. E*, 2010, **81**, 065302.

28. V. M. Fomin, M. Hippler, V. Magdanz, L. Soler, S. Sanchez and O. G. Schmidt, *IEEE Trans. Rob.*, 2014, **30**, 40–48.

29. S. Ebbens, M. H. Tu, J. R. Howse and R. Golestanian, *Phys. Rev. E*, 2012, **85**, 020401.

30. J. R. Howse, R. A. L. Jones, A. J. Ryan, T. Gough, R. Vafabakhsh and R. Golestanian, *Phys. Rev. Lett.*, 2007, **99**, 048102.

31. J. Palacci, B. Abecassis, C. Cottin-Bizonne, C. Ybert and L. Bocquet, *Phys. Rev. Lett.*, 2010, **104**, 138302.

32. J. P. Ebel, J. L. Anderson and D. C. Prieve, *Langmuir*, 1988, **4**, 396–406.

33. L. Li, J. Wang, T. Li, W. Song and G. Zhang, *Soft Matter*, 2014, **10**, 7511–7518.

34. V. Magdanz, G. Stoychev, L. Ionov, S. Sanchez and O. G. Schmidt, *Angew. Chem., Int. Ed.*, 2014, **53**, 2673–2677.

35. N. I. Kovtyukhova, *J. Phys. Chem. C*, 2008, **112**, 6049–6056.

36. A. A. Solovev, Y. Mei, E. B. Urena, G. Huang and O. G. Schmidt, *Small*, 2009, **5**, 1688–1692.

37. L. Restrepo-Perez, L. Soler, C. Martinez-Cisneros, S. Sanchez and O. G. Schmidt, *Lab Chip*, 2014, **14**, 2914–2917.

38. L. Soler, C. Martinez-Cisneros, A. Swiersy, S. Sanchez and O. G. Schmidt, *Lab Chip*, 2013, **13**, 4299–4303.

39. I. S. M. Khalil, V. Magdanz, S. Sanchez, O. G. Schmidt and S. Misra, *PLoS One*, 2014, **9**, e83053.

40. L. K. E. A. Abdelmohsen, F. Peng, Y. Tu and D. A. Wilson, *J. Mater. Chem. B*, 2014, **2**, 2395–2408.

41. W. Gao, A. Uygun and J. Wang, *J. Am. Chem. Soc.*, 2012, **134**, 897–900.

42. H. Wang, G. Zhao and M. Pumera, *J. Phys. Chem. C*, 2014, **118**, 5268–5274.

43. W. Gao, M. D'Agostino, V. Garcia-Gradilla, J. Orozco and J. Wang, *Small*, 2013, **9**, 467–471.

44. W. Gao, A. Pei and J. Wang, *ACS Nano*, 2012, **6**, 8432–8438.

45. D. A. Gregory, A. I. Campbell and S. J. Ebbens, *J. Phys. Chem. C*, 2015, **119**, 15339–15348.

46. J. Orozco, V. Garcia-Gradilla, M. D'Agostino, W. Gao, A. Cortes and J. Wang, *ACS Nano*, 2013, **7**, 818–824.

47. H. Wang, G. J. Zhao and M. Pumera, *J. Am. Chem. Soc.*, 2014, **136**, 2719–2722.

48. D. P. Manica, Y. Mitsumori and A. G. Ewing, *Anal. Chem.*, 2003, **75**, 4572–4577.

49. X. Wang, B.-T. Lee and A. Son, *Appl. Microbiol. Biotechnol.*, 2014, **98**, 8719–8728.

50. W. Gao, S. Sattayasamitsathit, J. Orozco and J. Wang, *J. Am. Chem. Soc.*, 2011, **133**, 11862–11864.

51. F. Kuralay, S. Sattayasamitsathit, W. Gao, A. Uygun, A. Katzenberg and J. Wang, *J. Am. Chem. Soc.*, 2012, **134**, 15217–15220.
52. S. Ghosh, S. Chakrabarty, D. Bhowmik, G. S. Kumar and N. Chattopadhyay, *J. Phys. Chem. B*, 2015, **119**, 2090–2102.
53. P. Nicholls, *Arch. Biochem. Biophys.*, 2012, **525**, 95–101.
54. S. A. Costa and R. L. Reis, *J. Mater. Sci.: Mater. Med.*, 2004, **15**, 335–342.
55. M. Koneracká, P. Kopčansky, M. Antalík, M. Timko, C. N. Ramchand, D. Lobo, R. V. Mehta and R. V. Upadhyay, *J. Magn. Magn. Mater.*, 1999, **201**, 427–430.
56. M. H. Liao and D. H. Chen, *Biotechnol. Lett.*, 2001, **23**, 1723–1727.
57. A. Prabhune and H. Sivaraman, *Appl. Biochem. Biotechnol.*, 1991, **30**, 265–272.
58. Y. Ren, J. G. Rivera, L. He, H. Kulkarni, D.-K. Lee and P. B. Messersmith, *BMC Biotechnol.*, 2011, **11**, 63.
59. Q. A. Feng, X. Xia, A. F. Wei, X. Q. Wang, Q. F. Wei, D. Y. Huo and A. J. Wei, *J. Appl. Polym. Sci.*, 2011, **120**, 3291–3296.
60. D. Pantarotto, W. R. Browne and B. L. Feringa, *Chem. Commun.*, 2008, 1533–1535.
61. J. Vicario, R. Eelkema, W. R. Browne, A. Meetsma, R. M. La Crois and B. L. Feringa, *Chem. Commun.*, 2005, 3936–3938.
62. C. Vepari and D. L. Kaplan, *Prog. Polym. Sci.*, 2007, **32**, 991–1007.
63. B. Kundu, N. E. Kurland, V. K. Yadavalli and S. C. Kundu, *Int. J. Biol. Macromol.*, 2014, **70**, 70–77.
64. L. S. Wray, X. Hu, J. Gallego, I. Georgakoudi, F. G. Omenetto, D. Schmidt and D. L. Kaplan, *J. Biomed. Mater. Res., Part B*, 2011, **99B**, 89–101.
65. Y. Q. Zhang, *Biotechnol. Adv.*, 1998, **16**, 961–971.
66. S. Z. Lu, X. Q. Wang, Q. Lu, X. Hu, N. Uppal, F. G. Omenetto and D. L. Kaplan, *Biomacromolecules*, 2009, **10**, 1032–1042.
67. P. Wang, C. Qi, Y. Yu, J. Yuan, L. Cui, G. Tang, Q. Wang and X. Fan, *Appl. Biochem. Biotechnol.*, 2015, **177**, 472–485.
68. E. M. Pritchard, P. B. Dennis, F. Omenetto, R. R. Naik and D. L. Kaplan, *Biopolymers*, 2012, **97**, 479–498.
69. K. Numata and D. L. Kaplan, *Adv. Drug Delivery Rev.*, 2010, **62**, 1497–1508.
70. X. Wang, T. Yucel, Q. Lu, X. Hu and D. L. Kaplan, *Biomaterials*, 2010, **31**, 1025–1035.
71. X. Wang, E. Wenk, A. Matsumoto, L. Meinel, C. Li and D. L. Kaplan, *J. Controlled Release*, 2007, **117**, 360–370.
72. E. Wenk, A. J. Wandrey, H. P. Merkle and L. Meinel, *J. Controlled Release*, 2008, **132**, 26–34.
73. S. Hofmann, C. T. Wong Po Foo, F. Rossetti, M. Textor, G. Vunjak-Novakovic, D. L. Kaplan, H. P. Merkle and L. Meinel, *J. Controlled Release*, 2006, **111**, 219–227.
74. E. Wenk, H. P. Merkle and L. Meinel, *J. Controlled Release*, 2011, **150**, 128–141.
75. B. B. Mandal, A. Grinberg, E. Seok Gil, B. Panilaitis and D. L. Kaplan, *Proc. Natl. Acad. Sci. U. S. A.*, 2012, **109**, 7699–7704.
76. B. Kundu, R. Rajkhowa, S. C. Kundu and X. Wang, *Adv. Drug Delivery Rev.*, 2013, **65**, 457–470.

77. H. J. Jin, J. Park, V. Karageorgiou, U. J. Kim, R. Valluzzi and D. L. Kaplan, *Adv. Funct. Mater.*, 2005, **15**, 1241–1247.
78. D. N. Rockwood, R. C. Preda, T. Yucel, X. Wang, M. L. Lovett and D. L. Kaplan, *Nat. Protoc.*, 2011, **6**, 1612–1631.
79. H. J. Jin and D. L. Kaplan, *Nature*, 2003, **424**, 1057–1061.
80. A. Motta, L. Fambri and C. Migliaresi, *Macromol. Chem. Phys.*, 2002, **203**, 1658–1665.
81. R. Suntivich, I. Drachuk, R. Calabrese, D. L. Kaplan and V. V. Tsukruk, *Biomacromolecules*, 2014, **15**, 1428–1435.
82. D. A. Gregory, Y. Zhang, P. J. Smith, S. J. Ebbens and X. Zhao, *Small*, 2016, **12**, 4048–4055.
83. A. I. Campbell and S. J. Ebbens, *Langmuir*, 2013, **29**, 14066–14073.
84. G. Dunderdale, S. Ebbens, P. Fairclough and J. Howse, *Langmuir*, 2012, **28**, 10997–11006.
85. S. J. Ebbens, *Curr. Opin. Colloid Interface Sci.*, 2015, **21**, 14–23.
86. R. Kumar, M. Kiristi, F. Soto, J. Li, V. V. Singh and J. Wang, *RSC Adv.*, 2015, **5**, 78986–78993.
87. W. Zhu, J. Li, Y. J. Leong, I. Rozen, X. Qu, R. Dong, Z. Wu, W. Gao, P. H. Chung, J. Wang and S. Chen, *Adv. Mater.*, 2015, **27**, 4411–4417.
88. X. Y. Zhang and Y. D. Zhang, *Cell Biochem. Biophys.*, 2015, **72**, 777–782.
89. A. M. J. van den Berg, P. J. Smith, J. Perelaer, W. Schrof, S. Koltzenburg and U. S. Schubert, *Soft Matter*, 2007, **3**, 238–243.
90. A. M. J. van den Berg, A. W. M. de laat, P. J. Smith, J. Perelaer and U. S. Schubert, *J. Mater. Chem.*, 2007, **17**, 677–683.
91. C. Mateo, J. M. Palomo, G. Fernandez-Lorente, J. M. Guisan and R. Fernandez-Lafuente, *Enzyme Microb. Technol.*, 2007, **40**, 1451–1463.
92. H. D. Chirra, T. Sexton, D. Biswal, L. B. Hersh and J. Z. Hilt, *Acta Biomater.*, 2011, **7**, 2865–2872.
93. H. Tao, B. Marelli, M. Yang, B. An, M. S. Onses, J. A. Rogers, D. L. Kaplan and F. G. Omenetto, *Adv. Mater.*, 2015, **27**, 4273–4279.
94. C. E. Hendriks, P. J. Smith, J. Perelaer, A. M. J. Van den Berg and U. S. Schubert, *Adv. Funct. Mater.*, 2008, **18**, 1031–1038.
95. H. Yamada, H. Nakao, Y. Takasu and K. Tsubouchi, *Mater. Sci. Eng., C: Biomimetic Supramol. Syst.*, 2001, **14**, 41–46.
96. A. Ajisawa, *J. Seric. Sci. Jpn.*, 1998, **67**, 91–94.
97. Y. Zhang, D. A. Gregory, P. J. Smith and X. Zhao, *presented in Part at the Printing for Fabrication (NIP)*, Manchester, 2016.
98. D. L. Nelson and M. M. Cox, *Lehninger Principles of Biochemistry 3rd Edn*, Worth, New York, 3rd edn, 2000.
99. D. A. Gregory, Y. Zhang, P. J. Smith, S. J. Ebbens and X. Zhao, *presented in Part at the Printing for Fabrication (NIP)*, Manchester, 2016.
100. Y. Liu, M. Tsai, Y. Pai and W. Hwang, *Appl. Phys. A*, 2013, **111**(2), 509–516.
101. W. Gao, S. Sattayasamitsathit, J. Orozco and J. Wang, *Nanoscale*, 2013, **5**, 8909–8914.
102. S. Sanchez, A. N. Ananth, V. M. Fomin, M. Viehrig and O. G. Schmidt, *J. Am. Chem. Soc.*, 2011, **133**, 14860–14863.

CHAPTER 9

Reactive Inkjet Printing for Additive Manufacturing

YINFENG HE, ALEKSANDRA FOERSTER, BELEN BEGINES, FAN ZHANG, RICKY WILDMAN, RICHARD HAGUE, PHILL DICKENS AND CHRISTOPHER TUCK*

University of Nottingham, UK
*E-mail: Christopher.Tuck@nottingham.ac.uk

9.1 Introduction

Chemistry is at the backbone of many Additive Manufacturing (AM), colloquially known as 3D Printing, systems and concepts. The conversion from one state to another is a basic premise of AM, wherein the layer by layer nature of the systems requires, in general, that material feedstocks are in a 'flowable' state prior to conversion into a continuous solid structure. Using the ASTM F42 definitions of AM technologies, this is especially true for both the Vat Polymerisation and Material Jetting methods which have traditionally relied on photo-curing techniques to produce solid components from liquid precursors. This chapter seeks to inform the reader on these materials and provide an insight into how chemistry with particular focus on reactive inkjet (RIJ) printing is being manipulated to increase the material palette for AM.

Smart Materials No. 32
Reactive Inkjet Printing: A Chemical Synthesis Tool
Edited by Patrick J. Smith and Aoife Morrin
© The Royal Society of Chemistry 2018
Published by the Royal Society of Chemistry, www.rsc.org

9.1.1 Photo-polymerisation in AM

Photo-polymerisation is one of the chemical reactions which has been widely used in Material Jetting and Vat Polymerisation processes. A phase change is induced in either an inkjet printed ink or vat of photo-polymer converting the material from liquid into solid through the use of electromagnetic radiation (*e.g.* Ultra-violet (UV) light) triggered polymerisation to form solid two- and three-dimensional (2D and 3D) structures.

Photo-polymerisation has been well established in the jetting and coating industry for a number of decades.[1-3] In recent years, this reaction has been applied in AM to produce freeform 3D objects through Material Jetting. Commercial AM machines supplied by 3D Systems and Stratasys are based on photo-polymerisation.[4] Both companies provide different inks to produce specimens with very different properties (*e.g.* rubbery, rigid, glassy transparent *etc.*), which help to expand the application of this technique. Figure 9.1 shows a generic schematic to describe how the system works, wherein inks for structure and support are deposited respectively and then exposed to UV light to trigger polymerisation reactions to solidify the liquid ink and stack up into a 3D object. In addition, there can be more than one model materials deposited within a single machine.

In this section, we first introduce the basic mechanisms of photo-polymerisation as well as the main side reaction during this process: oxygen inhibition, then present the commonly used ink formulations and finally the applications are highlighted at the end of this section.

9.1.2 Photo-polymerisation

Photo-polymerisation is a range of chemical reactions that are based on photo-reactive monomers or oligomers, which react and polymerise upon irradiation at a given wavelength (*e.g.* UV, visible light *etc.*) to become solid.

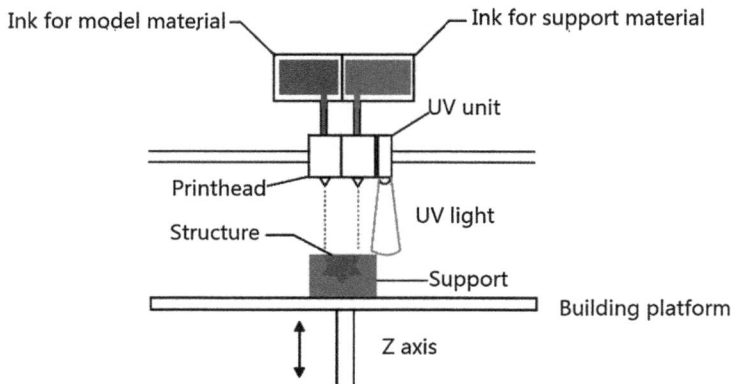

Figure 9.1 Example of Material Jetting System based on photo-polymerisation.

Generally speaking, the initiator will be triggered by the incident radiation and become reactive to the photo-reactive groups (*e.g.* acrylate, methyl acrylate, epoxy *etc.*) inside the monomers or oligomers to initiate chain propagation and form strong covalent bonds to connect the monomers or oligomers and form a crosslinked network. There are two kinds of photo-polymerisation mechanisms: free radical photo-polymerisation and Ionic (mainly in Cationic) photo-polymerisation. In the following sections, these two reactions will be discussed respectively.

9.1.2.1 Free Radical Polymerisation

Free radical based UV curing reactions normally require monomers or oligomers containing acrylate or methyl acrylate groups which can be triggered by free radical species generated by a photo-initiator.

The reaction typically includes three stages, which are initiation, propagation and termination (as shown in Figure 9.2). During the initiation stage, the photo-initiator (PI) will be activated by absorbing energy from the radiation source and generating free radical species ($I\cdot$). These free radical species will then attack photo-reactive acrylate groups ($CH_2=CHR$) and form free radicals ($ICH_2C\cdot HR$) which will then initialise chain propagation. During chain propagation, previously formed free radical species ($ICH_2C\cdot HR$) will attack other photo-reactive groups ($CH_2=CHR$) to form covalent bonds and new species. By repeating this procedure, the length of the polymer chain will increase. Through the whole of the polymerisation sequence, the free radicals always have the chance to terminate through combination or a radical disproportionation reaction. However, in the initial stage of reaction, the concentration of the photo-reactive material is much higher than that

Initiation

$$PI \xrightarrow{\ h\nu\ } I^{\bullet}$$

Propagation

Termination

Figure 9.2   Free radical based photo-polymerisation process.

of the free radicals, so the reaction tends to propagate instead of terminate. As the polymerisation reaction goes on, the free radical concentration keeps increasing while the concentration of unreacted groups reduces. Thus the possibility that a free radical terminates will increase and the reaction will then gradually enter the termination stage.

One of the key issues for free radical based UV curing polymerisation is the presence of oxygen which inhibits the curing reaction by scavenging free radical species. This will be further discussed in Section 9.1.2.3.

9.1.2.2   Cationic Polymerisation

Cationic polymerisation is another reaction which is applied in photo-polymerisation. Similar to free radical polymerisation, this reaction also goes through initiation, propagation and termination stages. However the mono-mers or oligomers used need to contain epoxy or vinyl ether groups instead of acrylate or methyl acrylate. Also, the initiator used to trigger the reaction is different from free radical reaction.

Compared to free radical based polymerisation, cationic polymerisation does not suffer oxygen inhibition and the cured sample has very low shrinkage.[4–6] However, it is sensitive to moisture and once the reaction is triggered, it is hard to terminate. This can lead to better conversion of the reactive groups in the ink but also brings challenges in ink preparation, transport and stor-age. Typical cationic initiators, Lewis or Brønsted acids can result in almost instant gelation which makes it hard to use in practice.[7] Instead, initiators based on diaryliodonium and triarylsulfonium salts are more stable in the absence of light and can be applied to material jetting inks and coatings.

Another disadvantage resides in the curing speed which is not as fast as free radical based photo-polymerisation and the cured specimens are usu-ally brittle. Therefore, in practice, acrylate based inks are used together with epoxy to increase the curing speed as well as reduce the brittleness of the final product.[8]

9.1.2.3   Oxygen Inhibition

As mentioned in Section 9.1.2.1, an oxygen inhibition reaction is one of the key issues for free radical based photo-polymerisation. It has been found that in an ambient environment, the oxygen diffusion coefficient can reach more than 2.5 cm^2 s^{-1} for a 15 μm depth from the environment contact surface and then sharply reduced to less than 0.5 cm^2 s^{-1} afterward, which results in 3 orders of magnitude initiation rate differences.[9,10] This may only lead to surface problems for coating applications, but it can cause specimen fail-ure in photo-polymerisation based material jetting processes. For example, a typical droplet from an inkjet printhead is around 45 μm in diameter[11] and the thickness of a deposited droplet can be as little as 8 μm (depending on the print head and substrate surface energy). As shown in Figure 9.3, oxygen can diffuse into all the jetted and deposited ink and prevent solidification.

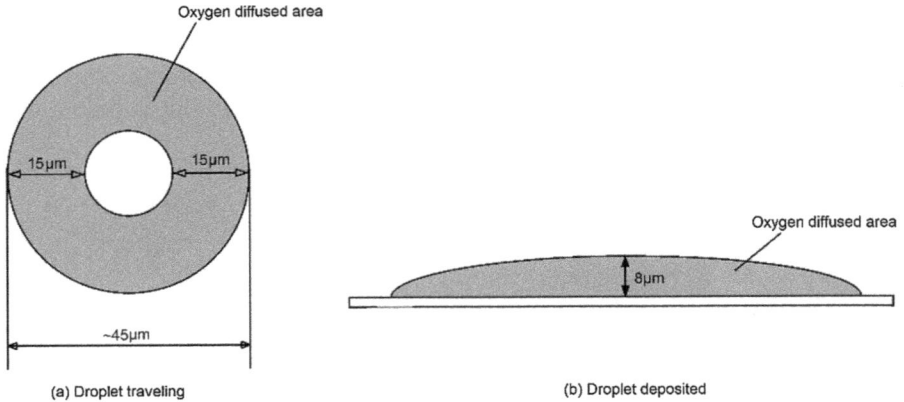

(a) Droplet traveling   (b) Droplet deposited

Figure 9.3   Example of oxygen diffusion and potential inhibition during material jetting, (a) x–y plane of a single droplet and (b) x–z plane view.

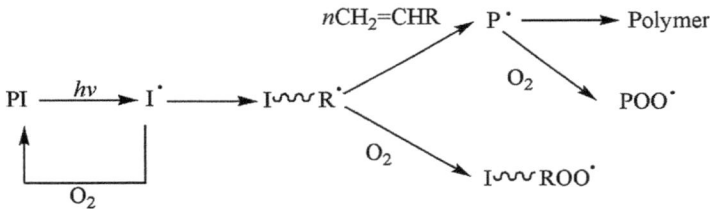

Figure 9.4   Oxygen inhibition process during free radical based photo-polymerisation.

Therefore the oxygen inhibition reaction is described in an independent section in this chapter.

Oxygen can inhibit the polymerisation reaction through all the polymerisation stages (shown in Figure 9.4). During the initiation stage, oxygen can react with the excited initiators ($I\bullet$) and transform them into an unreactive state. At the chain propagation stage, oxygen can react with the free radical species ($R\bullet$ or $P\bullet$) and form stable peroxide free radicals ($ROO\bullet$ or $POO\bullet$), hence inhibiting the chain growth. Therefore, the amount of oxygen diffused into the ink droplet and deposited ink layer can inhibit the polymerisation reaction considerably.

There have been numerous studies on how to overcome the oxygen inhibition effect[12–15] the basic principle being to increase the concentration of active free radical species. The commonly used methods include:

- Increase photo-initiator concentration.
- Increase radiation intensity.
- Use an inert atmosphere.
- Use type II photo-initiator.

Both of the first two methods aim to overcome the inhibition reaction by increasing the number of reactive free radical species in a defined period of time. Utilising an inert atmosphere is the most effective way to minimise oxygen inhibition by reducing the oxygen concentration in the material jetting environment. Commonly used inert environments include nitrogen or carbon dioxide, which can significantly increase the curing rate of photo-polymerisation.[12,13] However, for the practicalities of production in a manufacturing environment the use of inert atmospheres is not always practicable. The final method is to use a type II photo-initiator system. Compared with the usual type I photo-initiator, type II initiator systems are less sensitive to oxygen. Other methods such as modifying the monomer structure to increase the reactivity and reduce its sensitivity on oxygen, using oxygen scavengers can also be used if necessary.[16]

9.1.3 Ink

The recent rise in popularity observed for the Material Jetting technique has consequently resulted in the investigation of new components for UV-curable inks, a summary of a typical formulation for these inks is shown in Table 9.1.

Monomer/Oligomer constitutes the ink's main body and determines the general properties of the final material. Currently, mono- or multi- acrylates/methacrylates[17] are most commonly used precursors, however other types of UV-initiated polymers can be utilised such as vinyl or epoxy derivatives or thiol–ene engenderers. The number of functionalities in the monomer/oligomer is an important aspect. The higher the number of reactive points, the higher the cross-linking density in the final material. This will entail an increase in the reactivity, chemical resistance, stiffness but also in brittleness. An option to broaden the variety of polymers able to be synthesised by photo-polymerisation is modifying precursors of said polymers to obtain their corresponding UV-curable derivatives.[18-20]

Photo-initiators are molecules that generate reactive species after being exposed to UV-Visible radiation to start the polymerisation. As mentioned previously, two different types of photo-initiators may be found depending on the photo-polymerisation process: radical and ionic photo-initiators.[21]

Table 9.1 Typical UV-curable ink formulation.

Name	Function	Concentration
Monomer/Oligomer	Main body of the ink. Defines the properties of the cured material	60–90 wt%
Photo-initiator	Absorbs UV light and initialises polymerisation	<10 wt%
Diluent	Participates in the UV curing reaction, adjusts ink viscosity	0–40 wt%
Other additives: Surfactant, dye, cross-linker, *etc.*	Modifies ink properties	0–3 wt%

The basic mechanism of a radical photo-initiator consists in the absorption of radiation to produce free radical species. The active radicals will then add to the monomer to induce a process of chain-growth polymerisation. Typically, the rate of a photo-induced free radical polymerisation is proportional to the square root of the light intensity absorbed, therefore two of the most important points when selecting a certain photo-initiator are its absorbance peak and the radiation source to be used. Although most photo-initiators are strong initiators in the far-UV, the nature of the chromophore (substituent R in the chemical structures shown in Figures 9.5 and 9.6) will influence the initiator absorbance peak, increasing the wavelength up to the visible zone. This chromophore may also affect other properties such as solubility, reactivity, oxygen-inhibition resistance, *etc.* There are basically two types of radical photo-initiators.

- Type I initiators generate two radical species through photo-fragmentation by a homolytic bond cleavage. An example of this cleavage is the photo-fragmentation of α-alkoxydeoxybenzoins (Figure 9.5). Both created radicals can initiate the polymerisation process.
- Type II initiators undergo a bimolecular process[22] of hydrogen abstraction from the environment, which may be the monomer itself, a solvent or a hydrogen-donor species, generally called a co-initiator. Benzophenone is used in Figure 9.6 as an example to display this bimolecular reaction. Tertiary amines are one of the most commonly used hydrogen

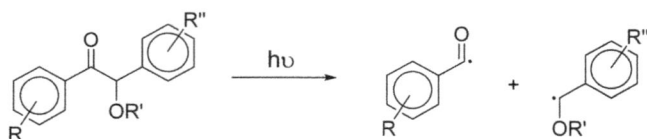

Figure 9.5   Cleavage of α-alkoxydeoxybenzoins.

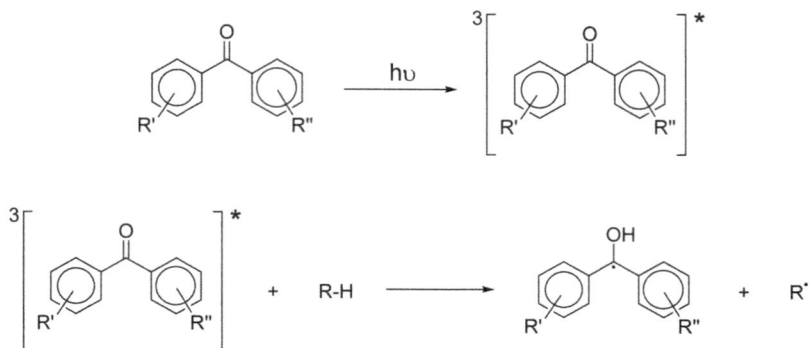

Figure 9.6   Bimolecular radical generation of benzophenones.

donors (or synergist) due to their availability and reactivity regarding the carbonyl group from the initiator. In this case, the active radical initiating the polymerisation is the one coming from the co-initiator, while the species generated from the photo-initiator stay unreactive against monomers. Alcohols may also be used as co-initiators. However, these functional groups are less effective than amines *versus* the photo-initiator's excited state, therefore the reaction will be slower. In this case, both radical species could initiate the polymerisation.

As referred to in Section 9.1.2.3, oxygen inhibition is a particularly pronounced problem in radical photo-polymerisation, mainly when the conditions lead to the generation of a low amount of radicals,[23] but it is even more pronounced in inkjet printing due to the potential for significant oxygen diffusion through the ejected drop or the thin layer of deposited material. This inhibition produces shorter polymer chains and lower crosslinking density, resulting in the production of a soft or tacky material or, in extreme cases a complete failure of curing. The more straightforward methodology to avoid oxygen inhibition is through nitrogen blanketing, but obviously this is an expensive and time-consuming option. The use of tertiary amine co-synergists also impairs the quenching effect of oxygen due to the ability of the alkylamino radicals to react with oxygen molecules. Other options more recently explored are additions of thiol and phosphite compounds, the use of oxygen-insensitive initiators[24] or a combination of type I and type II initiators.[25]

Ionic photo-polymerisation is divided into cationic and anionic polymerisation according to the ionic species generated after initiator photo-fragmentation. From an industrial point of view, cationic reactions are more significant than anionic, although none of them are as notable as radical photo-polymerisation. This is basically due to the lower reactivity and higher toxicity and price of ionic photo-initiators. Cationic initiators have been developed for the polymerisation of vinyl ether- and epoxy-based monomers. Most commonly used cationic photo-initiators are onium salts.[26] These compounds are formed by an organic cation paired with an inorganic anion. Upon photolysis, these initiators undergo irreversible homolytic photo-fragmentation to generate cation radicals and Brønsted acids. The organic cation is the light-absorbing specie, so it determines the UV absorption characteristics and photo-sensitivity. The inorganic anion and its stability affects the strength of the Brønsted acid formed and its initiation efficiency.

Diluents may be solvents or reactive species, however, they are always used to modify a specific ink property, mainly viscosity and/or surface tension to meet the high rheological requirements of inkjet printing. If a certain monomer is used as a diluent, it will form part of the polymer structure, affecting the properties of the final material. If the diluent is a solvent, unless a hydrogel is aimed to be prepared, it will need to be removed by a post-deposition process.

Other type of components may be added to the ink composition. Some examples are:

- Dyes, usually to give colour to the ink.
- Surfactants, to adjust surface energy.
- Cross-linkers, to increase the crosslinking density of the polymer. This will entail a modification of the mechanical properties of the final material.

9.1.4 Applications

Currently, photo-polymers are widely utilised in different fields such as the food industry, dentistry and biomedical industry, coatings or screen printing among others. In AM, stereolithography (a Vat Polymerisation method) is the technique with the largest commercial use of photo-polymers, mainly for prototyping purposes. One of the most common applications of stereolithography, in addition to prototyping, is the fabrication of microfluidic devices.[27] As well as the use of photo-polymers in stereolithography, the utilisation of these polymeric materials is gaining huge relevance in the field of inkjet printing due to the advantages this technique offers against the former technology. This advantage mainly resides in the potential application of multiple material deposition within the same printing process, which could enable the fabrication of multi-functional devices in just one single manufacturing step.

Although the range of materials that are currently commercially available is relatively wide for AM, many of the material formulations are patented, with their corresponding uncertainty regarding composition and high prices. Due to this, the number of photo-polymers-based applications is relatively limited in the literature. In the biomedical field, these materials have been used to help plan medical procedures, such as orthopaedic,[28] cardiac[29] and intracranial aneurysm[30] surgeries. They have also been used to test the dimensional error generated by the technique in the reproduction of anatomic details of the maxillofacial region.[31,32] These have always been applications in simulation and improvement of procedures, but never as final products due to the reasons previously discussed. However, currently an increase in the development of new biocompatible and/or biodegradable materials is occurring. Thus, He *et al.*[33] have demonstrated the inkjet printing of a methacrylate-derived polycaprolactone. These polymers have been widely used in Laser Sintering, however the possibility of utilising it in Material Jetting opens a door to the fabrication of biodegradable multi-material devices. Begines *et al.*[34] built 3D structures with photo-polymers resistant to bacterial attachment, which may result in a reduction in nosocomial infections caused by bacterial biofilms on the surfaces of biomedical devices. Gradient refractive index (GRIN) lenses have been reported by Check *et al.*[35] to be inkjet printable using diethylene-glycol diacrylate containing ZrO_2 nanoparticles.

(Note: Binder jetting is considered as or named by some authors inkjet printing, however this technique is out of the scope of this work and its corresponding applications were not included. For the same reasons 2D applications of this technology have not been included.)

9.2 Heat-assisted Reactive Ink Jetting

As discussed RIJ is a potential AM process during which the droplet solidifies through reaction under the initiation of a catalyst, heat or high energy radiation such as UV light.[36] Currently, UV has been the most used initiation method due to its simplicity in reaction and equipment accessories.[37] However, the materials that can be initiated using UV are limited—mostly photo-polymers. In order to broaden the range of polymers that can be used for 3D printing, a series of new methods were developed in recent years, which enable the AM of two important engineering polymers—polyamide and polyimide.

9.2.1 Nylon 6 (Polyamide)

Nylon 6 (polyamide) is an important engineering polymer that has a wide range of applications.[38] Developing an AM technique that can produce Nylon 6 with high precision would have a great commercialisation potential. Nylon 6 can be made from caprolactam ($C_6H_{11}NO$) through a ring opening process in the presence of an activator *N*-acetylcaprolactam and the consequent anionic polymerisation[39] (Figure 9.7), which is much faster (a few minutes) than classical hydrolytic polymerisation (about 12–24 hours).[40] Therefore, anionic polymerisation has been employed to produce inkjet printable nylon 6.

Initial work done by Fouchal and Dickens showed that using the correct mixtures of caprolactam, catalyst and activator and processing conditions, the solidification time of Nylon 6 was reduced to about 1 minute for large quantities (30–50 ml).[41] Based on these results, Fathi and co-workers established an inkjet additive manufacturing process to jet two low viscosity reactive mixtures of molten caprolactam (monomer of Nylon 6) containing catalyst and activator, respectively.[42] During printing, two mixtures were heated at 80 °C and deposited on top of each other to mix and polymerise with the aid of thermal radiation heating.[43] A Xaar XJ126 piezoelectric DoD printhead with 126 nozzles of 50 μm in size was employed to systematically study the jetting process (Figure 9.8). According to Fathi *et al.* jetting voltage and vacuum level were the main parameters influencing the droplet generation process.[44] The activator and catalyst mixtures showed similar droplet characteristics, suggesting that the melt supply level did not influence the droplet shape in trials with the catalyst mixture.[44]

According Khodabakhshi *et al.*,[40] there are three main parameters affecting the anionic polymerisation of caprolactam, which are catalyst, activator,

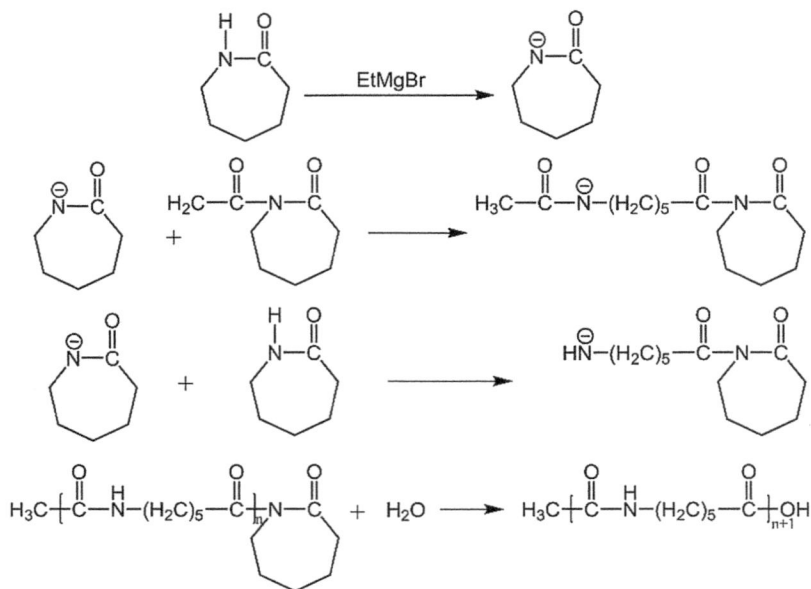

Figure 9.7   Reaction mechanisms of the anionic polymerisation of caprolactam.

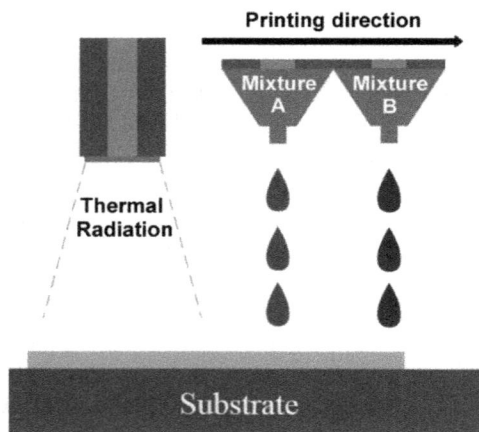

Figure 9.8   Two mixtures were jetted on top of each other to mix and polymerise
with the aid of thermal radiation heating.

and initial polymerisation temperature. After the jetting of molten caprolactam, Khodabakhshi *et al.* carried out a series of work to investigate its
anionic polymerisation process. Ethyl magnesium bromide (EtMgBr) was
confirmed to be a more efficient catalyst compared with sodium hydride
(NaH), which shortened the $t_{1/2}$ and improved the final properties of S-PA6,

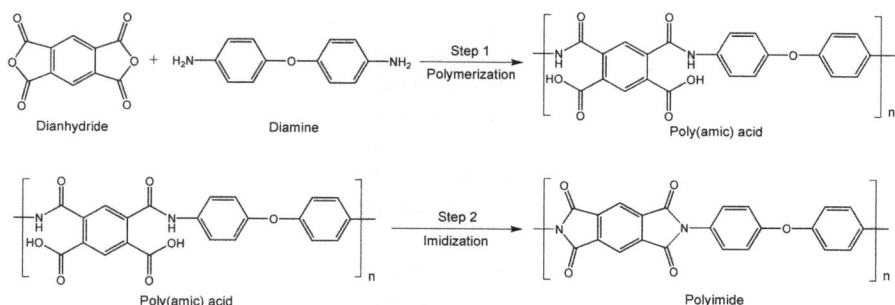

Figure 9.9 Polyimide synthesis in the two-step PAA method. Step 1: Polymerisation of PAA; Step 2: PAA cyclized into polyimide through an imidization process.

while *N*-acetylcaprolactam (ACL) was used as an effective activator.[40] The polymerisation conditions, including the catalyst and activator concentration and initial polymerisation temperature, were optimised to balance the polymerisation speed and final product properties.[40] The anionic polymerisation of caprolactam was investigated in small scale using differential scanning calorimetry (DSC) by addition of the required amounts of ink mixtures into aluminium DSC pans. It was found that the monomer conversion increased with the increase of polymerisation temperature from 140 to 180 °C, then decreased from 180 to 200 °C.[45] At slow heating rates, the produced polymer underwent crystallisation during the polymerisation process.[46] The crystallinity of the polymer decreased with the increase of polymerisation temperature and the increase in the heating rate.[45,46] Although no Nylon 6 product has yet been jetted, the current results suggest that the "Jetting of Nylon" is not just a concept, but a promising approach to compete with traditional methods.

9.2.2 Polyimide

Polyimides are a class of polymers well-known for their excellent thermal stability, chemical resistance, good mechanical and electrical properties.[47,48] Due to these outstanding properties, polyimides are particularly suitable for microelectronic applications.[49] However, these excellent properties of polyimides also increase the difficulty in manufacturing. In this case, developing an AM technique that can produce polyimide parts with high precision would be a great addition to the current polymer industry.

The most widely used procedure in polyimide synthesis is a two-step poly(amic) acid (PAA) process (Figure 9.9), which involves polymerising a dianhydride and a diamine to yield the corresponding PAA, followed by cyclizing into polyimide upon heating or chemical treatment.[47,50,51] In conventional methods, PAA solution is usually cast or spin-coated onto a substrate, followed by two heating processes. The first heating process involves the evaporation of solvent at lower temperatures (100–150 °C), while the second one enables the PAA molecules to lose water and convert to polyimide, which is called thermal imidization (150–450 °C).[51] Strong base catalysts have been developed to reduce the temperature and shorten the time required for thermal imidization, such as 1,4-diazabicyclo[2.2.2]octane (DABCO),[49,52] 1,8-diazabicyclo[5.4.0]undec-7-ene (DBU)[49,53] and 4-hydroxy-pyridine (4HP),[49] which would greatly reduce the difficulty in polyimide printing.

Jensen and co-workers have reported the inkjet printing of gold nanoparticle arrays for electrochemical detection applications. In their work, PAA was diluted to 1 wt% using *N*-methyl-2-pyrrolidone (NMP), and was subsequently printed (using a Fujifilm Dimatix DMP 2800 printer) over sintered gold nanoparticle arrays to insulate the leads.[54] The printed PAA was then thermally imidized at 200 °C for 30 minutes to form polyimide, which served well as a stable and flexible protective coating.

Inspired by these results, Zhang *et al.* systematically studied the inkjet printing of polyimide using a similar PAA ink, and evaluated its applications in microelectronics.[55] In their work, PAA was diluted by NMP and was printed onto a preheated substrate (100–160 °C) to enable the thermal imidization to occur whilst printing. During the PAA ink droplet ejection process, droplet pinning and solvent evaporation finished within 16 ms and 1400 ms, respectively, while the thermal imidization process took much longer. FT-IR results showed that the tiny PAA ink droplets could convert to polyimide by heating at above 160 °C for more than 15 minutes, allowing a continuous printing process. According to Zhang *et al.* the morphology of the printed samples was greatly affected by three factors, the droplet velocity, substrate temperature and droplet spacing. A lower droplet velocity and a higher substrate temperature resulted in smaller droplet diameter, according to which a suitable droplet spacing should be chosen. With the increase of substrate temperature, the film surface tended to show regular waviness, which was compensated by introducing 30% overlapping into the pattern. The insulating property of printed polyimide film was examined by sandwiching between two conductive layers to form parallel plate capacitors (Figure 9.10), the capacitance of which was measured to be 2.82 ± 0.64 nF, and the corresponding relative permittivity to be 2.9 ± 0.02. These results showed that the inkjet printing technique can be used to fabricate polyimide dielectric layers, suggesting its wide applications in microelectronic devices.[55]

9.3   High Viscosity Ink Deposition

9.3.1   Introduction

High viscosity fluids are important materials in printing technology. In general the functionality of the end products improves with increasing viscosity of the used fluids, as such there is growing interest in processing

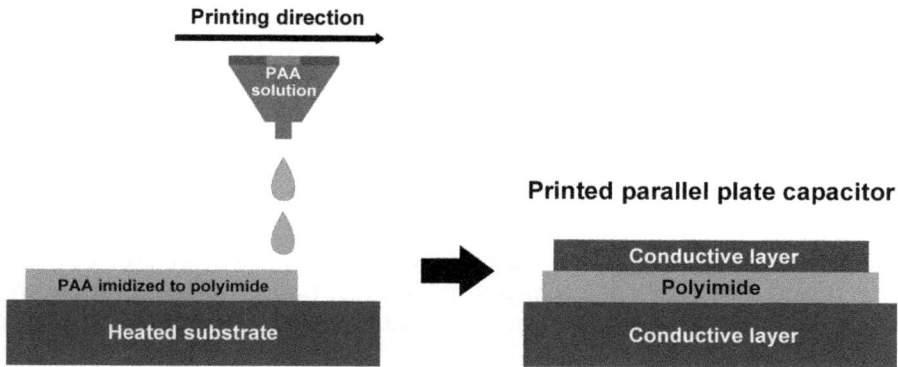

Figure 9.10   PAA ink was printed onto a preheated substrate to form polyimide, which was sandwiched between two conductive layers to act as a parallel plate capacitor.

these fluids with most of the studies focused on using high viscosity fluids with direct ink writing[56] or extrusion-based 3D printing methods.[57] In these examples architectures manufactured from a variety of materials including polymers,[58] metal,[59] ceramics[60] and composites[61] have been achieved. The ability to process such fluids using a reactive inkjet printing approach would enable better control over the process and broaden the application of printing technology.

The viscosity of fluids is among their most important properties and at the same time is the main limiting factor in the common inkjet printing methods that drive Material Jetting. Since there is a possibility to improve functional properties of the end product by using high viscosity fluids, including high molecular polymers[58] and highly concentrated particle dispersions,[62] researchers are focused on developing printing mechanisms which could enable their deposition, and have started to investigate the possibility of applying the reactive inkjet printing (RIJ) approach to high viscous fluids.[36,63]

9.3.2   Influence of Viscosity

The RIJ approach to printing technology makes it particularly suited for polymers since one reactive component can be deposited on the top or to the side of another in a controlled manner and then be allowed to polymerise on the substrate. Nonetheless, to jet a drop of highly viscous polymer is more difficult than a low viscosity one. The presence of high molecular weight (above 500 000 Mw) hence high viscosity polymers in ink formulations introduces the viscoelastic properties characteristic for non-Newtonian fluids.[41] These properties influence the droplet formation during printing and effect a tail disintegration from the droplet.[64,65] From fluid dynamics it is known that for the successful drop ejection the ratio Re/We, that takes into account relationship and effect between internal, viscous, and surface tension importance on fluid flow, in general has to have a value of 1–10:

$$\frac{\text{Re}}{\text{We}} = \frac{(\gamma \rho a)^{1/2}}{\eta} \tag{9.1}$$

where Re is Reynold number, We is the Weber number, a is the characteristic length, γ, ρ, η are fluid surface tension, density, and viscosity respectively.[66] The higher the viscosity, the more difficult it becomes to detach the droplet from the nozzle as the viscous forces are dominant,[67] as the Re/We ratio lowers so the required energy to eject the fluid increases and a large pressure drop is required.[68,69]

9.3.3 High Viscosity Printing Methods

Many researchers have focused on developing printing mechanisms which could support printing of high viscosity fluids. Houben[70] developed a continuous inkjet based system to facilitate jetting of high viscous fluids, where the energy required for highly viscous droplet ejection was provided by using higher pressure delivery by the fluid supply and increased mechanism of the actuation element. In search of different printing mechanisms allowing high viscosity fluids to be jetted, other non-contact, data-driven high viscosity fluid dispensing technology derived from inkjet technology have been established with some pneumatically systems driven being introduced.[71-74] When these systems are used with shear thinning fluids the overall operating pressure can be reduced. As the viscosity of these fluids reduces with increasing shear rate, the pressure to begin fluid flow can stay the same as when the fluid flows through the nozzle.

Most studies have focused on the RIJ approach with low viscosity fluids and small attention has been given to high viscosity fluids despite their great functional properties. There are only a few studies investigating RIJ approaches towards high viscous fluids, in the literature.[36,75-78]

Kröber *et al.*[77] demonstrated reactive inkjet printing of 3D polyutherane features using high viscosity components. Two printheads were employed to deposit two different inks, one containing isophorone diisocyanate and the other included a mixture of poly (propylene glycol), catalyst and crosslinking agent. Under the influence of temperature polymerisation took place on the substrate and polyurethane was formed.

Yang *et al.*[36] have produced silicone films using the RIJ approach. Silicone polymers are interesting materials used in many fields.[79] The solid silicone lines were fabricated on a substrate by jetting two separate inks consecutively from two separate printheads, with one ink containing vinyl terminated polydimethylsiloxane (PDMS) with catalyst, and the other consisting of hydro terminated PDMS. A number of inks with various viscosities from non-Newtonian to very high viscosity, thixotropic materials have been investigated using PicoDot jet valves. The results indicated that by varying process parameters, such as temperature and pressure the volume of the jetted drop could be optimised.

Shepherd *et al.*[78] have produced 3D features of poly (2-hydroxyethyl methacrylate) using RIJ. The photo-polymerised hydrogel inks were designed and consisted of high molecular pHEMA chains, photo-curable HEMA monomer solution, photo-initiator and water. Printed inks were exposed to UV light, resulting in the formation of a hydrogel network. As mentioned, the RIJ approach has a great benefit for polymers providing that good mixing of the components on the substrate takes place.

9.3.4 Droplet Mixing

Studies of drop impact on the substrate has attracted many researchers. As a result extensive work has been reported describing the phenomenon of drop impact on both, solid and liquid surfaces.[80-85] Many parameters have been reported that influence the outcome of a drop impact, and they include drop size, velocity, density, viscosity, and viscoelasticity of the fluid. The roughness and wettability of the substrate and interfacial tension between the drop and the substrate[83] is also a factor. To assure good mixing and avoid any undesired behavior such as floating, bouncing or splashing of one drop on the other, the droplet kinetics should be lower than a value defined based on dimensionless numbers.[81,82] The wettability gradient[85] and difference in deposited size of droplets has been reported to enhance droplet mixing.[81] When two drops (unequal in size) coalesce, the smaller falling drop merges smoothly into the bigger sessile droplet forming a vortex and consequent mixing takes place.[81,83] The energy enabling mixing is provided from the difference in internal pressure of the droplets.[81]

Several techniques have been introduced to investigate and visualise internal mixing inside deposited droplets during or after coalescence. High speed cameras, particle image velocimetry and confocal microscopy are all useful tools to observe and better understand the mixing between two deposited droplets.

Castrejon-Pita *et al.*,[85] investigated impact and coalescence of the sessile drop with a consequently deposited second droplet using particle image velocimetry, focusing on the velocity field during early time coalescence. The same technique has been used by Verdied and Brizard[86] to investigate internal mixing within deposited droplets of the same size. Lai *et al.*,[82] using a high-speed camera, have observed that the time for mixing between two droplets as a result of the release of surface energy is very short (a few hundred milliseconds) and the following dynamics are ruled by diffusion.

Confocal laser scanning microscopy and fluorescence dye have been explored to investigate the mixing between two consequently printed droplets on the substrate by different research groups.[87] Kröber *et al.*[77] used this technique to study the polymerisation reaction that forms solid polyutherane while Fathi and Dickens[87] explored the technique to investigate the synthesis of polyamide. The results from both works showed a high degree of droplet mixing on the substrate. Complete mixing was achieved by Kröber and more than 80% by Fathi and Dickens.

9.4 Summary

In summary, RIJ approaches offer the capability to form functional features using a range of techniques, from the commonly available photo-reactive chemistries common in today's Material Jetting and Vat Polymerisation systems, to experimental multicomponent RIJ with low and high viscosity fluids. What is clear is that there is an ongoing drive towards better understanding of mixing between two droplets and the development of new printing mechanisms which could be suitable for higher throughput and higher viscosity fluids. Given their fast progress, reactive inkjet printing approaches should emerge to bring new AM materials for a broad range of applications.

References

1. C. E. Hoyle, in *Radiation Curing of Polymeric Materials*, ed. C. E Hoyle and J. F. Kinstle, American Chemical Society, Washington, 1990, Photocurable coatings, p. 1.
2. D. C. Neckers and W. Jager, *Photoinitiation for Polymerisation: UV & EB at the Millenium*, Wiley, London, 1998.
3. A. Hancock and L. Lin, *Pigm. Resin Technol.*, 2004, **33**(5), 280.
4. B. Golaz, V. Michaud, Y. Leterrier and J. A. E. Manson, *Polymer*, 2004, **53**(10), 2038.
5. E. W. Nelson, J. L. Jacobs, A. B. Scranton, K. S. Anseth and C. N. Bowman, *Polymer*, 1995, **36**(24), 4651.
6. V. Sipani and A. B. Scranton, *J. Photochem. Photobiol., A*, 2003, **159**(2), 189.
7. J. V. Crivello, *Polym. Chem.*, 1999, **37**(23), 4241.
8. L. Lu, J. Y. H. Fuh, A. Y. C. Nee, E. T. Kang, T. Miyazawa and C. M. Cheah, *Mater. Res. Bull.*, 1995, **30**(12), 1561.
9. A. K. O'Brien and C. N. Bowman, *Macromol. Theory Simul.*, 2006, **15**(2), 176.
10. A. K. O'Brien and C. N. Bowman, *Macromolecules*, 2006, **39**, 2501.
11. C. D. Meinhar and H. S. Zhang, *J. Microelectromech. Syst.*, 2000, **9**(1), 67.
12. K. Studer, C. Decker, E. Beck and R. Schwalm, *Prog. Org. Coat.*, 2003, **48**(1), 92.
13. K. Studer, C. Decker, E. Beck and R. Schwalm, *Prog. Org. Coat.*, 2003, **48**(1), 101.
14. C. Decker, T. V. T. Nguyen, D. Decker and E. Weber-Koehl, *Polymer*, 2001, **42**(13), 5531.
15. N. Craiger and S. Herlihy, *presented at International Conference on Digital Printing Technologies*, Orlando, Florida, 1999.

16. S. C. Ligon, B. Husar, H. Wutzel, R. Holman and R. Loska, *Chem. Rev.*, 2014, **114**, 557.
17. J. V. Crivello and E. Reichmanis, *Chem. Mater.*, 2014, **26**, 533.
18. Y. He, C. J. Tuck, E. Prina, S. Kilsby, S. D. R. Christie, S. Edmondson, R. J. M. Hague, F. R. A. Rose and R. D. Wildman, *Biomed. Mater. Res., Part B*, 2016, 1.
19. C. Yang and Z. G. Yang, *J. Appl. Polym. Sci.*, 2013, **129**(1), 187.
20. M. N. Kirikova, E. V. Agina, A. A. Bessonov, A. S. Sizov, O. V. Borshchev, A. A. Trul and S. A. Ponomarenko, *J. Mater. Chem. C*, 2016, **4**(11), 2211.
21. N. S. Allen, *J. Photochem. Photobiol., A*, 1996, **100**, 101.
22. J. P. Fouassier, *Photochemistry and UV Curing: New Trends*, Research Signport, Kerala, 2006.
23. P. Glöckner, T. Jung, S. Struck and K. Studer, *Radiation Curing: Coatings and Printing Inks*, Vincentz Network, Hannover, 2008.
24. N. S. Allen, F. Catalina, K. O. Fatinikun, W. Chen, P. N. Green and W. A. Green, *J. Oil Colour Chem. Assoc.*, 1987, **11**, 332.
25. H. J. Hageman and L. G. H. Jansen, *Makromol. Chem.*, 1988, **189**, 2781.
26. M. Sangermano, N. Razza and J. V. Crivello, *Macromol. Mater. Eng.*, 2014, **299**, 775.
27. A. Au, W. Huynh, L. F. Horowitz and A. Folch, *Angew. Chem., Int. Ed. Engl.*, 2016, **55**(12), 3862.
28. D. G. Ahn, J. Y. Lee and D. Y. Yang, *J. Mech. Sci. Technol.*, 2006, **20**, 19.
29. Y. L. Cheng and S. J. Chen, *Mater. Sci. Forum*, 2006, **505–507**, 1063.
30. B. O. Erbano, A. C. Opolski, M. J. Olandoski, A. Foggiatto, L. F. Kubrusly, U. A. Dietz, C. Zini, M. M. Makita Arantes Marinho, A. G. Leal and R. Ramina, *Acta Cir. Bras.*, 2013, **28**, 756.
31. D. Ibrahim, T. L. Broilo, M. G. De Oliveira, H. W. De Oliveira, S. M. W. Nobre, J. H. G. Dos Santos Filho and D. N. Silva, *J. CranioMaxillofac. Surg.*, 2009, **37**, 167.
32. M. Salmi, K. S. Paloheimo, J. Tuomi, J. Wolff and A. Mäkitie, *J. CranioMaxillofac. Surg.*, 2013, **41**, 603.
33. Y. He, C. J. Tuck, E. Prina, S. Kilsby, S. D. R. Christie, S. Edmondson, R. Hague, F. R. A. J. Rose and R. Wildman, *J. Biomed. Mater. Res., Part A*, 2017, 1645–1657.
34. B. Begines, A. L. Hook, M. R. Alexander, C. J. Tuck and R. D. Wildman, *Rapid Prototyping J.*, 2016, **22**(5), 835.
35. C. Check, R. Chartoff and S. Chang, *React. Funct. Polym.*, 2015, **97**, 116.
36. H. Yang, Y. He, C. Tuck, R. Wildman, I. Ashcroft, P. Dickens and R. Hague, *presented at Solid Freeform Fabrication Symposium*, Texas, 2013.
37. R. Pandey, *Photopolymers in 3D Printing Applications*, Arcade University of Applied Sciences, 2014.
38. M. I. Kohan, *Nylon Plastics Handbook*, Hanser Gardner Publications, Cincinnati, Ohio, 1995.
39. K. Ueda, K. Yamada, M. Nakai, T. Matsuda and M. Hosoda, *Polym. J.*, 1996, **28**, 446.

40. K. Khodabakhshi, M. Gilbert, P. Dickens and R. Hague, *Adv. Polym. Technol.*, 2010, **29**(4), 226.
41. S. Fathi, *Fundamental Investigation on Inkjet Printing of Reactive Nylon Materials*, Loughborough University, 2011.
42. S. Fathi, P. Dickens and R. Hague, *Rapid Prototyping J.*, 2013, **19**, 189.
43. S. Fathi and P. Dickens, *J. Mater. Process. Technol.*, 2013, 84.
44. S. Fathi and P. Dickens, *Int. J. Adv. Manuf. Technol.*, 2013, **69**(1), 269.
45. S. Fathi and P. Dickens, *J. Therm. Anal. Calorim.*, 2014, **115**(1), 383.
46. K. Khodabakhshi, M. Gilbert and P. Dickens, *Polymer. Adv. Techol.*, 2013, **24**(5), 503.
47. J. H. Lai, R. B. Douglas and K. Donohoe, *Ind. Eng. Chem. Prod. Res. Dev.*, 1986, **25**, 38.
48. T. Ogura, T. Higashihara and M. Ueda, *J. Polym. Sci., Part A: Polym. Chem.*, 2009, **47**, 3362.
49. K. Fukukawa, Y. Shibasaki and M. Ueda, *Chem. Lett.*, 2004, **33**, 1156.
50. F. W. Harris, in *Polyimides*, ed. D. Wilson, H. D. Stenzenberger and M. Hergenrother, Springer, Netherlands, 1990, Synthesis of aromatic polyimides from dianhydrides and diamines, p. 1.
51. S. Diaham, M. Locatelli and R. Khazaka, in *High Performance Polymers – Polyimides Based*, ed. M. J. Abadie, InTech, Croatia, 2012, From Chemistry to Applications.
52. K. Fukukawa, Y. Shibasaki and M. Ueda, *Polym. Adv. Technol.*, 2006, **17**(2), 71.
53. T. Ahn, Y. Choi, M. Jung and M. Yi, *Org. Electron. Phys., Mater. Appl.*, 2009, **10**, 12.
54. G. C. Jensen, C. E. Krause, G. A. Sotzing and J. F. Rusling, *Phys. Chem. Chem. Phys.*, 2011, **13**, 4888.
55. F. Zhang, C. Tuck, R. Hague, Y. He, E. Saleh, C. Sturgess and R. Wildman, *J. Appl. Polym. Sci.*, 2016, **133**(18), 1.
56. B. G. Compton and J. Lewis, *Adv. Mater.*, 2014, **26**(34), 5930.
57. E. B. Duoss, T. H. Weisgraber, K. Hearon, C. Zhu, W. Small, T. R. Metz and T. S. Wilson, *Adv. Funct. Mater.*, 2014, **24**(31), 4905.
58. G. M. Gratson, M. Xu and J. A. Lewis, *Nature*, 2004, **428**(6981), 386.
59. B. Y. Ahn, E. B. Duoss, M. J. Motala, X. Guo, S. I. Park, Y. Xiong and J. A. Lewis, *Science*, 2009, **323**(5921), 1590.
60. K. Sun, T. S. Wei, B. Y. Ahn, J. Y. Seo, S. J. Dillonb and J. A. Lewis, *Adv. Mater.*, 2013, **25**(33), 4539.
61. B. G. Compton and J. A. Lewis, *Adv. Mater.*, 2014, **26**(34), 5930.
62. A. Kosmala, A. R. Wright, Q. Zhang and P. Kirby, *Mater. Chem. Phys.*, 2011, **129**(3), 1075.
63. S. Fathi and P. Dickens, *J. Manuf. Process.*, 2012, **14**(3), 403.
64. B. J. de Gans, P. C. Duineveld and U. S. Schubert, *Adv. Mater.*, 2014, **16**(3), 203.
65. N. Reis, C. Ainsley and B. Derby, *J. Appl. Phys.*, 2005, **97**(9), 094903.
66. K. Seerden, N. Reis, J. Evans, P. Grant, J. Halloran and B. Derby, *J. Am. Ceram. Soc.*, 2001, **84**(11), 2514.

67. D. Xu, V. Sanchez-Romaguera, S. Barbosa, W. Travis, J. de Wit, P. Swan and S. G. Yeates, *J. Mater. Chem.*, 2007, **17**(46), 4902.
68. B. Derby, *Annu. Rev. Mater. Res.*, 2010, **40**(1), 395.
69. D. Jang, D. Kim and J. Moon, *Langmuir*, 2009, **25**(5), 2629.
70. R. J. Houben, Pat., 2004018212 A1, 2004.
71. X. Shu, H. Zhang, H. Liu, D. Xie and J. Xiao, *Sci. China, Ser. E: Technol. Sci.*, 2010, **53**(1), 182.
72. J. Ledesma-Fernandez, C. Tuck and R. Hague, *presented in Part at Solid Freeform Fabrication Symposium*, Texas, 2015.
73. D. W. Rosen, L. Margolin and S. Vohra, *Presented at 19th Annual International Solid Freeform Fabrication Symposium*, Texas, 2008.
74. I. H. Choi and J. Kim, *Micro Nano Lett.*, 2016, **4**(1), 4.
75. E. Uhlmann and P. C. Elsner, *presented at 16th Solid Freeform Fabrication Symposium*, Texas, 2005.
76. M. Muller, Q. U. Huynh, E. Uhlmann and M. H. Wagner, *Prod. Eng.*, 2004, **8**(1–2), 25.
77. P. Kröber, J. Perelaer, J. T. Delaney and U. S. Schubert, *presented at NIP & Digital Fabrication Conference*, Manchester, 2009.
78. J. N. H. Shepherd, S. T. Parker, R. F. Shepherd, U. Martha and J. A. Lewis, *Adv. Funct. Mater.*, 2011, **21**(1), 47.
79. M. J. Owen, in *Advances in Silicones and Silicone-modified Materials*, ed. S. J. Clarson, M. J. Owen, S.D. Smith and M.E. Van Dyke, American Chemical Society, 2010, Foam control, Washington, p. 269.
80. J. R. Adam, N. R. Lindblad and C. D. Hendricks, *J. Appl. Phys.*, 1968, **39**(11), 5173.
81. D. Liu, P. Zhang, C. K. Law and Y. Guo, *Int. J. Heat Mass Transfer*, 2013, **57**(1), 421.
82. Y. H. Lai, M. H. Hsu and J. T. Yang, *Lab Chip*, 2010, **10**(22), 3149.
83. M. Rien, *Fluid Dyn. Res.*, 1993, **12**, 61.
84. J. R. Castrejón-Pita, E. S. Betton, K. J. Kubiak, M. C. Wilson and I. M. Hutchings, *Biomicrofluidics*, 2011, **5**(1), 1–13.
85. J. R. Castrejón-Pita, K. J. Kubiak, M. C. T. Wilson and I. M. Hutchings, *Phys. Rev. E*, 2013, **88**(2), 023023.
86. C. Verdier, M. Brizard, C. Verdier and M. Brizard, *Rheol. Acta*, 2002, **41**, 514–523.
87. S. Fathi, P. Dickens, K. Khodabakhshi and M. Gilbert, *J. Manuf. Sci. Eng.*, 2013, **135**(1), 011009.

CHAPTER 10

Reactive Inkjet Printing of Metals

PAUL CALVERT

New Mexico Tech., Socorro, NM 87181, USA
*E-mail: paul.calvert@nmt.edu

10.1 The Need for Printed Metals

10.1.1 Conventional Technologies

The classic example of patterned metal conductors is the printed circuit board. A sheet of copper 20–80 microns thick is laminated to a composite board, coated in photopolymer, exposed to crosslink the polymer and then the unexposed region is washed off and the underlying copper removed by etching with iron trichloride. Alternatively, the pattern can be silk-screen printed. With conventional methods, linewidths down to 150 μm are achieved. A typical line resistance would be 0.02 Ω cm^{-1}.

Flexible printed circuits are used on small electronics, in automobile instrument panels and where parts move relative to one another, as in a laptop hinge or printer head. A typical flexible printed circuit would be produced from copper foil bonded to polyimide film and etched.

Direct printing of the conductors is a cheaper approach suitable for large quantities of circuits. A roll-to-roll process can be used to print conductors and dielectrics so that connectors, resistors and capacitors can be formed. Conventional printing processes such as gravure, offset and flexographic can

---

Smart Materials No. 32
Reactive Inkjet Printing: A Chemical Synthesis Tool
Edited by Patrick J. Smith and Aoife Morrin
© The Royal Society of Chemistry 2018
Published by the Royal Society of Chemistry, www.rsc.org

be used for printing onto polymeric films such as polyethylene terephthalate, polyethylene naphthenate or polyimide. In addition, organic semiconductors can be printed to make active circuits.

For lower volumes silk-screen printing can be used. Silver and carbon "thick film" inks are readily available with high loadings of conducting particles in a paste. Solvent loss during drying leaves contacting particles in a tough matrix polymer. Typical resistance values are 3–50 mΩ square^{-1} mil^{-1} thickness for silver and 30–100 Ω sq^{-1} mil^{-1} for carbon inks. This translates as roughly 1–10 × 10^{-6} Ω m resistivity for silver and 8–25 × 10^{-3} Ω m for carbon. This compares to bulk silver and copper at about 1.6 × 10^{-8} Ω m. For a line 25 microns thick and 100 microns wide, this would give us at best 4 Ω cm^{-1} for silver and 32 kΩ for carbon.

To put these line resistance numbers into context we can consider two simple functions, one involving carrying power and the other just carrying a signal. In the first case, assume the printed conducting line is carrying current to illuminate a small light emitting diode requiring 15 mA. To do this with a power supply running at 12 V, we can tolerate line resistance of a few hundred ohms and printed silver is suitable but carbon is not. Alternatively, assume we have a sensor, sensing a key-press for example to give a signal into a microprocessor. The impedance of the circuit when the connection is closed could be up to about 100 kΩ and printed carbon lines should be just feasible. These questions are usually unimportant with laminated copper circuits but do become critical as we consider using printed metals or conducting polymers. Clearly the suitability of the conductor depends on the material and on the width, thickness and length of the line.

10.1.2 Non-contact Printing

A major target for metal printing has been inkjet printing of nanoparticulate silver. The ideal ink must have nanoparticle metal to prevent settling before printing and ideally should have a high volume fraction of silver but a low viscosity. A typical volume fraction of 10 nm particles is 12%, with a viscosity of about 10 cPs in an organic carrier. Compared to silk-screen pastes, the carrier liquid can evaporate or wick away after deposition, to leave a denser silver layer. There are many more interfaces with the small particles which reduces the conductivity of the layers as printed. However, unlike the larger silk-screened particles, the nanoparticles can partly sinter at quite low temperatures to improve conductivity. After sintering at 200 °C a typical resistivity is about 2 times that of bulk silver.

How much sintering can be achieved depends on the maximum temperature, which in turn depends on the substrate material, any other components on the printed foil and whether sintering is in a furnace or by laser or microwave heating. One clear target is to achieve good conductivity for line on a cheap polyethylene terephthalate substrate, which has a maximum distortion-free temperature of about 100 °C. Laser sintering has the disadvantage that it is serial and therefore slow, an area treatment is preferable.

Silver nanoparticle inks can be difficult to print due to aggregation and clogging the nozzles. An alternative is to use a silver precursor such as silver nitrate or silver acetate that decomposes to form silver after printing. One problem with this approach is to find a system that has a high loading of dissolved silver but there are a number of promising systems.[1-3] As an examples Schubert reports a conductivity for microwave sintered silver on polyimide of 5% of bulk, with a resistivity of 3×10^{-7} Ω m.[4,5] An alternative strategy is to add a silver-precursor salt such as silver acetate to a particle suspension to form a hybrid ink. The precursor can be decomposed after printing to provide inter-particle bridges.[6]

An alternative technology to inkjet printing is aerosol jet printing, principally developed by Optomec Inc. which will also deposit lines on a similar scale to inkjet printing, which will then sinter to about 2× the resistivity of bulk silver.[7]

10.1.3 Reactive Inkjet Printing

Set against the difficulties of inkjet printing heavily particle-loaded inks, we could turn to reactive inkjet printing where two inks are deposited simultaneously and react to form a solid deposit on the substrate. Reactive inkjet printing has been reviewed by Smith and Morrin.[8] In addition to metals as discussed here it has been applied to polyurethanes,[9] silk,[10,11] epoxies and composites[12] and polyelectrolytes.[13,14]

A number of targets could be seen for reactive inkjet printing of metals. A system of fast reel-to-reel printing with a linewidth of 100 microns or less and a conductivity of about 1/5 that of copper, heat-treated at 100 °C or less would revolutionize printed electronics. The more modest goal of producing 100 micron lines with a conductivity of about 1/100 that of copper on cheap substrates for connection of sensors and 1 mm lines for connection of current devices would allow a form of customizable printed electronics that could find widespread use parallel to the applications of inkjet printing in textiles, labelling and displays. The major advantages could be simpler, more stable inks and greater versatility of materials.

10.2 Printing Technologies for Reactive ijp

There are many different inkjet technologies available.[12,15,16] Most of our experience has been with thermal inkjet printers using refilled consumer printheads. These have the advantages that thermal printing, which depends on expulsion of the drop by a bubble of vaporized solvent is quite robust where piezoelectric printing requires careful tuning of the pulse conditions to the ink properties. In addition, thermal heads are cheap to replace. We have found it particularly easy to use the larger drop systems such as the MSSC which uses a commercial HP cartridge with a larger nozzle size.[18] On the other hand, when development and commercialization follows exploration, it is probably better to switch to a piezoelectric system where the printing system and conditions can be better controlled.[17,19]

Two aspects of the printing system do deserve special consideration. A standard inkjet ink is a dilute solution of dye in an aqueous solution and most dyes are anionic. In contrast metal precursor inks are mainly solutions of metal salts which can be quite corrosive. Chloride ions, in particular, can corrode gold and so slowly destroy any exposed electrodes. In addition these solutions are very conductive and may short-circuit any exposed leads. There are patents on vapor treatment to deposit a parylene film on inkjet heads to improve corrosion resistance.

A second problem which afflicts most inkjet processes is drying of ink in the nozzles. Humectants such as glycerol are usually added to ink so that it will dry to a soft paste of dye or pigment in humectant and this can be mobilized with a little pressure or suction. In laboratory printing some thought needs to be given to the balance between the dry conditions desirable for rapid drying on the substrate and humid conditions to prevent drying in the nozzle.

10.3 Reactive Inks

Here we will include all non-metal-particle inks under the general reactive inks umbrella, since we are then dealing with approaches which rely on chemistry instead of purely physical processes such as drying and sintering.

10.3.1 Jetting

There have been many studies of the jetting process in inkjet printers. The importance of surface tension and viscosity are well known.[20,21] Suspensions and polymer solutions can be quite non-Newtonian and this will be expected to have a significant effect on the jetting process.[20] In terms of droplet formation and jetting, there are no special circumstances relating to reactive inkjet printing that do not apply to any other inkjet process other than the need for two drops to be expelled from two nozzles in short succession. For commercial 3-color printheads the nozzles for the 3 colors are not necessarily in line. On our system the traverse direction of the stage is adjusted in order to line up the nozzles.

10.3.2 Reaction Stoichiometry

In any chemical reaction, a first consideration is the stoichiometry. For instance, in the case of the reduction of copper sulfate by sodium borohydride, the reaction is:

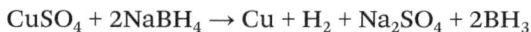

$$CuSO_4 + 2NaBH_4 \rightarrow Cu + H_2 + Na_2SO_4 + 2BH_3$$

The borane BH_3, will either hydrolyze to boric acid and hydrogen or bubble off as gas depending on the substrate layer.

The solubility of copper sulfate in water is about 1.3 molar, (208 g l^{-1}) and that of sodium borohydride is about 9.25 molar (350g l^{-1}) though it also slowly decomposes to release hydrogen. The decomposition is much slower in dilute base.

Thus, to get the right stoichiometry if we are delivering equal numbers of equally sized drops, we need a 1M solution of copper sulfate (allowing a little leeway to avoid precipitation that would block the nozzles) and a 2M solution of borohydride. We can also adjust this by delivering different numbers of drops. At these concentrations, we need to deliver 280 cc of solution per cc of copper formed. Clearly, removal of excess liquid will be a key step in the process.

Aside from the specific issues relating to copper salts and reducing agents, this example reminds us that reactive inks bring many issues that do not arise with conventional solution and suspension inks.

10.3.3   Kinetics of Deposition and Reaction

The kinetics of fast chemical reactions can be hard to study because the normal time taken to mix a few ml of two solutions is a few seconds and many instrumental measurements take a few seconds for each data point. As a result, most chemical kinetics are measured on a timescale from minutes to hours. In principle, the mixing of two small drops on a surface using an inkjet printer seems like an ideal way of measuring kinetics.

Studies of the reaction of copper and nickel salts with borohydride give reaction times of tens of minutes at mmol concentrations.[22,23] If we assume the reaction is first order in copper and borohydride and we have molar concentrations, we can assume the chemistry will be several hundred times faster and the reaction half-time will be seconds.

Within this timeframe the reaction is likely to be influenced by or controlled by diffusion and mixing processes. Mixing of drops is discussed elsewhere in this volume. Drop behavior on impact has been studied by several groups.[20,24,25] In the absence of evaporation and on a smooth surface one could expect dynamic processes during impact to lead to a settled sessile drop determined by the surface tension. Inkjet papers are designed as multilayers of particles, fibers and binder which can absorb different components of the ink. Such a paper should rapidly soak up the first drops but can be flooded if too much ink is printed quickly.[26]

For simplicity, we can consider two cases. In one a drop containing some polymeric binder, lands on a flat surface and dries in less than a second to a flat pancake a few nm thick and with a radius comparable to the original drop. When the second drop lands on top of this one, the water rapidly diffuses and remobilizes the first drop, then diffusion of reactants occurs between two pancakes. Assuming 100 nm pancake thicknesses and small diffusants with a diffusion coefficient of about 10^{-6} cm^2 s^{-1}, the time for this process would be much less than a second. Hence, whether we consider diffusion control or reaction control, the process should occur with a few seconds or less. However, this does depend on solvent loss being sufficiently rapid or printing being sufficiently slow that we do not build up big puddles of reagents on the substrate.

There are standard calculators for water evaporation times from pools as a function of humidity and wind speed. Using this to estimate a drying time gives water loss rates of the order of 10 microns (depth) per second. Since a flattened but undried drop will have a thickness of about 10 microns also, this gives us an estimate of droplet deposition rates expressed in terms of depths/unit area of about 10^{-3} ml cm^{-2} sec^{-1}. These rates are for water. The ink will also contain salts and humectant which will ultimately concentrate to the point where they prevent further drying. Experience with photoprinting does tell us that it is quite easy to flood the paper and clearly this does depend strongly on the substrate morphology.

Paper flooding may be an inconvenience in photoprinting but printed metals will need to reach a substantial thickness in order to achieve high conductivity and the 1M printing solutions suggested above are very dilute (about 1%) when viewed in terms of volume of solid copper per volume of ink. Thus the real kinetic barrier to reactive printing appears to be water removal, either through drying or soaking into a porous substrate.

The copper films shown in Figures 10.1 and 10.2 were printed with 250 passes at a slow droplet rate. In this case the inks quickly dry by evaporation and absorption of the water into the paper. As a result a coherent metal layer forms on the top surface.

10.4 Other Deposition Chemistries

Paquet *et al.* have described the use of a commercial ink of copper oxide particles formulated with a reducing agent such that copper forms when the deposit is reduced and sintered using a flashlamp.[27] Black *et al.* have described printing an organometallic ink with a low decomposition temperature (120 °C)

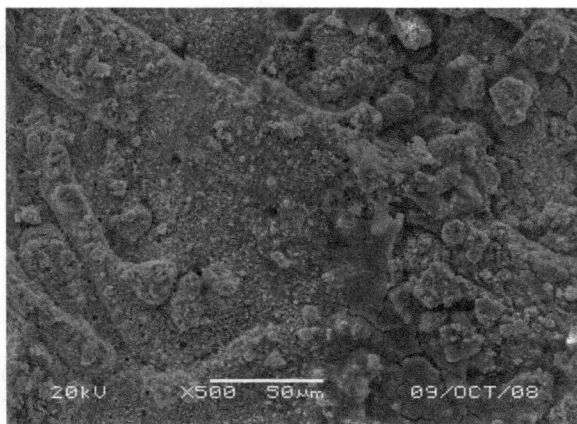

Figure 10.1   Surface of filter paper coated with copper. 250 print cycles of copper sulfate and sodium borohydride solutions.

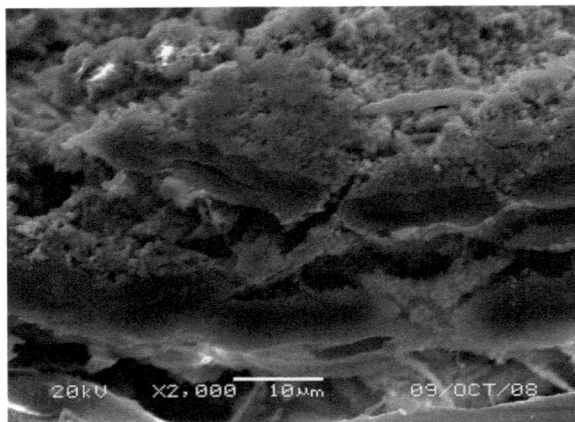

Figure 10.2   Cross section of paper from Figure 10.1 showing that the metal is mainly on the surface but has partly penetrated the paper.

and a reducing agent to form silver films on glass.[28] There are many such options for inks containing silver with a reducing agent since the salts are readily reduced.[29] Laser heating has been used to deposit copper from an ink containing copper salt and a reducing agent.[30] A similar approach has been used for nickel.[31]

Various workers have used inkjet printing to print a seed layer of metal, such as silver or palladium, followed by electroless plating to develop thicker conductors.[32,33]

10.5   Substrates

Thick film technology depends on the silk-screen printing of thick inks onto ceramic or polymeric substrates followed by burn-off or evaporation of the carrier to leave a largely metal film. This same approach has largely been followed for inkjet printing of nanoparticle metal inks. On the other hand inkjet photoprinting commonly uses special layered papers that are designed to immobilize the color and allow the aqueous carrier to wick away. Where the ink is delivered too rapidly, it is common to see that the paper is flooded and the image starts to lose resolution through mixing. On the other hand, a desire for intense colors means that a relatively large volume of ink must be delivered per unit area.

10.5.1   Porous Substrates

Inkjet inks were initially dye solutions but dyes tend to be vulnerable to color-bleaching through photo-oxidation. Thus, carbon nanoparticle black inks replaced black azo-dyes and provided better stability and better resolution. Subsequently many manufacturers also replaced colored dyes with nanoparticle pigments. In either case, photopapers can be specifically designed with layers that adsorbed particular inks. Anionic dyes or pigments with anionic surface charge will adsorb strongly to cationic particles in the paper and *vice versa*. It is also possible to add more specific chemistry to the pigment and to the paper to promote bonding. Many photopapers have complex layered structures that both localize the pigment or dye and wick away the carrier fluid.[34] Ohlund *et al.* have discussed a modified paper specifically for photoreduction of printed copper oxide to copper.[35]

In considering the physical requirements for a good colored image *versus* those for a conducting metal pattern, it is clear that it should be possible to design a paper substrate especially for nanoparticle metal inks or, differently, for reactive inkjet printing of metals. So far, this does not seem to have been done.

10.5.2 Porous Substrates for RIJP of Metals

Ideally a thick porous substrate could wick away excess water while allowing a thick metal layer to build up on the surface. Inkjet photopapers function by having layers that will absorb individual colors based on charge or dye chemistry while allowing the excess liquid to penetrate deep into the paper and ultimately dry.[24] In other work on printed ceramic slurries we have found that a porous plaster substrate will remove excess water while leaving a thick layer of tightly-packed ceramic powder. As the part thickens the ceramic layer itself conducts away the liquid.[36]

In our experiments we use filter papers as substrates for printing metals because they rapidly conducted away the liquid. As the metal layer developed it was not a smooth film but, as seen by electron microscopy, resembled a cauliflower with small particles stuck together into larger aggregates. Clearly the morphology of the metal should be porous so that it can continue to conduct away the excess liquid.

The formation of particles in solution will be a nucleation and growth process. Many studies have addressed this for the formation of silica particles by the hydrolysis of silicon alkoxides.[37] Initially the reduction process should be homogeneous, with isolated metal atoms aggregating to form small clusters. Following this, particles may grow by further addition of atoms from solution, by catalytic reduction of ions on the particle surface or by aggregation of clusters onto larger particles. Which of these is dominant will depend on the relative rates of the surface catalyzed and homogeneous reactions and on the concentrations of the various species in solution. Clearly these rates will also change as the substrate surface becomes coated with metal.

The balance of these processes will also affect the morphology of the resulting metal film. If the rate of the surface reaction is slow, this will be the rate-determining step and the film will be expected to be quite dense. If the surface reaction is fast, the process will be limited by diffusion to the surface and the surface will be rough as regions projecting further through the diffusion boundary layer will grow faster than regions where the layer is thinner. These issues have been discussed in the context of biomimetic mineralization, where the interest was in preparing dense inorganic films on a functionalized organic surface by chemical precipitation.[38]

10.5.3 Modeling of Inks on Substrates

The processes which occur as an ink drop collides with a substrate have been studied by several groups[20] and are discussed elsewhere in this volume. At slow printing rates each drop can dry in a few seconds to form a pancake. If a drop of the second reagent lands on top of the first, the development of the metal involves competing processes of diffusion and reaction as illustrated in Figure 10.3. In many of the cases considered here diffusion is probably limiting. At higher printing speeds the second drop may combine with the first while it is still liquid to produce a mixed solution.

In this context Figures 10.1 and 10.2 can be interpreted as showing that the water of the ink is rapidly absorbed by the coarse filter paper, causing the metal to form as a surface layer on the paper.

The flow of liquid into a porous substrate will be driven by capillarity with adsorption of the solutes onto the surface of the pores plus swelling of the pore walls by the solvent. Figure 10.4 shows a PEDOT ink printed onto a woven textile to form a line at the top of the image. The fluid wicks away from the line down the bundles of fibers making up the individual yarns with more limited cross-over between yarns. As this process continues the water is taken up by the cellulose fibers. A model for this process has been developed and shown to fit experiment.[39] Similar models would be useful in the design of substrates for reactive inkjet printing.

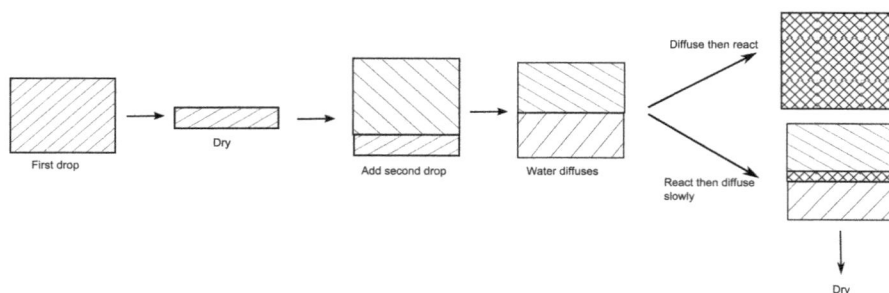

Figure 10.3   A simple model showing possible sequences of reaction and drying when drops of two reacting solutions are printed.

Figure 10.4   Woven textile with a line of printed PEDOT solution going left to right above the top of the image. Excess solution percolates down from the main line and partly transfers to crossing yarns.

10.5.4   Other Substrates

One step to tailoring the substrate to the ink is to modify the surface of a polymer foil. Kirikova printed an alkane thiol layer onto various polymers before printing a nanoparticulate silver. The adhesion of the silver to the polymer film was improved giving better retention of conductivity during stretching and bending.[40]

Tobjork and Osterbacka have reviewed paper substrates for printed electronics.[41] Conventional photopapers have been widely used for the printing of antennas using nanoparticle silver inks.[42,43]

Sawhney *et al.* printed silver nitrate onto textiles and then reduced this in a bath of glucose solution to conducting silver lines. Microscopy showed that the individual textile fibers in the woven yarns were coated with a layer of silver. This is significant because parallel studies on resistance changes in inkjet printed PEDOT showed that this formed both a thick layer on the surface of the fabric and a deeper coating around the fibers. On stretching the surface layer cracks irreversibly causing a resistance increase, the deeper layer does not crack but shows a resistance decrease on stretching. This decrease was attributed to the individual twisted fibers in the yarn rotating and being pulled closer together on stretching.[44,45]

Stempien *et al.* printed silver onto textiles using Tollen's reagent, an ammoniacal solution of silver nitrate that is readily reduced to metal by aldehyde. Metal conductors were printed onto various fabrics and characterized for tolerance to bending and washing.[46] Stempien *et al.* also printed conducting PEDOT onto textiles by inkjet printing solutions of iron chloride and solutions of EDOT monomer which then polymerized *in situ*.[47]

Li *et al.* used a nozzle dispensing system to deposit copper onto textiles *via* sequential deposition of a copper salt and a reducing agent.[48]

10.6   Properties of Printed Patterns

10.6.1   Electrical Properties

Two key properties of metal films on soft substrates are conductivity and strain to break. For conductivity, consider two possible examples. In one case consider a sensor on a flexible patch 1 cm square which is somehow to be connected into a wearable system. Sensor resistances vary but 10 kΩ is typical, so our conducting lines need to be considerably less than this, say 10 Ω. For a 1 cm printed connection length, with a linewidth of 0.1 mm and a thickness of 10 microns, this would require a conductivity of 10^4 S cm^{-1}. This compares with a conductivity for copper of 6×10^5 S cm^{-1} and requires a printed film with 1/60 of the conductivity of bulk copper, which is quite feasible.

As an alternative, consider a similar line delivering power to an LED running at 3 V and 0.5 amps. In this case we would want the connection to consume less than 10% of the power and so would need a resistance of less than 0.6 Ω, possibly even less because the power loss could lead to overheating on a soft organic substrate. This gives us a conductivity requirement of about 1/3 that of copper. Clearly it is possible to use thicker or wider lines but then there may be no advantage of inkjet printing over conventional paste extrusion methods. There are also many other considerations for connection of other devices but much of current electronics is based on the assumption that connectors will have properties similar to bulk copper and it requires something of a cultural shift to consider using poorer conductors.

10.6.2   Mechanical Properties

For a metal film deposited onto a dense flexible substrate, such as a polyester sheet, there are two considerations, flexing and extension.

A thin metal sheet will typically strain by a few % before tearing. On a surface, this means that the film will crack and its resistance will rapidly increase but then decrease again when the sheet is unbent. The resistance changes can give spurious signals if the metal lines are connecting leads for a sensor, for instance.[49]

Similar arguments apply to stretching of the film, where a strain over a few % can cause the metal to crack. Printed films tend to be more brittle because they are essentially granular with limited particle to particle bonding and so limited strain to break. In the case of printed PEDOT films we have found that the addition of carbon nanotubes reduces the conductivity loss as the film is stretched because the nanotubes bridge the cracks that form.[50]

When the metal is deposited into a porous, fibrous substrate such as paper or textile, this behavior is modified because much of the linear strain can occur by rotation of the fibers rather than their extension. In our studies of silver printed onto textiles, we found that there were two structures. The silver layer on the surface of the yarn cracked under stress and caused a resistance increase. Silver which had penetrated into the twisted-fiber yarn showed a resistance decrease on stretching as the individual fibers were straightened and pulled closer together.[44]

Another approach to provide printed connections resistant to extension is to deposit a wavy or zig-zag line rather than a straight line. This converts much of the extension to in-plane bending.[51]

In many examples of flexible electronics, bending rather than extension will be the main form of deformation. For a simple film in bending, the tensile strain in the convex surface can be directly calculated from the radius of curvature and the thickness of the film. For a composite sheet, such as a metal layer on a plastic sheet the calculation is more complicated and depends on the thickness and elastic moduli of the two layers. Since the metal has a modulus about 20× that of the plastic, a thin metal layer on a soft substrate can experience a much reduced strain in the metal layer when compared to the tensile strain on the convex surface of a simple sheet of one material.

The response of metal connectors to repeated loading may also include failure of the bonding to the substrate, work-hardening of the copper followed by fracture after many cycles or crack growth through the sintered metals. These issues will become progressively more important as wearable and soft electronics become more widely used.

10.7 Comparison Between Different Metals

Copper is the standard choice for electrical connections. It is highly conductive when pure but probably as important, it is very malleable so it can easily be shaped into wires and foils and it is quite resistant to corrosion because it forms a dense oxide layer. Laminated copper foil has long been the standard conductor for rigid and flexible printed circuit boards. Against this background there has been little need for electronic engineers to design for less conductive metal connectors.

Copper has one significant drawback and that is its rapid work hardening when repeatedly flexed. The resulting brittleness and the concurrent increase in resistance can cause serious problems in current-carrying wires. This is a major issue in flexible and wearable electronics. One recent project on metal-cored textile fibers for wearable electronics selected indium as a conductor, both for its low resistivity and its good fatigue properties.[52]

For particulate inks, the high surface area makes corrosion resistance especially important. Hence, where silk-screen printed paste inks are used, silver and carbon are the standard materials depending on the conductivity needed because both are quite resistant to oxidation compared to copper.

Thus, in discussing conductors deposited by reactive methods, the deposition chemistry, the mechanical properties of the layers and the corrosion resistance of these high surface area structures may be more important that maximum conductivity and other metals or combinations of metals may perform better than copper or silver.

Nickel is a much poorer conductor than copper but does print more easily because it is quite resistant to oxidation. Where fine copper particles oxidize rapidly[53] so that relatively thick layers must be deposited if conductivity is to be retained, nickel remains conducting even in thin layers of sintered particles.[54] It may thus be preferable to use a metal conductor or alloy that is less conductive than copper and cheaper than silver, depending on the detailed requirements for current or signal transmission.

Mohl *et al.* used reactive inkjet printing to deposit palladium or silver in circles formed on drying ("coffee-rings"). These interlinked circles could then be electroless-plated to make conducting but transparent layers.[33] Bhuvana *et al.* have inkjet printed palladium by depositing thiolates and then converting these to metal by thermolysis at 230 °C.[55]

Conducting polymers generally have too low a conductivity for power transmission in electronics but could be used for signal transmission. Many groups have inkjet printed PEDOT lines from aqueous suspensions. Stempien *et al.* have printed PEDOT using reactive inkjet printing of EDOT and ferric nitrate.[47] Jeon *et al.* have formed the conducting polymer polyphenylenevinylene (PPV) *in situ* by reactive inkjet printing.[56] A similar process can be used for gold.[57]

Carbon is widely used for printed electronics. Kim *et al.* have printed conducting graphene lines by printing graphene oxide suspensions and reducing them to graphene.[58] Zhang *et al.* have inkjet printed inks of graphene oxide with attached silver nanoparticles.[59]

10.8   Comparison with Nanoparticle Printed Metals

Most research on direct printing of metals has involved the use of nanoparticulate silver inks. Inkjet printing is more problematic than other contact printing methods such as offset printing, or jet printing such as aerosol jet printing[7] or nozzle extrusion such as the Micropen, nScript and other systems.[60] Kheawom *et al.* have compared reactive printing methods with reactive sintering in a hot vapor. In their study the reactive sintering approach gave better conductivity.[61]

The most favored metal is silver and high conductivity is obtained by sintering of the particles after deposition. The key to this is the lower sintering temperature of nanoparticles, which allows the use of polymeric substrates. A major effort has been devoted to achieving lower sintering temperatures. Polyimide substrates can withstand 230 °C. Polyester sheet substrates would be more economical but can withstand only 130 °C while thermal sintering of silver nanoparticles is very limited below 170 °C.[62] One approach, relevant to reactive printing, is to include soluble silver salts with low decomposition temperatures that decompose on heating and form conducting silver bridges between the particles.[6] Laser[63] or microwave sintering[5] or other chemical treatments[64] can also produce good conductivity at low temperatures.

Nanoparticle inks of other metals are available but copper, the obvious conductor, does oxidize on the surface and this can result in rapid, complete loss of conductivity in nanoparticles.[53,65] Forming the particles with an excess of reducing agent can protect them from oxidation.[66]

In consumer inkjet printers, the ink is carefully designed to function well in the printer. Nanoparticle suspensions can be inkjet printed but this adds extra constraints to an already difficult process. Most inkjet printers have a viscosity limit which means that concentrated suspensions cannot be used. Particles do tend to settle which may require some method to re-suspend them during printing. Thermal inkjet printers rely of rapidly vaporizing the fluid to form a bubble and piezoelectric printers rely on a pressure pulse transmitted through the fluid. Both processes may be affected by a dense, thermally conducting suspension of silver nanoparticles.

Setting up an inkjet printer to work with a new ink is rarely routine and usually requires careful formulation. Dilute suspensions or solutions with low viscosity generally work best but industrial processes tend to need concentrated suspensions and viscous, polymeric, solutions to work best. Thus, one reason to explore reactive inkjet printing is that, by using two separate liquids that form the desired solid *in situ* on the substrate, one can relax some of the constraints that normally limit single-ink printing.

10.9 Future Prospects

10.9.1 Industrial Inkjet Processes

Much of this chapter is dedicated to the problems of inkjet printing of concentrated solutions. It should be emphasized that these are problems relating especially to laboratory experiments with changing inks and substrates. Once the right system and conditions are established inkjet printing can be very reliable and there are many industrial processes which use inkjet. Examples include high speed labeling of cardboard boxes, dye masks for LCD displays, dosing pharmaceuticals into tablets, and reel-to-reel printing in flexible electronics, textiles.

10.9.2 Other Chemistries and Processes

As mentioned above, one stimulant for our interest in reactive methods for material deposition is the fact that most biological growth occurs by room temperature chemical processes whereas most materials shaping is done by heating and cooling. In this sense, chemistry is much more versatile than physics and there are many more possible ways of producing a metal film by chemistry than there are by particle deposition and sintering.

These means that we have barely begun to explore the possibilities and many novel approaches are being suggested.

10.9.3 Outstanding Problems

One key issue which needs to be addressed in developing the process is to identify more concentrated inks which are still printable in order to reduce the excess water that must be absorbed by the substrate. Viewed as a volume fraction of metal, aqueous copper sulfate is only about 0.3% at saturation. Modification of the solvent and salt can almost certainly improve on this significantly. The same argument applies to the reducing solution.

Similarly, there is considerable room to develop substrates that will wick away the water or solvent while retaining the metal at the surface in order to allow fast printing without puddling. The substrate may be a thick layered paper or a thin porous peelable layer over a spongy support.

A further issue is to explore ways to form more coherent films with less included void space for higher conductivity and better mechanical properties. One would expect that surface-active additives in the printing solutions should be able to suppress particle nucleation and promote dense film formation.

References

1. S. B. Walker and J. A. Lewis, *J. Am. Chem. Soc.*, 2012, **134**, 1419–1421.
2. G. McKerricher, M. Vaseem and A. Shamim, *Microsyst. Nanoeng.*, 2017, **3**, 16075.
3. K. S. Bhat, R. Ahmad, Y. Wang and Y.-B. Hahn, *J. Mater. Chem. C*, 2016, **4**, 8522–8527.
4. J. Perelaer and U. S. Schubert, *J. Mater. Res.*, 2013, **28**, 564–573.
5. J. Perelaer, B.-J. de Gans and U. S. Schubert, *Adv. Mater.*, 2006, **18**, 2101–2104.
6. J.-T. Wu, S. Lien-Chung Hsu, M.-H. Tsai, Y.-F. Liu and W.-S. Hwang, *J. Mater. Chem.*, 2012, **22**, 15599.
7. A. Mahajan, C. D. Frisbie and L. F. Francis, *ACS Appl. Mater. Interfaces*, 2013, **5**, 4856–4864.
8. P. J. Smith and A. Morrin, *J. Mater. Chem.*, 2012, **22**, 10965.
9. P. Kröber, J. T. Delaney, J. Perelaer and U. S. Schubert, *J. Mater. Chem.*, 2009, **19**, 5234.
10. P. Rider, Y. Zhang, C. Tse, Y. Zhang, D. Jayawardane, J. Stringer, J. Callaghan, I. M. Brook, C. A. Miller, X. Zhao and P. J. Smith, *J. Mater. Sci.*, 2016, **51**, 8625–8630.
11. D. A. Gregory, Y. Zhang, P. J. Smith, X. Zhao and S. J. Ebbens, *Small*, 2016, **12**, 4048–4055.
12. P. Calvert, *Chem. Mater.*, 2001, **13**, 3299–3305.

13. Y. Yoshioka, G. E. Jabbour and P. D. Calvert, *Proc. SPIE- Int. Soc. Opt. Eng. Nanoscale Opt. Appl.*, 2002, **4809**, 164–169.

14. P. Calvert, G. Jabbour and Y. Yoshioka, *Polym. Prepr.*, 2002, **43**(2), 358–359.

15. E. Tekin, P. J. Smith and U. S. Schubert, *Soft Matter*, 2008, **4**, 703–713.

16. B. R. Ringeisen, C. M. Othon, J. A. Barron, D. Young and B. J. Spargo, *Biotechnol. J.*, 2006, **1**, 930–948.

17. D. McCallum, C. Ferris, P. Calvert, G. G. Wallace and M. in het Panhuis, *ICONN 2010 Int. Conf. Nanosci. Nanotechnol.*, 2010, pp. 257–260.

18. S. Limem, D. P. Li, S. Iyengar and P. Calvert, *J. Macromol. Sci., Part A: Pure Appl. Chem.*, 2009, **46**, 1205–1212.

19. C. J. Ferris, K. J. Gilmore, S. Beirne, D. McCallum, G. G. Wallace and M. in het Panhuis, *Biomater. Sci.*, 2013, **1**, 224–230.

20. B. Derby, *Annu. Rev. Mater. Res.*, 2010, **40**, 395–414.

21. P. Calvert and T. Boland, in *Inkjet Technology for Digital Fabrication*, ed. I. M. Hutchings and G. D. Martin, John Wiley & Sons, Ltd, Chichester, UK, 2014, pp. 275–305.

22. M. Alsawafta, S. Badilescu, M. Packirisamy and V.-V. Truong, *React. Kinet., Mech. Catal.*, 2011, **104**, 437–450.

23. G. N. Glavee, K. J. Klabunde, C. M. Sorensen and G. C. Hadjipanayis, *Langmuir*, 1994, **10**, 4726–4730.

24. K. Mielonee, S.-S. Ovaska, T. Laukala and K. Backfolk, *J. Imaging Sci. Technol.*, 2016, **60**, 30502.

25. G. Desie, G. Deroover, F. De Voeght and A. Soucemarianadin, *J. Imaging Sci. Technol.*, 2004, **48**, 389–397.

26. S. Limem, D. McCallum, G. G. Wallace, M. in het Panhuis and P. Calvert, *Soft Matter*, 2011, **7**, 3818.

27. C. Paquet, R. James, A. J. Kell, O. Mozenson, J. Ferrigno, S. Lafrenière and P. R. L. Malenfant, *Org. Electron.*, 2014, **15**, 1836–1842.

28. S. N. Black, L. A. Bromley, D. Cottier, R. J. Davey, B. Dobbs and J. E. Rout, *J. Chem. Soc., Faraday Trans.*, 1991, **87**, 3409–3414.

29. C. Lefky, A. Mamidanna, Y. Huang and O. Hildreth, *Phys. Status Solidi A*, 2016, **213**, 2751–2758.

30. M. S. Panov, I. I. Tumkin, A. V. Smikhovskaia, E. M. Khairullina, D. I. Gordeychuk and V. A. Kochemirovsky, *Microelectron. Eng.*, 2016, **157**, 13–18.

31. Y. Rho, K.-T. Kang and D. Lee, *Nanoscale*, 2016, **8**, 8976–8985.

32. D. I. Petukhov, M. N. Kirikova, A. A. Bessonov and M. J. A. Bailey, *Mater. Lett.*, 2014, **132**, 302–306.

33. M. Mohl, A. Dombovari, R. Vajtai, P. M. Ajayan and K. Kordas, *Sci. Rep.*, 2015, **5**, 13710.

34. D. W. Donigian, P. C. Wernett, M. G. McFadden and J. J. McKay, *Tappi J.*, 1999, **82**, 175–182.

35. T. Öhlund, A. K. Schuppert, M. Hummelgård, J. Bäckström, H.-E. Nilsson and H. Olin, *ACS Appl. Mater. Interfaces*, 2015, **7**, 18273–18282.

36. J. Cesarano, Solid Freeform and Additive Fabrication, *Mater. Res. Soc. Symp. Proc.*, 1999, **542**, 133–139.
37. C. J. Brinker and G. W. Scherer, *Sol-gel Science: The Physics and Chemistry of Sol-gel Processing*, Academic Press, Boston, 1990.
38. P. Calvert and P. Rieke, *Chem. Mater.*, 1996, **8**, 1715–1727.
39. P. Calvert and P. Chitnis, *Res. J. Text. Apparel*, 2009, **13**, 46–52.
40. M. N. Kirikova, E. V. Agina, A. A. Bessonov, A. S. Sizov, O. V. Borshchev, A. A. Trul, A. M. Muzafarov and S. A. Ponomarenko, *J. Mater. Chem. C*, 2016, **4**, 2211–2218.
41. D. Tobjörk and R. Österbacka, *Adv. Mater.*, 2011, **23**, 1935–1961.
42. S. Kim, Y. Kawahara, A. Georgiadis, A. Collado and M. M. Tentzeris, *IEEE Sens. J.*, 2015, **15**, 3135–3145.
43. T. Zhang, X. Wang, T. Li, Q. Guo and J. Yang, *J. Mater. Chem. C*, 2014, **2**, 286–294.
44. A. Sawhney, A. Agrawal, T.-C. Lo, P. K. Patra, C. H. Chen and P. Calvert, *AATCC Rev.*, 2007, **7**, 42–48.
45. P. Calvert, D. Duggal, P. Patra, A. Agrawal and A. Sawhney, *Mol. Cryst. Liq. Cryst.*, 2008, **484**, 657–668.
46. Z. Stempien, E. Rybicki, T. Rybicki and J. Lesnikowski, *Sens. Actuators, B*, 2016, **224**, 714–725.
47. Z. Stempien, E. Rybicki, T. Rybicki and M. Kozanecki, *Synth. Met.*, 2016, **217**, 276–287.
48. B. Li, D. Li and J. Wang, *Text. Res. J.*, 2014, **84**, 2026–2035.
49. P. Manandhar, P. D. Calvert and J. R. Buck, *IEEE Sens. J.*, 2012, **12**, 2052–2061.
50. B. Hu, D. Li, P. Manandharm, Q. Fan, D. Kasilingam and P. Calvert, *J. Mater. Chem.*, 2012, **22**, 1598–1605.
51. D.-H. Kim, J. Xiao, J. Song, Y. Huang and J. A. Rogers, *Adv. Mater.*, 2010, **22**, 2108–2124.
52. W. R. Perera and G. J. Mauretti, *US Patent*, 8800136, 2014.
53. J. H. Kim, S. H. Ehrman and T. A. Germer, *Appl. Phys. Lett.*, 2004, **84**, 1278–1280.
54. D. Li, D. Sutton, A. Burgess, D. Graham and P. D. Calvert, *J. Mater. Chem.*, 2009, **19**, 3719–3724.
55. T. Bhuvana, W. Boley, B. Radha, B. D. Dolash, G. Chiu, D. Bergstrom, R. Reifenberger, T. S. Fisher and G. U. Kulkarni, *Micro Nano Lett.*, 2010, **5**, 296.
56. S. Jeon, S. Park, J. Nam, Y. Kang and J.-M. Kim, *ACS Appl. Mater. Interfaces*, 2016, **8**, 1813–1818.
57. M. Abulikemu, E. H. Da'as, H. Haverinen, D. Cha, M. A. Malik and G. E. Jabbour, *Angew. Chem., Int. Ed.*, 2014, **53**, 420–423.
58. K. Kim, S. I. Ahn and K. C. Choi, *Carbon*, 2014, **66**, 172–177.
59. W. Zhang, E. Bi, M. Li and L. Gao, *Colloids Surf., A*, 2016, **490**, 232–240.
60. P. Sarobol, A. Cook, P. G. Clem, D. Keicher, D. Hirschfeld, A. C. Hall and N. S. Bell, *Annu. Rev. Mater. Res.*, 2016, **46**, 41–62.

61. S. Kheawhom and K. Foithong, *Jpn. J. Appl. Phys.*, 2013, **52**, 05DB14.
62. S. K. Volkman, S. Yin, T. Bakhishev, K. Puntambekar, V. Subramanian and M. F. Toney, *Chem. Mater.*, 2011, **23**, 4634–4640.
63. K. Maekawa, K. Yamasaki, T. Niizeki, M. Mita, Y. Matsuba, N. Terada and H. Saito, *IEEE Trans. Compon., Packag., Manuf. Technol.*, 2012, **2**, 868–877.
64. S. Magdassi, M. Grouchko, O. Berezin and A. Kamyshny, *ACS Nano*, 2010, **4**, 1943–1948.
65. M. R. Pinnel, H. G. Tompkins and D. E. Heath, *Appl. Surf. Sci.*, 1979, **2**, 558–577.
66. S. Magdassi, M. Grouchko and A. Kamyshny, *Materials*, 2010, **3**, 4626–4638.

CHAPTER 11

The Use of Reactive Inkjet Printing in Tissue Engineering

CHRISTOPHER TSE*, YI ZHANG AND PATRICK J. SMITH

Department of Mechanical Engineering, University of Sheffield, Sheffield, United Kingdom
*E-mail: christopher.tse@sheffield.ac.uk

11.1   Introduction

This chapter discusses the use of reactive inkjet printing in the field of tissue engineering. The field of tissue engineering is briefly introduced with an explanation offered as to its appeal, before a short discussion of the four main areas in which tissue engineering research is being performed; these are cell printing, creation of biocompatible scaffolds, vascularisation and drug delivery methods. The discussion then moves on to how reactive-inkjet printing can address these four areas, before moving on to a survey of specific examples with the focus being on alginate-based and fibrin-based systems. The chapter concludes with an overview of work that employed reactive inkjet printing and a gelatin-based system.

11.2   Tissue Engineering

Tissue engineering is defined as the placement of fabricated structures into a living organism (predominantly humans) that restore, maintain and improve tissue function that has been lost due to injury and disease. Tissue engineering

Smart Materials No. 32
Reactive Inkjet Printing: A Chemical Synthesis Tool
Edited by Patrick J. Smith and Aoife Morrin
© The Royal Society of Chemistry 2018
Published by the Royal Society of Chemistry, www.rsc.org

is the fabrication, or transplantation, of biocompatible material that can be placed into the patient with minimal harmful side effects. Beneficiaries of tissue engineering are individuals who may experience a decreased quality of life, through disease and injury that limits their movement and interaction with others, or even have a life-threatening condition. Examples of conditions where tissue engineering can provide assistance include the debilitating effects of physical injuries and burns, nerve damage and genetic diseases. Tissue engineering aims to alleviate such ailments and restore an individual's quality of life, which may be permanently reduced without such intervention.

Currently, there are four main challenges in tissue engineering. These challenges are: cell sourcing and refinement, biomaterial creation and implementation, angiogenesis of engineered tissues and drug delivery systems.

1. Cell sources are a significant issue as they need to be cultured from the patient. Due to immunohistochemical markers that are unique to each individual, only cells derived from the patient will be compatible (unless other measures are introduced such as chemotherapy). The acquirement of the cells can also be problematic; cells ideally are extracted from the patient without injuring them in the process. After extraction, sorting the cells and selecting the right ones can be difficult and time-consuming.
2. The creation of biocompatible scaffolds is an active area of research. Biomaterials, which can originate from different sources, are required to repair damaged tissue and can be injectable or implanted, and all need to be safe upon introduction into the patient.
3. Angiogenesis, which is the growth of new blood vessels, is essential if the implanted tissue is going to be viable. Alternatively, the creation of an artificial vasculature implanted directly within a thick artificial biological tissue is required, if diffusion is insufficient for the transport of nutrients and waste. Without a steady blood supply to the new tissue, the cells will die due to insufficient nutrient uptake and accumulation of waste in the local area will build up, causing necrosis of the surrounding tissue. If necrosis occurs, the end result for the patient is an area of tissue that will then have to be removed. Vascularisation strategies that promote the delivery of nutrients and removal of wastes are an active area of research.
4. Drug delivery systems, some of which direct cell behaviour, need not be directly implanted into the patient, but can serve to facilitate tissue engineering applications. Delivery systems can be inhaled, swallowed, inserted or applied topologically *via* the skin or *via* a mucosal membrane.

Reactive inkjet printing selectively deposits reagents onto desired spatial locations and can provide varying degrees of response to address three of the above-mentioned challenges. Cell sourcing is outside the capabilities of reactive inkjet printing, as it requires a completely different set of skills, such as the ability to harvest, dice and extract required cells from host tissue. The creation of biocompatible scaffolds is the foundation of what makes reactive inkjet-printing (RIJ) appealing. Through tailoring the scaffolds' properties, not only can RIJ be used in the replacement of damaged tissue (challenge 2), but RIJ can also be employed to create artificial vasculature that mimic vessels to supply blood (challenge 3) and create drug delivery systems that tailor the release of drugs and growth factors into the host (challenge 4). As such, reactive inkjet printing makes a strong contribution to tissue engineering.

In the following sections the most prominent biomaterials used with reactive inkjet printing will be discussed, which include alginate, fibrin and gelatin. Each section will provide an explanation on formation, application and potential future applications.

11.3   Alginate-based Systems

Alginate, mainly sodium alginate, has been increasingly used to react with aqueous calcium chloride ($CaCl_2$) to form 3D alginate hydrogel as scaffolds for tissue engineering. This is because the formed alginate hydrogel is a network of hydrophilic polymer chains containing a large amount of water, and the porous internal structure plays an important role in the diffusion of nutrients for living cells and of the resultant wastes. Alginate forms a hydrogel by ionic cross-linking in the presence of a divalent cation such as Ca^{2+}, Mg^{2+} and Ba^{2+}. In this chapter, the focus is on Ca^{2+} as the cross-linker.

Bioprinting has been commonly used for fabricating alginate hydrogel with various designs of outer and inner structures for targeted tissues. This hydrogel system meets the two general requirements for printing successful 3D structures: (1) droplets have the ability to (partially) retain their 3D profiles after being deposited on substrates; (2) droplets must turn into a sufficiently robust solid structure to prevent coalescence.[1] Usually, solutions of aqueous sodium alginate and calcium chloride are loaded separately into cartridges (as printing inks) and droplets of each are ejected onto the same location to form the alginate hydrogel, a process that is readily described as reactive inkjet printing.

Xu *et al.*[2] used a platform-assisted 3D inkjet bioprinting system to fabricate 3D alginate hydrogel zigzag tubes with fibroblast cells for vessel-like tissue culture. The post-printing cell viability of printed cellular tubes has been found above 82%. Although the ionic-cross-linking (gelation) of alginate hydrogel is a relatively fast process, it is still a challenge to print overhanging parts without support. Figure 11.1 schematically shows the printing process of cell loaded zigzag 3D alginate hydrogel tubes. Xu *et al.* mixed the prepared cell suspension with 2% (w/v) sodium alginate solution at a volume ratio of 1:1. The final solution/ink has a cell density of 3×10^6 cells ml^{-1}. Then the mixed solution was loaded into an ink reservoir ready for printing. A platform which can move in the z-direction was combined with a container filled

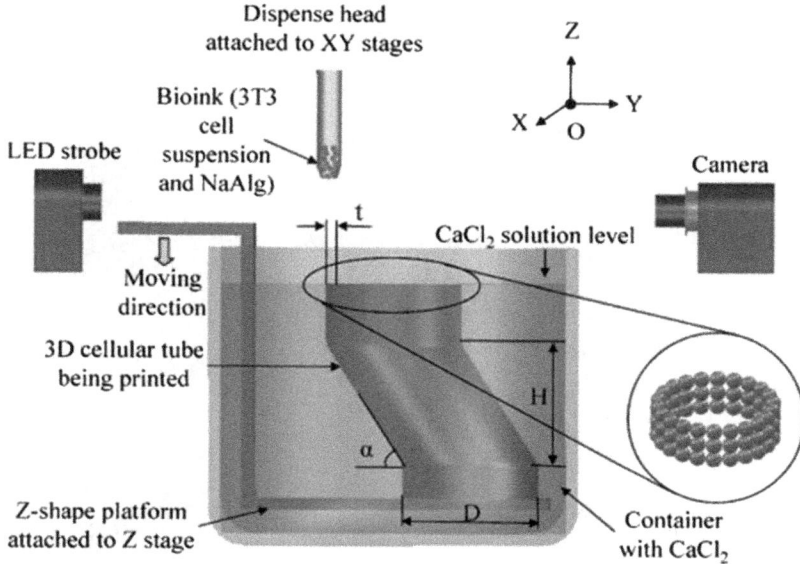

Figure 11.1   3D alginate hydrogel tubes fabricated by reactive inkjet printing. Droplets of the sodium alginate contacted with the bulk $CaCl_2$ solution to form calcium alginate. Reproduced from C. Xu, W. Chai, Y. Huang and R. R. Markwalk, *Biotechnology and Bioengineering*,[2] John Wiley and Sons, © 2012 Wiley Periodicals, Inc.

with 2% (w/v) $CaCl_2$ solution, therefore, once the sodium alginate solution is ejected from the dispensing device and contacts the $CaCl_2$ solution, an alginate hydrogel feature forms. The platform then moves down vertically to start the next layer and over time a 3D structure forms.

As mentioned earlier, suitable vascularization is essential for cell proliferation for tissue engineering. This work illustrated that it is feasible to fabricate complex 3D constructs by using reactive inkjet printing which makes printing of heterogeneous tubular structures possible. Based on this successful step mentioned by Xu *et al.*,[2] vascular-like cellular structures with bifurcations were fabricated using the same method and material system.

Printing overhanging structures without support material can be challenging for inkjet printing since the process involves solidifying of liquid droplets within a short time. Improving the quality of overhanging structures and fully addressing this challenge is vital to the success of vascular constructs.[3] Figure 11.2 schematically shows a variety of setup configurations for printing that can address the overhang challenges. Two printing directions were adopted to fabricate the same structure to evaluate the quality of the final products.

Figure 11.3 shows the horizontally and vertically printed bifurcated structures (without cells). Overhanging structures can be successfully printed

Horizontal Printing   Vertical Printing

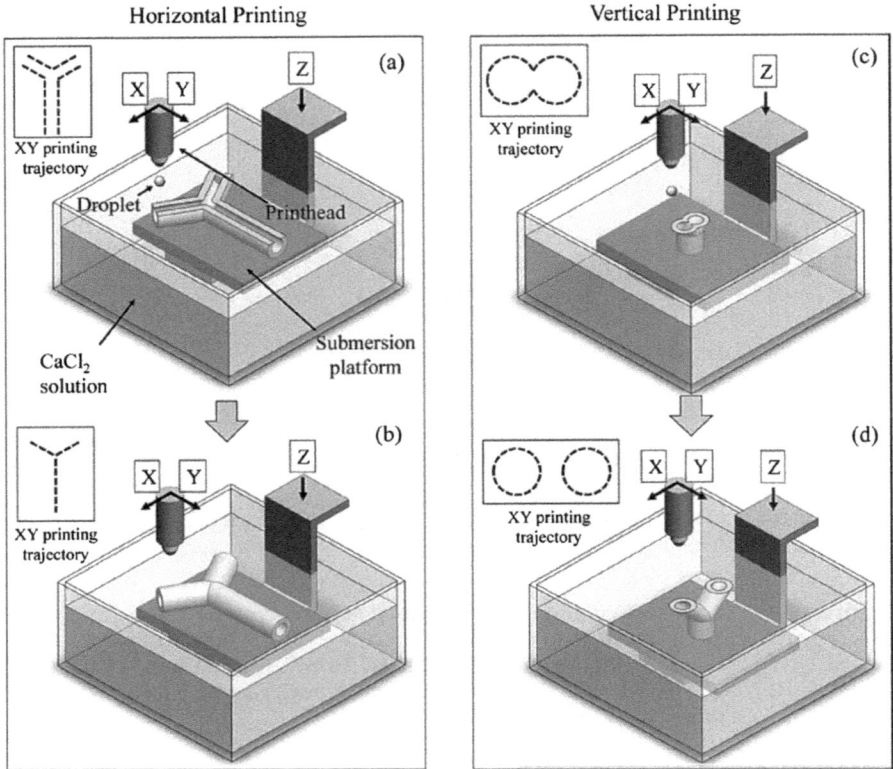

Figure 11.2   A schematic showing the system and the process with different printing directions; (a) and (b) printing in the horizontal direction, with (c)
and (d) printing in the vertical position. Reproduced from K. Christensen, C. Xu, W. Chai, Z. Zhang, J. Fu and Y. Huang, *Biotechnology and
Bioengineering*,[3] John Wiley and Sons, © 2014 Wiley Periodicals, Inc.

using both printing directions. Based on this success, Christensen *et al.*[3]
integrated both horizontal and vertical bifurcated structures at the same
time to have a close mould for real-world applications.

Figure 11.4 shows the printed vascular-like structures with, and without,
cells containing both horizontal and vertical bifurcations. It is noticed that
structures printed without cells have higher feature resolution than those
printed with cells. This difference in resolution is due to the presence of cells
in the printing ink causing a poor quality of droplet formation and trajectory
during printing.[4]

The cell viability of the printed cell-laden structures was tested to investigate the effect of the printing process on the living cells, *i.e.* NIH 3T3 mouse
fibroblasts. It was found that the post-printing cell viability of the printed
structure was 92.4% straight after printing and 90.8% after 24 hours of
incubation.[3] This result showed that the viability of cells after inkjet printing is barely affected by the printing process, indicating inkjet printing is

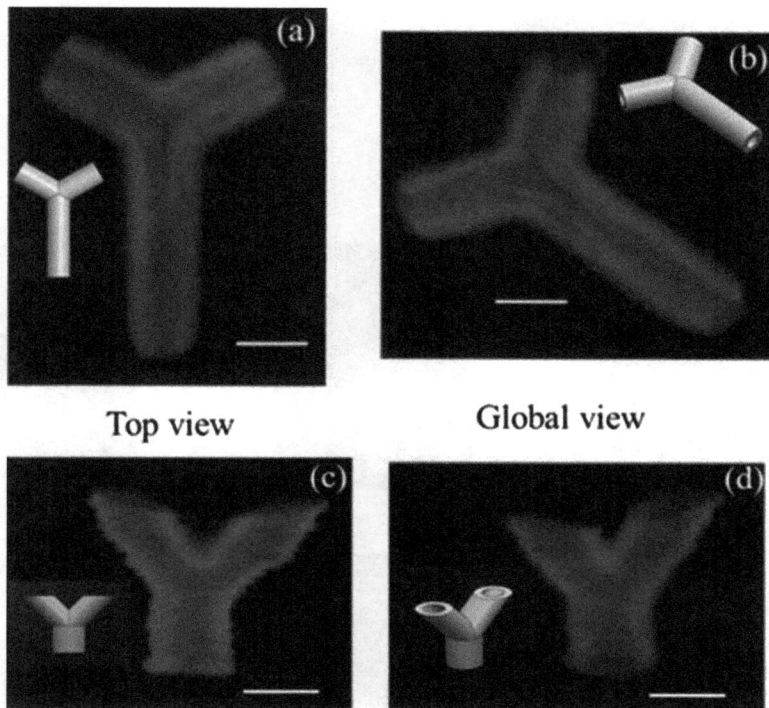

Top view   Global view

Figure 11.3   Printed alginate hydrogel bifurcation structures. (a) and (b) horizontal printing; (c) and (d) vertical printing. Scale bars are 3 mm unless specified. Reproduced from K. Christensen, C. Xu, W. Chai, Z. Zhang, J. Fu and Y. Huang, *Biotechnology and Bioengineering*,[3] John Wiley and Sons, © 2014 Wiley Periodicals, Inc.

suitable for fabricating 3D complex cell-laden structures as scaffolds for tissue engineering.

In order to verify the stability of the printed vascular-like structures whilst having fluid flowing/fluid transmission, Pataky *et al.*[1] furthered this work by adding actual fluid flow testing using printed bifurcated structures. A suspension of red fluorescent microbeads was aspirated through the capillary by syringe as shown in Figure 11.5G. Beads flowed through the channel without leaking outside of it, indicating a proper seal of printed vascular-like structures. By changing the driving pressure, bead velocities of several μm s^{-1} to tens of mm s^{-1} were observed in the centre of the channel.

Another similar work has been done by Xu *et al.*[5] who successfully fabricated a 'half heart' 3D structure using reactive inkjet printing also based on the same alginate hydrogel system. The difference of this work from the above printings is that they used the aqueous $CaCl_2$ solution as the printing ink instead of sodium alginate solution, conversely, sodium alginate solution was loaded in a container as substrate to form the alginate hydrogel as shown in Figure 11.6.

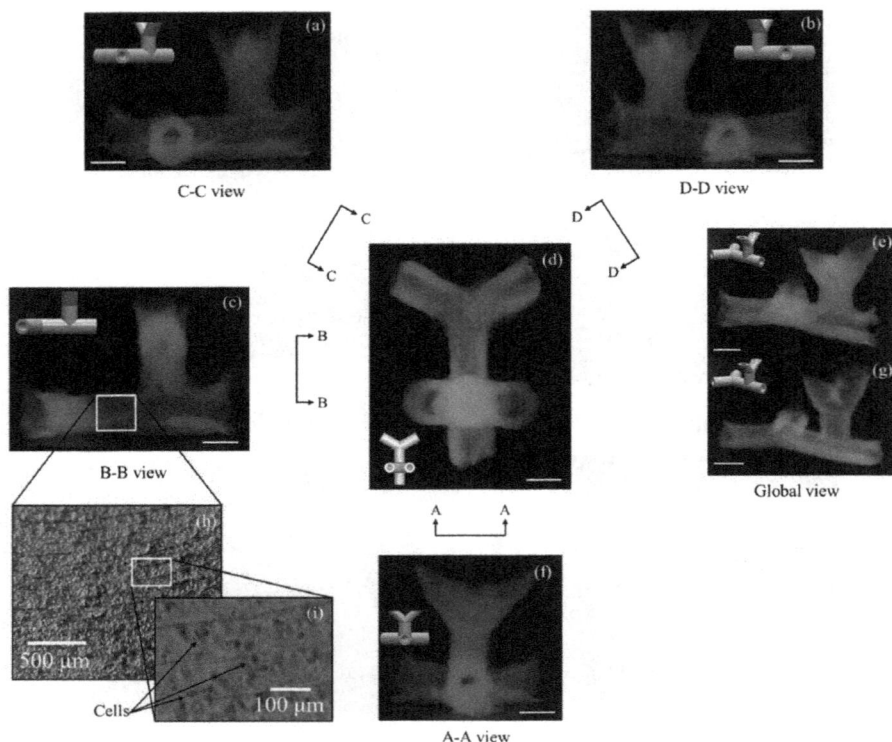

Figure 11.4 Different views of the printed vascular-like structures (with cells) (a–f), (g) structure printed without cells, (h, i) surface of printed structures with different magnifications. Scale bars are 3 mm unless specified. Reproduced from K. Christensen, C. Xu, W. Chai, Z. Zhang, J. Fu and Y. Huang, *Biotechnology and Bioengineering*,[3] John Wiley and Sons, © 2014 Wiley Periodicals, Inc.

In order to understand the mechanism of the gelation process, a cross-section of printed alginate hydrogel was examined using SEM. Figure 11.7a shows the final reactively printed 'half heart' 3D structure as designed. The SEM image of the cross-section of printed hydrogel shows when the $CaCl_2$ droplets contact with sodium alginate solution, the outside of the $CaCl_2$ droplet quickly forms a layer of hydrogel, leaving the inside as a hollow structure as shown in Figure 11.7b.

This unique hollow inner structure affects the mechanical properties of the final product. In order to find out how it affects the mechanical properties of printed alginate hydrogel, a conventional one was made by manually mixing sodium alginate solution with $CaCl_2$ solution. The comparison of the mechanical test is shown in Table 11.1.

It can be seen that the printed hydrogel has only about 30% of the tensile properties of the manually made hydrogel, which is due to the hollow

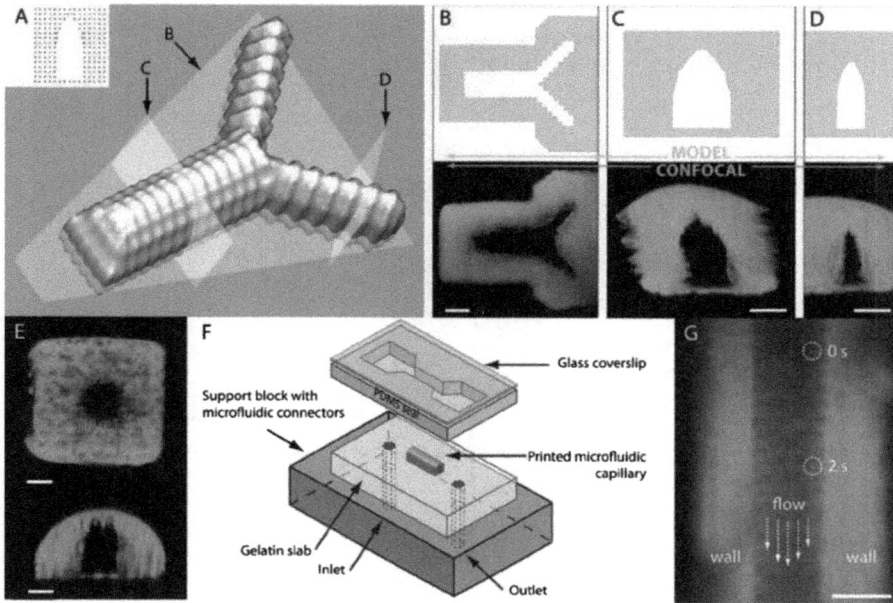

Figure 11.5 Different views of the bifurcated structure (A–D). Top view and cross-section of closed lumen with cells (E). Scheme showing the setup of the fluid flow test (F). Fluid flow through the printed microchannel. All scale bars are 200 μm. Reproduced from K. Pataky, T. Braschler, A. Negro, P. Renaud, M. P. Lutolf and J. Brugger, *Advanced Materials*,[1] John Wiley and Sons, Copyright © 2012 WILEY-VCH Verlag GmbH & Co. KGaA, Weinheim.

structure. It is understandable that the higher content of voids results in poorer mechanical properties. Therefore, the mechanical properties need to be improved before being used in clinical applications.

Alginate hydrogels are formed through the strength of ionic bonds between the respective ions. When a chelating agent is introduced to the mixture, the ionic interactions can be replaced and break up the hydrogel. Delaney *et al.*[6] took advantage of reactive inkjet printing to print uniform droplets of sodium alginate into $CaCl_2$ solution to form uniform hydrogel beads, which they then re-suspended in a different cross-linkable matrix. Once the matrix cross-linked, ethylenediaminetetraacetic acid (EDTA) was added to chelate the calcium ions in the alginate hydrogel beads within the matrix, leaving uniform sized pores in the matrix. Through this research, Delaney *et al.* showed: (1) the direct control of the macropore size can be controlled *via* inkjet printing on a scale relevant to cell culture and tissue engineering. (2) By packing the material with a sufficiently high concentration of hydrogel beads in the macroporous matrix, the beads will come into contact with one another, yielding a continuous 3D network of channels once they are removed.

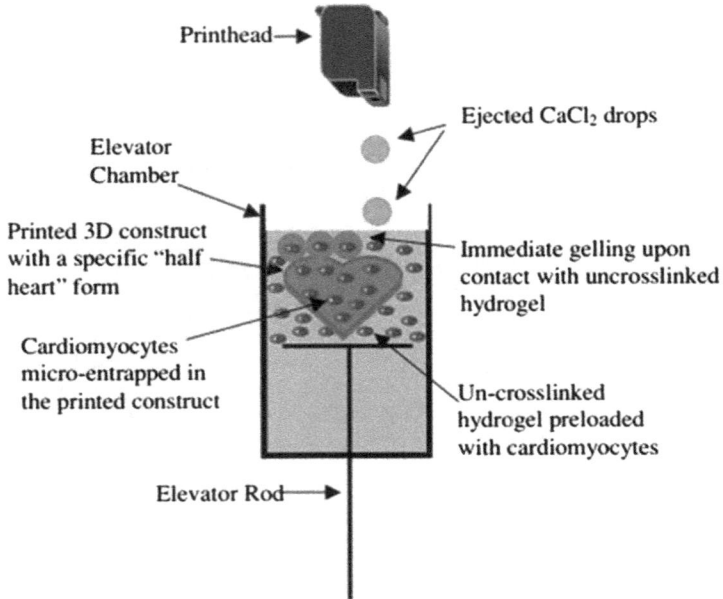

Figure 11.6 CaCl$_2$ solution was loaded in the printhead as ink, gelation happened when the CaCl$_2$ droplet made contact with the sodium alginate. Once the gelation of the upper layer was finished, the platform moved downwards to start the next layer. Reproduced from ref. 5, DOI: 10.1088/1758-5082/1/3/035001. © IOP Publishing. Reproduced with permission. All rights reserved.

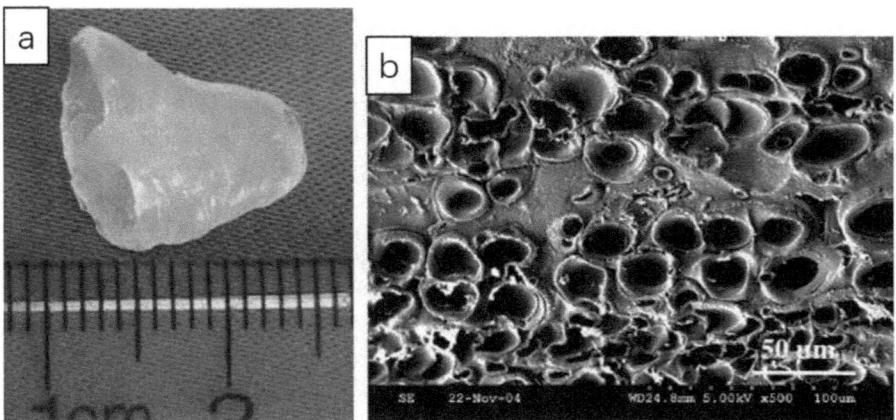

Figure 11.7 (a) Reactively inkjet printed 3D 'half heart' structure. (b) SEM image of cross-section of printed alginate hydrogel. Reproduced from ref. 5, DOI: 10.1088/1758-5082/1/3/035001. © IOP Publishing. Reproduced with permission. All rights reserved.

Table 11.1 Tensile properties of printed and manually made alginate hydrogel scaffolds.

	Printed	Manual
Modulus (kPa)	113.3 ± 17.5	380.0 ± 118.5
UTS (kPa)	43.9 ± 12.2	232.3 ± 88.9
n	5	6

11.4 Fibrin-based Systems

Fibrin gel has been increasingly used for bioprinting area. Fibrin is a bio-polymeric gel that plays an important role in real world applications, *e.g.* autologous scaffolds;[7] skin grafts and tissue engineered skin replacements.[8] The fibrin gel is formed by the enzymatic polymerization of fibrinogen. The protease thrombin first symmetrically cleaves fibrinogen molecules at two sites. After the fibrinogen is cleaved, the fibrinogen monomers self-assemble to form a non-covalently cross-linked polymer gel *via* the proteolytic exposure of binding sites.[9,10] Since the whole process of gelling is very fast, it may present an alternative material for inkjet printing as a gel can be made fast and less dissipation occurs. Printing of thrombin into fibrinogen will cause geometry-specific cross-linking, thus enabling the rapid construction of 3D fibrin scaffolds with specific structures and forms.

Another advantage of using fibrin for tissue engineering is that it can be degraded and remodelled by enzymatic activity during cell migration and wound healing. The *in-vitro* studies showed that fibrin gels can potentially help cell migration and proliferation. A promising area of research is the recovery of neuronal damage. Fibrin-based neural scaffolds could potentially provide suitable cell or tissue sources with *in-vivo* affinity for clinical treatments of serious neural injuries and degenerative diseases, such as spinal cord injury and Parkinson's disease.[9]

Cui *et al.*[11] successfully adopted this gel system for fabricating microvascular structures with living cells. Figure 11.8 schematically shows the printing setup. In order to assure prompt and optimum polymerization of fibrin gel after printing, they conducted fibrin gel polymerisations with various combinations of fibrinogen, thrombin, and Ca^{2+} concentrations.

Xu *et al.*[10] reported they also used this fibrin-based system to print complex cellular patterns and structures for neural research. The difference from Cui's work was that they alternatively printed gel material and cells layer-by-layer instead of mixing cells within the printing inks. Their methodology can be summarised as follows: a thin layer of fibrinogen was plated onto a clean, sterile coverslip, thrombin was loaded into a cartridge as one of the printing inks, another cartridge was loaded with NT2 neurons as the second ink. Three layers of thrombin were printed onto the pre-plated fibrinogen layer. Fibrin gel formation was observed immediately after thrombin contacting with fibrinogen. After gelation for 3 to 5 minutes, NT2 neuronal cells were printed. This process was repeated until a scaffold was created.

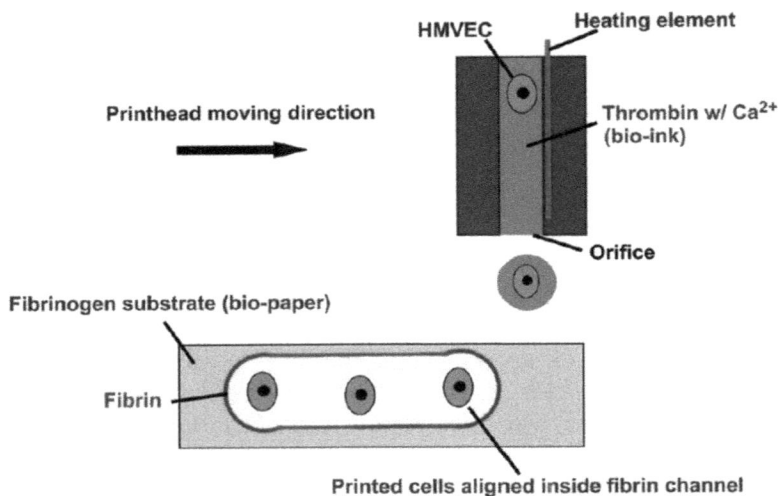

Figure 11.8   Schematic showing the fibrin printing process. Living cells were mixed with thrombin as ink and fibrinogen was used as substrate.

11.5   Gelatin-based Systems

Gelatin is one of the most commonly used hydrogels for mammalian cell growth[12,13] and biomedical applications. Such biomedical applications include drug delivery through hard and soft capsules, wound dressings, cell encapsulation and *in vitro* tissues.[14,15] Gelatin hydrogels can maintain a physiologically moist environment. Its precursor prior to chemical modification—collagen, is an essential major component of the extracellular matrix, and through its similarities to collagen, mammalian cells grow very well in gelatin's presence. Gelatin is a biocompatible denatured protein of collagen that is biodegradable, non-immunogenic and its degredation rate can be tailored. Cross-linking of the biopolymer is required as virgin gelatin turns into its sol state at biological temperatures (37 °C), which upon cooling returns to a thermos-reversible hydrogel. Proteolytic enzymes released by adhered cells help facilitate the virgin gelatin to dissolve and break down the gelatin into the cell culture media. There are many ways to cross-link gelatin; through using UV-light,[16] a cross-linking enzyme[17–19] or chemical cross-linkers. Such chemical cross-linkers include carbodiimide, diphenylphosphoryl azide, and glutaraldehyde.[13] However, glutaraldehyde provides the most stable cross-linking agent compared to others mentioned.

Cross-linking allows gelatin to remain as a hydrogel in biological conditions, whilst also improving thermal and mechanical stability. Glutaraldehyde cross-linking involves the reaction of free amino groups of lysine or hydroxylysine (lysine oxidised by lysyl hydroxylase enzymes) amino acid residues of gelatin and collagen with the aldehyde groups of glutaraldehyde.[20] High concentrations have proved to be toxic as residual glutaraldehyde is

Figure 11.9   Gelatin scaffolds used for tissue engineering cultured in well plates.

toxic to cells in cell culture. As a rule of thumb, glutaraldehyde concentrations of less than 2.5% are suitable for biomedical applications. 1 wt% glutaraldehyde is sufficient to cross-link 100% of exposed gelatin and increase its Young's modulus by 20 times. Once concentrations are above 2.5%, cell toxicity becomes a significant problem as glutaraldehyde can be slowly released from the gelatin over time.[21] Gelatin has been used in tissue engineering in bone,[22] cartilage,[23] wound healing of skin[24] and adipose.[25] Figure 11.9 shows an example of cross-linked gelatin.

This section discusses the work that was undertaken to investigate how glutaraldehyde could be inkjet printed onto a bed of gelatin to create cross-linked-gelatin patterned environments suitable for cell seeding. Cross-linking was required on gelatin hydrogels as virgin gelatin melts at biological temperatures (37 °C). Transglutaminase,[17,24,26] genipin,[18,27] carbodiimides,[28] and glutaraldehyde[29,30-32] have been researched extensively as cross-linking agents for gelatin. All cross-linking agents offer a degree of toxicity, and although glutaraldehyde has been shown to cause the most cell death as a cross-linking agent, it also provides the best mechanical strength. The cause of toxicity for glutaraldehyde is caused by residual glutaraldehyde leaching out and causing cell death. To minimise this toxicity research focused on using lower concentrations.

Cross-linking of gelatin by glutaraldehyde occurs through the aldehyde groups reacting to the free amino groups of lysine or hydroxylysine amino acid residues on the gelatin molecules,[21] Primary amines and secondary amines react with the aldehyde group in glutaraldehyde through nucleophilic addition to form carbinolamines, which can then dehydrate to give substituted imines and enamines respectively. A range of possible resultant products are outlined in Figure 11.10, which shows collagen reacting with glutaraldehyde. Many of the reactions involve the formation of a Schiff base intermediate (Structure III) that is able to form a plethora of products. It is through this reaction that the mechanical properties of gelatin and its mechanical properties increase; through the generation of larger complex compounds within the gelatin.

11.6   Optimal Printing Conditions of Glutaraldehyde Printing

To create inkjet printed scaffolds, the optimal printing conditions were required to be identified for the successful creation of biologically relevant cell-friendly scaffolds. Ideally, the viscosity of the ink should lie between 1–20 centipoise. Table 11.2 shows that within the range of temperatures tested, glutaraldehyde should be printable between 0–40 °C. This temperature range is ideal as this means that glutaraldehyde solution can be printed at room temperature.

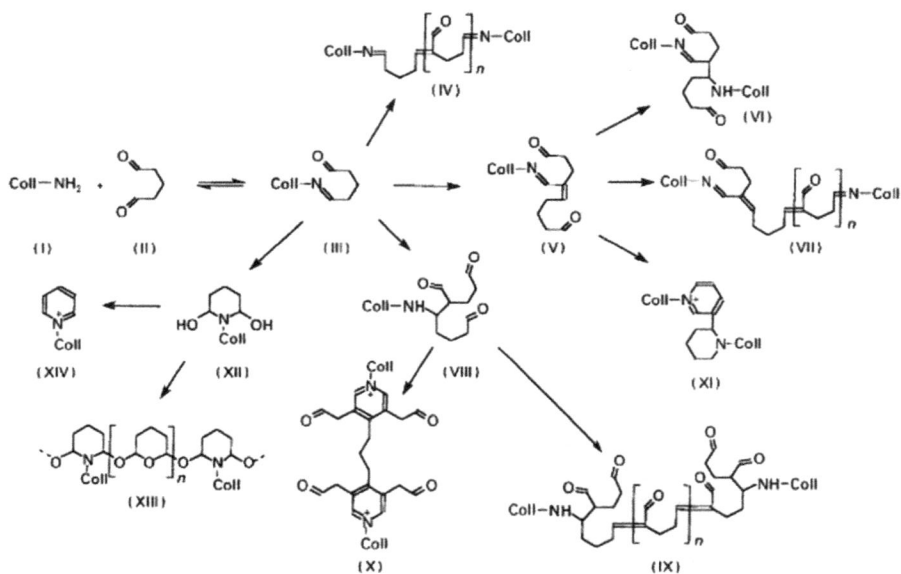

Figure 11.10   Gelatin is denatured collagen, and the various reactions pathways shown here can be related to the cross-linking of gelatin (*i.e.* Coll-NH$_2$ = Collagen-amine group). Reproduced from *Journal of Materials Science: Materials in Medicine*, Glutaraldehyde as a cross-linking agent for collagen-based biomaterials, 6, 1995, 460–472, L. H. H. Olde Damink, P. J. Dijkstra, M. J. A. Van Luyn, P. B. Van Wachem and P. Nieuwenhuis,[20] © Chapman & Hall 1995. With permission of Springer.

Table 11.2   Viscosity behaviour of 25% glutaraldehyde (w/w in water) at different temperatures.

Temperature/°C	Viscosity/centipoise (mPa s)
0	8.4
20	3.4
40	1.7

11.7 Inkjet Printing Glutaraldehyde to Cross-link Gelatin

To determine the best glutaraldehyde concentration to print onto a bed of gelatin, several concentrations were tested to evaluate the fidelity of the resultant patterns. 25% glutaraldehyde (v/v in water) was selected. Concentrations lower than this created thin printed tubular structures with diameters less than 500 microns thick and it was clear that the scaffolds were too weak, would curl up and collapse on themselves due to surface tension when being manipulated. If the scaffolds were too difficult to handle, then they would not be suitable as they would most likely break during the cell seeding and handling of the scaffolds during analysis.

3 mL of warm (~25 °C) gelatin solution (4 w/v %, in water) was pipetted onto glass slides to cover the whole surface area. The surface tension of the glass slide prevented the gelatin from flowing off the glass slide, and this was put into a fridge to cool prior to printing. Once the gelatin became solid, the samples were loaded into the printing system. 25% glutaraldehyde (v/v in water) was loaded into the printer, and the parameters optimised. A droplet spacing of 100 μm was used, and a cross-hatch pattern was printed to emulate a connected network (Figure 11.11).

After printing, the samples were left for different periods of time to allow cross-linking to occur (Figure 11.12), ranging from 5 minutes to 24 hours (overnight).

After the allocated time, a scalpel was used to cut the scaffold into smaller segments and placed into each well of a 6 well plate (Figure 11.13). Hot water (around 50 °C) was gently pipetted into each well to wash off uncross-linked gelatin, five times. Scaffolds were inspected and assessed to determine the most suitable scaffolds for cell seeding.

Figure 11.14 shows examples of the successful samples created.

11.8 Cell Seeding onto Inkjet-printed-glutaraldehyde-cross-linked-gelatin

1 mL of fibroblasts were seeded at 40 000 cells mL^{-1} into each well, and topped off with 2 mL of cell media, before being stored in an incubator. The samples were experimented in triplicate, with light and confocal images taken on day 1, 2 and 3. Confocal images were stained with FITC-phalloidin and DAPI, with the aim to determine cell viability of these scaffolds. Gelatin scaffolds made with 25% glutaraldehyde were handled successfully and cells were able to grow on the patterned environments on all days that they were observed (day 1, 2 and 3 after cell seeding).

Figure 11.15 shows an example of images taken near an intersection where several gelatin beams join together. The structure appears "furry" due to the surface of the hydrogel being populated by fibroblasts growing on the surface of the cross-linked gelatin scaffolds. The scaffolds did not show any signs of

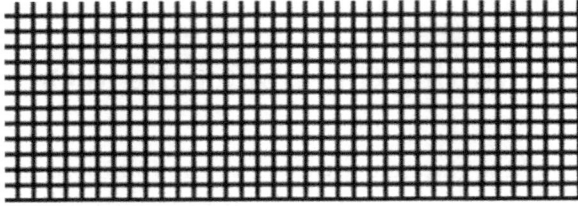

Figure 11.11 Image showing the hatch pattern designed for printing onto gelatin. A droplet spacing of 100 µm was used, and the distance between each connection within the pattern was 2.5 mm across.

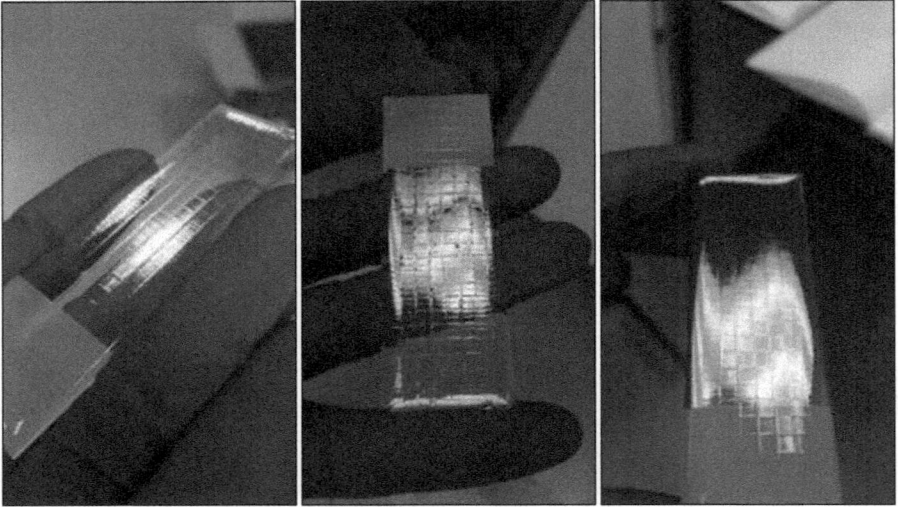

Figure 11.12 Photographs of inkjet printed glutaraldehyde onto a bed of 4% gelatin on a glass slide. The samples were cut out into segments that would fit into a well in a 6-well plate, using a scalpel.

Figure 11.13 Image showing an experimental procedure where the cross-linked samples were put into labelled 12 well plates prior to washing away the uncross-linked gelatin.

Figure 11.14   A collection of images showing cross-linked gelatin scaffolds being washed with hot water in well plates. The samples shown here are samples that were cross-linked with 25% glutaraldehyde and left overnight for 24 hours.

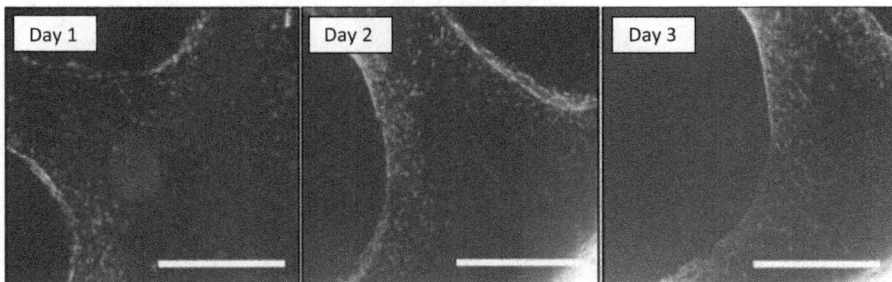

Figure 11.15   Microscopy images of fibroblasts seeded onto cross-linked gelatin through inkjet printing glutaraldehyde onto a bed of gelatin. Images taken 1, 2 and 3 days after cell seeding respectively. Scale bar = 300 μm.

toxicity, and cells proliferated at a rate that is typical for healthy cells. If the scaffolds did release glutaraldehyde, or it was detrimental to cell health, the population of cells would have decreased on the scaffold, or more cells would have adopted a spherical shape to indicate unfavourable culture conditions. These types of physical cues were not seen in any of the samples. Enlarged images of these sections are shown in Figure 11.16.

This showed that cells were able to grow on gelatin that had been cross-linked with up to 25% glutaraldehyde. No residual glutaraldehyde was expected to be released into the cell culture because there was no aggregation of glutaraldehyde during the cross-linking process; this was because the glutaraldehyde was able to diffuse freely in all three directions and was added in picolitre amounts.

The gelatin scaffolds that were created had a range of thicknesses, ranging from 250 to 100 μm, with the thickest at the intersections. The beams had a cylindrical shape, which was caused due to the way in which the glutaraldehyde diffused uniformly from the area of deposition. Through the use of reactive inkjet printing in forming cross-linked gelatin scaffolds, it would not be possible to create sharp edged scaffolds, unless the diffusion of the glutaraldehyde were to be inhibited by restricting its movement with a physical barrier. If this restriction were to be performed, care would need to be taken concerning the excess glutaraldehyde within the scaffold which would leak out over time and make the scaffold increasingly toxic.

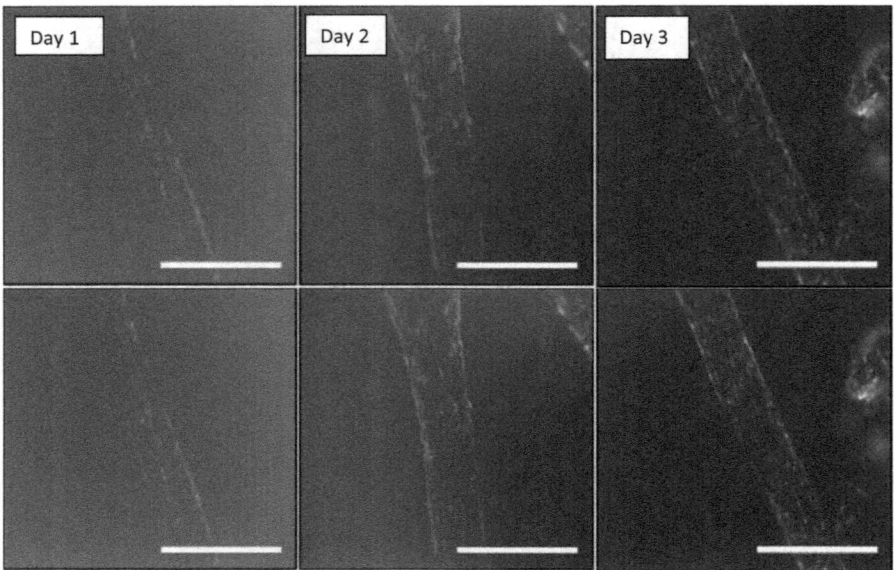

Figure 11.16 Microscopy images of fibroblasts seeded onto cross-linked gelatin created with inkjet printing. Images taken 1, 2 and 3 days after cell seeding respectively. Scale bar = 300 μm.

Figure 11.17 shows enlarged light microscopy images of the cell seeded gelatin scaffolds. On day 1 after cell seeding, cells manage to adhere and begin to proliferate on the scaffold. There is a larger population of cells proliferating on the side of the scaffold that was facing the cell seeding process, but it is interesting to see cells have also managed to grow on the periphery and underside of the scaffold. By day 2 and 3 (Figure 11.17) there appear to be no detrimental effects of cross-linking with glutaraldehyde, as the gelatin scaffold has not melted under biological temperatures, and cells have proliferated along the whole structure. More alignment of the cells can be seen on day 2, that are running in the same direction as the beams. The alignment was lost once the population of fibroblasts became over-confluent.

Confocal images were recorded to analyse the cell seeded gelatin scaffolds. Due to the 3D nature of the scaffold, the confocal images capturing a plane of the scaffold show a limited amount of data. However, the details that can be seen by analysing each pane individually make it clear that fibroblasts were able to infiltrate into the scaffold and proliferate within. Figure 11.18 shows a cross-section of a gelatin scaffold seeded with fibroblasts after three days of cell seeding. The periphery of the image shows a shell of cells on the surface of the gelatin scaffold; stained for actin (phalloidin-TRITC) and nuclei (DAPI), and also note the cells situated in the middle of the image. These cells are proliferating within the gelatin scaffold, and it must be assumed that the fibroblasts are able to not just proliferate on the surface of the gelatin, but also infiltrate the hydrogel and grow within it.

Several areas were analysed, and a z-stack was created of each area. By combining the z-stack images with the Zeiss LSM Image Browser software, a 3D model of the scaffold was created. Figure 11.19 (day 1) and Figure 11.20 (day 2) are such models, and a series of images are shown, with each image being rotated by 30°.

The majority of cells that proliferate on the scaffolds were proliferating on the surface. Initially, only a small percentage of the cells grew within the gelatin hydrogel. However, this internal population increased over time, most likely caused by the cells preferentially differentiating to areas with the least

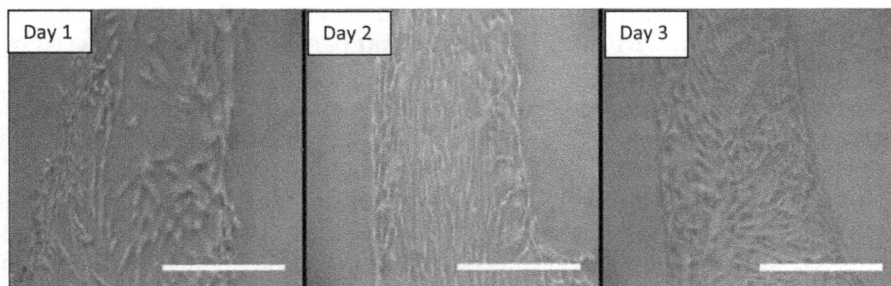

Figure 11.17 Enlarged microscopy images of fibroblasts seeded onto cross-linked gelatin through inkjet printing. Images taken 1, 2 and 3 days after cell seeding respectively. Scale bar = 150 μm.

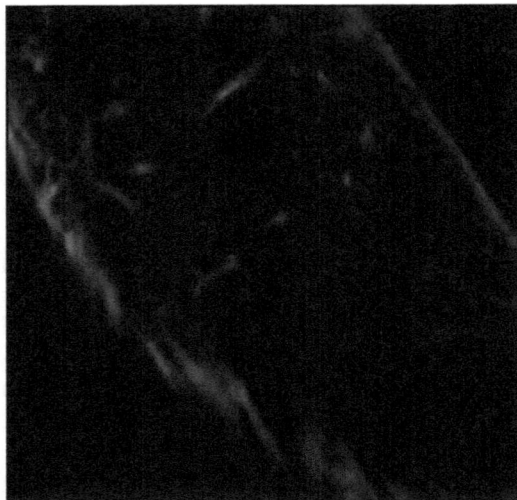

Figure 11.18   Confocal image of a cross-section of the middle of a gelatin scaffold
beam stained for actin.

resistance. It could be seen (Figure 11.20) that there were more cells infil-
trating the hydrogel during the second and third day after cell seeding; as
the surface of the gelatin became more populated, the action of proliferating into the gelatin hydrogel became a better alternative than competing for
space on the outer surface.

The morphology of the gelatin scaffolds varied over time; as originally,
the branches were typically cylindrical in shape. With the addition of cells,
which were able to eat away at the gelatin slowly, when the cell population
became confluent, the scaffolds became more elongated and flat. Figure
11.20 shows the flattening of the gelatin occurring 3 days after cell seeding.
In the target area, cells had fully covered the surface facing the cell seeding
side, and cells had begun migrating into and enveloping the scaffold. The
shape of the gelatin scaffold can be determined through the outlines given
by the population of cells adhering onto and into the scaffold. More cells can
be seen proliferating within the hydrogel compared to samples analysed on
day 1 after cell seeding.

With the cross-linked gelatin being a hydrogel (*i.e.* being mostly water),
cells were able to adhere and survive within the gelatin scaffold. The high
concentration of glutaraldehyde that was used did not cause the gelatin scaffold to become toxic, so it can be assumed that either all glutaraldehyde molecules fully reacted with the gelatin or all of the residual glutaraldehyde was
removed during washing. Thinner scaffolds could be created with a lower
concertation of glutaraldehyde, however to prevent damage to such scaffolds, the design and handling of the scaffolds needs attention. Thicker scaffolds could be created by the printing of multiple layers of glutaraldehyde

Figure 11.19 A sequence of 2D images of a 3D projected cross-linked gelatin scaffold with seeded fibroblasts after 1 day. Scale bar = 250 μm.

onto the bed of gelatin; however, like all other tissue engineered scaffolds, only scaffolds that do not exceed a few millimetres in thickness can have cells proliferate on/in them, as simple diffusion is a sufficient method to transport nutrients into cells from the cell media and remove waste. A limit to inkjet printing would be that scaffolds may not be more than a few millimetres thick, unless a vasculature network is designed within the scaffold to allow directed transport of nutrients and cell waste throughout the scaffold. It could be postulated that with the simple cross hatch pattern that was used to create these scaffolds, once cells have been seeded, the scaffold could be

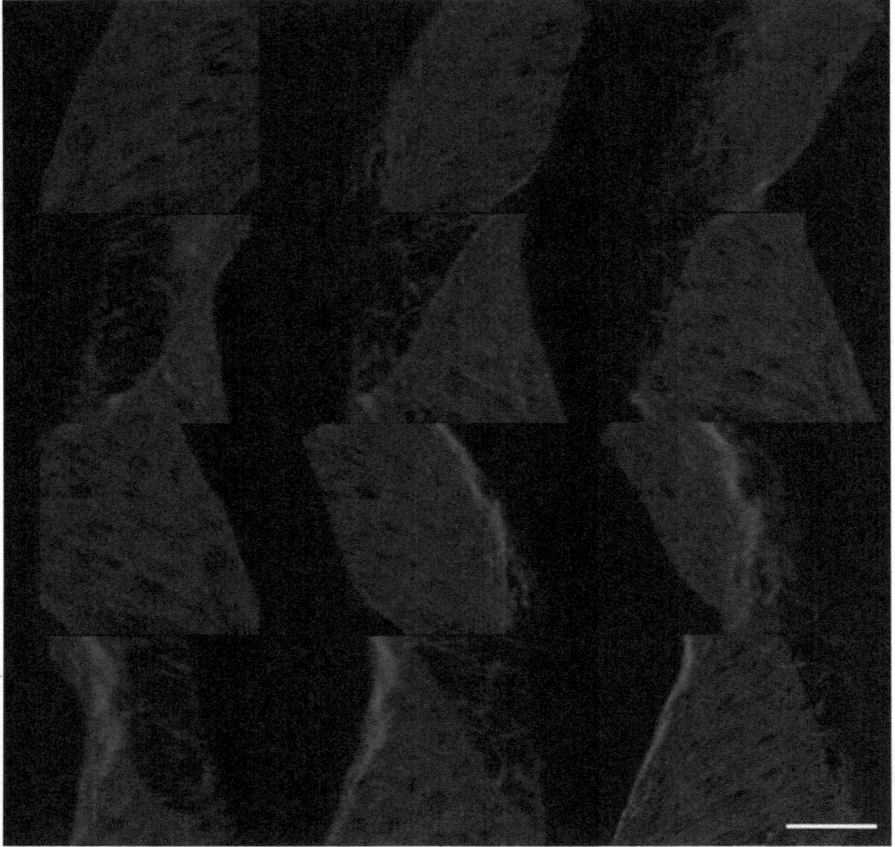

Figure 11.20   A sequence of 2D images of a 3D scaffold that shows fibroblasts grown on glutaraldehyde-cross-linked gelatin. A 3D image was generated through the stacking of multiple confocal images to create a z-stack. Stained for actin (phalloidin-TRITC) and nuclei (DAPI), day 3 after cell seeding. Cells proliferated significantly more on the surface side where cell seeding took place. Cells can be seen to infiltrate within the gelatin and grow within the scaffold. Scale bar = 250 μm.

collapsed and rolled together to form a larger, thicker scaffold of significant size, and with the porosity of the design of the scaffold, cells would survive at a better rate than a solid block of hydrogel of the same material.

11.9   Conclusions

The purpose of this chapter was to demonstrate the attractiveness of reactive inkjet printing (RIJ) for tissue engineering. Indeed, it is apparent that much of the tissue engineering research that has been performed using inkjet printing has employed an RIJ approach. The chapter discussed the preparation of fibrin gel in which thrombin is brought into contact with fibrinogen causing cleavage and forming fibrinogen monomers that subsequently self-assemble to form fibrin. RIJ has also be used in the formation of alginate hydrogels. Typically, sodium alginate is reacted with calcium chloride to form the hydrogel which can be used as a scaffold for tissue engineering. Finally, the preparation of cross-linked gelatin structures by the addition of glutaraldehyde was presented.

In all three of these systems RIJ plays a role that is vital. Inkjet allows the targeted placement of material, in these three systems the formed gel can be patterned due to the pre-determined deposition of the second reactant. This enhanced patterning ability, allowing a researcher to create the shape of choice from a biologically relevant material, underlines RIJ's appeal in that systems can be formed and shaped at the same time.

References

1. K. Pataky, T. Braschler and A. Negro, *et al.*, *Adv. Mater.*, 2012, **24**(3), 391–396.
2. C. Xu, W. Chai and Y. Huang, *et al.*, *Biotechnol. Bioeng.*, 2012, **109**(12), 3152–3160.
3. K. Christensen, C. Xu and W. Chai, *et al.*, *Biotechnol. Bioeng.*, 2015, **112**(5), 1047–1055.
4. C. Xu, M. Zhang and Y. Huang, *et al.*, *Langmuir*, 2014, **30**(30), 9130–9138.
5. T. Xu, C. Baicu and M. Aho, *et al.*, *Biofabrication*, 2009, **1**(3), 035001.
6. J. T. Delaney, A. R. Liberski and J. Perelaer, *et al.*, *Soft Matter*, 2010, **6**(5), 866–869.
7. M. Fussenegger, J. Meinhart, W. Hobling, W. Kullich, S. Funk and G. Bernatzky, *Ann. Plast. Surg.*, 2003, **51**, 493–498.
8. L. J. Currie, J. R. Sharpe and R. Martin, *Plast. Reconstr. Surg.*, 2001, **108**, 1713–1726.
9. M. C. Chiti, M. M. Dolmans and J. Donnez, *et al.*, *Ann. Biomed. Eng.*, 2017, 1–14.
10. T. Xu, C. A. Gregory and P. Molnar, *et al.*, *Biomaterials*, 2006, **27**(19), 3580–3588.
11. X. Cui and T. Boland, *Biomaterials*, 2009, **31**, 6221–6227.
12. R. Langer and J. P. Vacanti, *Science*, 1993, **260**, 920–926.
13. J. L. Drury and D. J. Mooney, *Biomaterials*, 2003, **24**, 4337–4351.
14. E. Esposito, R. Cortesi and C. Nastruzzi, *Biomaterials*, 1996, **17**, 2009–2020.
15. K. W. Wissemann and B. S. Jacobson, *In Vitro Cell. Dev. Biol.*, 1985, **21**, 391–401.
16. J. Elisseeff, K. Anseth, D. Sims, W. McIntosh, M. Randolph and R. Langer, *Proc. Natl. Acad. Sci. U. S. A.*, 1999, **96**, 3104–3107.
17. C. W. Yung, L. Q. Wu, J. A. Tullman, G. F. Payne, W. E. Bentley and T. A. Barbari, *J. Biomed. Mater. Res., Part A*, 2007, **83**, 1039–1046.

18. A. Bigi, G. Cojazzi, S. Panzavolta, N. Roveri and K. Rubini, *Biomaterials*, 2002, **23**, 4827–4832.
19. Y. Otani, Y. Tabata and Y. Ikada, *Biomaterials*, 1998, **19**, 2091–2098.
20. L. H. H. Olde Damink, P. J. Dijkstra, M. J. A. Luyn, P. B. Wachem, P. Nieuwenhuis and J. Feijen, *J. Mater. Sci.: Mater. Med.*, 1995, **6**, 460–472.
21. A. Bigi, G. Cojazzi, S. Panzavolta, K. Rubini and N. Roveri, *Biomaterials*, 2001, **22**, 763–768.
22. Y. Tabata and Y. Ikada, *Adv. Drug Delivery Rev.*, 1998, **31**, 287–301.
23. T. A. Holland, Y. Tabata and A. G. Mikos, *J. Controlled Release*, 2005, **101**, 111–125.
24. A. Ito, A. Mase, Y. Takizawa, M. Shinkai, H. Honda, K. -I. Hata, M. Ueda and T. Kobayashi, *J. Biosci. Bioeng.*, 2003, **95**, 196–199.
25. T. Masuda, M. Furue and T. Matsuda, *Tissue Eng.*, 2004, **10**, 523–535.
26. A. Paguirigan and D. J. Beebe, *Lab Chip*, 2006, **6**, 407–413.
27. C. H. Yao, B. S. Liu, C. J. Chang, S. H. Hsu and Y. S. Chen, *Mater. Chem. Phys.*, 2004, **83**, 204–208.
28. H. W. Sung, D. M. Huang, W. H. Chang, R. N. Huang and J. C. Hsu, *J. Biomed. Mater. Res.*, 1999, **46**, 520–530.
29. F. H. Lin, C. H. Yao, J. S. Sun, H. C. Liu and C. W. Huang, *Biomaterials*, 1998, **19**, 905–917.
30. Yu-C. Ou, C.-W. Hsu, L.-J. Yang, H.-C. Han, Yi-W. Liu and C.-Y. Chen, *Sensors*, 2008, **20**, 435–446.
31. A. P. McGuigan and M. V. Sefton, *Tissue Eng. Regener. Med.*, 2007, **1**, 136–145.
32. L.-J. Yang and Y.-C. Ou, *Lab Chip*, 2005, **5**, 979–984.

Subject Index